Boston Studies in the Philosophy and History of Science

Volume 348

The series *Boston Studies in the Philosophy and History of Science* was conceived in the broadest framework of interdisciplinary and international concerns. Natural scientists, mathematicians, social scientists and philosophers have contributed to the series, as have historians and sociologists of science, linguists, psychologists, physicians, and literary critics.

The series has been able to include works by authors from many other countries around the world.

The editors believe that the history and philosophy of science should itself be scientific, self-consciously critical, humane as well as rational, sceptical and undogmatic while also receptive to discussion of first principles. One of the aims of Boston Studies, therefore, is to develop collaboration among scientists, historians and philosophers.

Boston Studies in the Philosophy and History of Science looks into and reflects on interactions between epistemological and historical dimensions in an effort to understand the scientific enterprise from every viewpoint.

Vincenzo Politi

Thomas Kuhn's Philosophical Perspective

Communities, Languages, and Practices

Springer

Vincenzo Politi
Department of Philosophy, Logic, and
Philosophy of Science - Faculty of Philosophy
University of Seville
Seville, Spain

ISSN 0068-0346 ISSN 2214-7942 (electronic)
Boston Studies in the Philosophy and History of Science
ISBN 978-3-032-13539-1 ISBN 978-3-032-13540-7 (eBook)
https://doi.org/10.1007/978-3-032-13540-7

This Springer imprint is published by the registered company Springer Nature Switzerland AG
The registered company address is: Gewerbestrasse 11, 6330 Cham, Switzerland

Acknowledgements

A couple of years ago, at a conference, I shared my intent to write a book on Thomas Kuhn with a philosopher whom I have known for quite some time. On hearing of my plans, the philosopher asked: "Why would you want to do such a thing? A book on Kuhn would do nothing for your career! Besides, no one reads books anymore, articles are more read and more cited." Obviously that philosopher only wanted the best for me. Isn't it true that articles, instead of books, are read more and, in several instances, cited more? And shouldn't a researcher jumping from one fixed-term position to the other try to improve their chances to get a permanent position by, well, writing readable and citeable articles, instead of books (that apparently no one reads anymore)? The words of that philosopher made me think about many things. Indeed, *why* did I want to bring into the world yet another book on Kuhn? I came up with three reasons. First and foremost because, even though there are many books and articles about Thomas Kuhn, I believe that there are still several aspects of his philosophy that have not been explored in enough depth yet. If I manage, with this book, to give a small contribution to the study of one of the major philosophers of science of the twentieth century, then it would have been worth the effort. Second, to close a circle. I wrote my PhD thesis on the philosophy of Thomas Kuhn. I then expanded it in some of my published articles. As I steadily move on towards other research interests, I believe that a monograph could be a good way to take stock of my studies on Kuhn's philosophy. Last, but perhaps not least, because I want to. It is true: nowadays reading a philosophy book from cover to cover is rare. Articles are a far more valuable commodity in the increasingly competitive market of academia, especially for someone who, like me, still needs to secure a permanent position. Graduate students are sold the idea that academia, especially in the Humanities, is a place of unbounded freedom where scholars can explore exciting concepts and ideas. This is true to a point. Sometimes the reality may be more mundane and bleaker than that. Nevertheless, I am not ready to give up my intellectual freedom yet to chase a higher citation count. At the end of the day, I simply wanted to write this book. Therefore, I thank that philosopher of science for making me think about my choices, my desires, and, ultimately, my identity as a philosophy researcher.

Throughout the years, the study of the philosophy of Thomas Kuhn has put me in contact with a network of interesting philosophers who, in a way or another, have contributed to refine the ideas contained in this book. I especially wish to thank Alexander Bird, my PhD supervisor, for his continued support and encouragement. I also wish to thank Paul Hoyningen-Huene, Philip Kitcher, Hasok Chang, Vasso Kindi, Brad K. Wray, Yafeng Shan, and Jouni-Matti Kuukkanen: the present work has benefited from the invaluable exchanges with them, in ways that I cannot even begin to put into words.

I thank Chris Wilby for all his help and patience.

Very special thanks to Atocha Aliseda, for her support, encouragement, and friendship.

This book cannot have been written without the research period funded by Beatriu de Pinós/Marie Skłodowska–Curie Action Cofund programme (grant number: 2020-BP-00196), spent at the Institute for the History of Science at the Autonomous University of Barcelona (UAB). For their support and help, my thanks go to Xavier Roqué and all the colleagues and friends at the Institute. Special thanks also go to the unemployment benefits scheme of Catalunya (Spain), without which I could not have paid the rent, let alone concluded this book.

I thank Taylor & Francis and Routledge for granting me the permission to reproduce parts of my published articles in this book.

The paradox of untenured academics is that they often do not have a stable place to call home, even though they have lived in many places where they felt at home. I have visited many cities and nations, where I felt at home, because home is where friends are. To my philosophy and non-philosophy friends in Rome, London, Bristol, Mexico City, Paris, Oslo, and Barcelona: this book is dedicated to you all, who are my home.

Competing Interests The author has no competing interests to declare that are relevant to the content of this manuscript.

Contents

Part III Kuhnian Issues

About the Author

Vincenzo Politi earned a Phd in Philosophy at the University of Bristol, where he pursued a project about Thomas Kuhn's post- *Structure* philosophy. He has held research positions in Mexico City, Paris, Oslo. He has been a Beatriu de Pinós Fellow at the Universidad Autónoma de Barcelona. Currently, he is a Ramón y Cajal Fellow at the University of Seville, where he is currently working on issues about science, values, and society.

Abbreviations[1]

BB *The Black-Body Theory and the Quantum Discontinuity 1894-1912.* Oxford: Oxford University Press, 1978; second edition, with the afterword "Revisiting Planck": 1987

CR *The Copernican Revolution: Planetary Astronomy in the Development of Western Thought*, Cambridge: Harvard University Press, 1957

ET *The Essential Tension: Selected Studies in Scientific Tradition and Change.* Chicago: Chicago University Press, 1977

PW *The Plurality of Worlds: An evolutionary theory of scientific development*, in *The Last Writings of Thomas Kuhn*, 90–265, Chicago: Chicago University Press, 2022

QPT *The Quest for Physical Theory: problems in the methodology of scientific research - Thomas S. Kuhn's Lowell Lectures (1951)*, ed. George Reisch, Boston: The M.I.T. Libraries, Department of Distinctive Collections, 2021

RSS *The Road since Structure. Philosophical essays, 1970-1993, with an autobiographical interview*, ed. James Conant and John Haugeland, Chicago: Chicago University Press, 2000

SKHP "Scientific Knowledge as Historical Product", first published (in Japanese) in 1986; reprinted in *The Last Writings of Thomas Kuhn,* 1–22, Chicago: Chicago University Press, 2022

SML *The Shearman Memorial Lectures (23-25 November 1987)*, in *The Last Writings of Thomas Kuhn*, 20–89, Chicago: Chicago University Press, 2022

SSR The Structure of Scientific Revolutions, Chicago: Chicago University Press; first edition: 1962; second edition, with Postscript: 1970.

TL *The Thalheimer Lectures: Scientific Development and Lexical Change (12-19 November 1984)* published in Spanish (translated by Leandro Giri), ed. Pablo Melogno and Miguel Hernán, 2017.

[1] Abbreviations of Thomas Kuhn's Works

Chapter 1
Recovering Thomas Kuhn's Philosophy

Abstract In this introductory chapter, I argue for a revised interpretation on Thomas Kuhn's philosophy. I will begin with a general analysis of the contemporary debates around Kuhn, which reveals the existence of two different and non-communicating parties: the 'Kuhnians' and the 'post-Kuhnians'. Both parties share what I call a '*Structure*-centred' interpretation. Paying attention to Kuhn's extensive oeuvre, however, makes it possible to develop a '*Structure*-decentred' interpretation of his philosophy. Since it is virtually impossible to discuss Kuhn's philosophy without referring to his most famous book, I will then provide a concise overview of the central concepts of *The Structure of Scientific Revolutions*, which will be presupposed in the remainder of the book. Finally, I will present my interpretation, which will be elaborated upon in subsequent chapters.

Keywords Thomas Kuhn · *The Structure of Scientific Revolutions* · Post-Kuhnian philosophy of science · Scientific revolutions · Incommensurability

1.1 Introduction

The Structure of Scientific Revolutions (*SSR*) is one of the most cited books in philosophy of science and, also, one of the rare cases of philosophy best-sellers. His author, Thomas Kuhn, is credited with the creation of a historical approach to philosophy of science as well as for popularising expressions such as 'paradigm change', now widely used beyond philosophy. He is also blamed for defending, or at least inspiring, a rather demystifying conception of science as driven by propaganda, rhetoric, and social and political interests, rather than by genuinely epistemic factors. Through the years and decades, Kuhn's philosophy has been discussed, analysed, criticised and dissected. *SSR* being part of pretty much every syllabus of introductory courses of philosophy of science, it seems that everybody has a (more or less) authoritative opinion about Kuhn's philosophy and feels legitimated to express it. Since there already exists a vast if not immense literature about it, one

V. Politi, *Thomas Kuhn's Philosophical Perspective*, Boston Studies in the Philosophy and History of Science 348,
https://doi.org/10.1007/978-3-032-13540-7_1

could not help but wonder: what is the point of writing *yet another book* on Thomas Kuhn's philosophy? The question is legitimate. My answer is an indication of the path ahead.

Many philosophers have traditionally viewed Kuhn's contributions to philosophy as encapsulated in his famous history-based model of scientific progress presented in *SSR*. That book was published in 1962, with a second edition containing the extensive Postscript coming out in 1970. However, Kuhn begun his career *before* *SSR* and kept writing philosophy well *after* the publication of its second edition—see, for example, the writings contained in *The Essential Tension* (*ET*), published in 1977, and in *The Road since Structure* (*RSS*), published posthumously in 2000. Until 1996, the year of his premature death, he was working in his book *The Plurality of Worlds: An evolutionary theory of scientific development* (*PW*), which sadly remained unfinished. All these works contain Kuhn's responses to his critics, which led him to re-define and clarify some of the central elements of his history-based model of science. Focusing almost exclusively on the model presented in *SSR* means analysing a version that Kuhn himself considered in need of further refinement and clarification. Ironically, then, the '*Structure*-centred' approach to Kuhn ends up undermining itself: if one wants to really get the model originally discussed in *SSR*, one should not ready only *SSR*.

Things have been gradually changing. In the past couple of decades, several philosophers have finally begun to pay attention to Kuhn's post-*SSR* writings. Even so, it may be the case that the *Structure*-centred approach is not completely overcome yet. While many philosophers keep considering *SSR* as the defining framework of Kuhn's philosophy, those who analyse his later works frequently interpret them as a nuanced elaboration upon *SSR*, thereby ensuring its continued prominence in discussions of his thought.

This book proposes a '*Structure*-decentred' approach. I will not assume that the history-based model of scientific progress through revolution is Kuhn's most valuable contribution to philosophy of science. In my view, in fact, that place belongs to Kuhn's new *philosophical perspective* on science, which eventually led to the development the history-based model of science, and which may even clarify it.

In Sect. 1.2, I discuss how Kuhn's philosophy is debated by two separate and mostly non-communicating groups of philosophers—the 'Kuhnian' and the 'post-Kuhnians'—and how both share a *Structure*-centred approach. In Sect. 1.3, I explain how and why a '*Structure*-decentred' approach can be developed. Since the shadow casted by *SSR* looms large on every attempt to discuss Kuhn's philosophy, in Sect. 1.4 I provide a concise summary of some of its central concepts, which I will simply take for granted in the remainder of the book. Finally, in Sect. 1.5, I provide a concise overview of the rest of the book.

1.2 The Kuhnians and the Post-Kuhnians

According to Alexander Bird (2002), Thomas Kuhn's legacy presents a paradox. Bird asserts that while Kuhn is widely regarded as a highly influential philosopher of science due to the extraordinary impact of *SSR*, he appears not to have established a distinct philosophical tradition. On the one hand, even though some historians and sociologists of science regard themselves as carrying a sort of Kuhnian project, they do so in ways that Kuhn himself rejected. For instance, several Kuhn-inspired historical and sociological studies of science are developed from an externalist and constructionist perspective, whereas Kuhn clarified on more than one occasion that he regarded himself as an internalist and that he was even repulsed by the extreme versions of constructionism. On the other hand, when it comes to philosophers of science, there is no proper 'Kuhnian school of thought' in sight.

The situation may be even *more* paradoxical than that, however. It is true that, in some philosophical circles, some Kuhnian concepts, such as scientific revolutions and incommensurability, are perceived as rather old fashioned or simply irrelevant for contemporary debates. Yet, at the same time, the number of publications about Kuhn does not seem to decrease. One of the first book-length analyses of his philosophy is Hoyningen-Huene (1993), who interprets Kuhn as a neo-Kantian philosopher.[1] With the beginning of the new millennium, Kuhn has been enjoying renewed popularity among philosophers. This is due in part to the publication of Bird (2000), who provided a new interpretation of his philosophy. Other monographs followed, such as Andersen (2001), Sharrock and Read (2002), Marcum (2005), and Kuukkanen (2008). Andersen et al. (2006) develop a cognitivist reinterpretation of Kuhn's post-*SSR* philosophy, while Wray (2011) focuses on the evolutionary and social epistemological aspects of Kuhn's thought. More recently, Mladenović (2017) has proposed an interpretation of Kuhn through the lenses of the philosophical tradition of American Pragmatism. Furthermore, in the year 2012, which marked the 50th anniversary of the publication of *SSR*, The University of Chicago Press printed a fourth edition of the book, and many conferences were held around the world. As a result, many collective volumes about Kuhn appeared shortly—see, for example, Kindi and Arabatzis (2013), Devlin and Bokulich (2015), Richards and Daston (2016). Similarly, the year 2022 marked the 100th anniversary of Kuhn's birth, leading to, again, international conferences and new publications, such as Giri et al. (2023), Wray (2024), Shan (2024). Countless articles on Kuhn keeps appearing in major international philosophy journals. With the long-awaited publication of his last writings, including the unfinished manuscript of *PW*, the interest towards Kuhn can only increase. In short, it appears like the Kuhnian

[1] The original German edition of Hoyningen-Huene's book appeared in 1989 and it is mainly based on the material collected and examined by its author during the academic year 1984–85, when he visited the Department of Linguistics and Philosophy at M.I.T. The final draft of the book was completed in the year 1987–88 at the Centre for Philosophy of Science of the University of Pittsburgh. It is not too controversial to say, therefore, that Hoyningen-Huene's book, despite its enormous breadth, does not include an analysis of what Kuhn published in the 1990s.

scholarship is alive and well, despite Kuhn being a philosopher who has left no distinctive Kuhnian tradition in philosophy.

But *who* has this interest towards Kuhn philosophy? That the 'Kuhnian scholarship' is becoming a separate specialty in the philosophy of science may cause perplexities. Do the Kuhnians bring something new and valuable to contemporary philosophy of science, or do their work mainly aim at exegesis? Is the Kuhn reconstructed and interpreted by the Kuhnians a towering philosopher to be understood and contemplated in his magnificent isolation, or could he be put in conversation with contemporary philosophers? And the Kuhnians themselves, are they interested in engaging with the 'non-Kuhnians' (or with the 'a-Kuhnians', who neither defend nor dismiss Kuhn's view) or do they prefer commenting only on each other work?

Although I am giving an exaggerated and not entirely charitable reading of the 'contemporary Kuhnians', these questions appear even more legitimate when one considers yet another paradox with the contemporary reception of Kuhn's thought. While the Kuhnians keep on publishing their Kuhnian works, outside of their restricted (and, one may fear, increasingly isolated) network, the number of self-defined *post*-Kuhnians is rising. At the same time, however, it is not entirely clear what being a 'post-Kuhnian' means or implies exactly. For example, Michael Friedman (2001) proposes a post-Kuhnian historiography of science, in which paradigm shifts are made possible, and are rationally justified, by changes in the so-called 'relativised' a priori principles of science. Drawing from the history of chemistry, Eric Scerri (2016) defends a self-proclaimed post-Kuhnian model of science, which regards scientific progress as a process carried forward not only by 'big' revolutionary theoreticians, but 'smaller' scientists in traditional historical narratives. Ankeny and Leonelli (2019) develop a post-Kuhnian philosophy of science, that accounts for some social and institutional elements of contemporary research, thus overcoming Kuhn's (alleged) focus on dramatic and wide-ranging revolutionary changes. Moti Mizrahi (2015) invokes a decisive transformation in the Kuhnian image of science on the ground that Kuhn's incommensurability thesis, and especially its more mature 'taxonomic' version, runs into considerable difficulties.

I want to point out three things about the post-Kuhnians. First, it may be the case that many of them just misunderstand Kuhn. Friedman aims at neutralising some of the problems associated with incommensurability, such as incommunicability and the impossibility of a rational comparison between paradigms. However, in his post-*SSR* works, Kuhn made it clear that incommensurability does *not* imply incommunicability or incomparability. Furthermore, Friedman is focussed on the analysis of the central concepts of some theories in physics, thus overlooking Kuhn's account of the scientific community and of the practical aspects of research. Such an account is completely neglected by Scerri, who attributes to Kuhn the view for which scientific progress depends on the work of revolutionary individuals. Kuhn said exactly the opposite, by putting a major emphasis on the problem-solving activity of the scientific community; and even his scientific revolutions, although they may be initiated by individuals, are completed and resolved only at the community level. Ankeny and Leonelli seem to criticise a rather 'textbook' version of Kuhn, by considering paradigms as inflexible monoliths and scientific revolutions as big theoretical changes (again, this is not what Kuhn said). Mizrahi seems to consider taxonomic incommensurability as constitutive of a 'Kuhnian image' of science, which maybe he perceives as dominant and in need of being discarded. His argument is that, since taxonomic incommensurability faces difficulties, then the Kuhnian image of science must be overthrown. I find some of Mizrahi's argument against taxonomic incommensurability compelling and I too criticise Kuhn's replacement of paradigm with conceptual taxonomies (see Chap. 6). However, since there is much more to its philosophy than taxonomies, Kuhn's philosophy may survive the demise of taxonomic incommensurability.

Second, and quite ironically, it appears that post-Kuhnian philosophers live the very Kuhnian situation of talking at cross-purposes—assuming that they are talking to one another—because of the incommensurability surrounding the very expression 'post-Kuhnian', which appears to have a different meaning for each of them. The problems, however, begin with the meaning of 'Kuhnian'. Kuhn said many things and, therefore, he could be criticised for different reasons. The point is that it is not clear whether the criticism and rejection of one or some of the many elements of Kuhn's philosophy correspond to a rejection of the 'Kuhnian image of science'. Nor is it clear whether accepting some Kuhnian concepts is enough for being a Kuhnian. In short, while every post-Kuhnian philosopher is post-Kuhnian in their own way, it is not the case that all Kuhnian philosophers, like all happy families, are alike.

Third, one should wonder: *why* do some philosophers have decided to define themselves as 'post-Kuhnians'? When philosophers call themselves in such a way, they do so to differentiate themselves from a 'Kuhnian image' that they perceive as dominant. But if, as Bird (2002) says, Kuhn did not leave any distinctive 'Kuhnian school of thought' and if, indeed, there are more philosophers who criticise the model of *SSR* than philosophers who defend it, then it is not clear what the post-Kuhnians are reacting against. Perhaps, 'post-Kuhnian' should be interpreted in a chronological sense. 'Post-Kuhnian philosophers' are simply the 'philosophers who came after Kuhn'. However, by choosing Kuhn as a chronological reference, post-Kuhnians implicitly recognizes and restates his importance. They are not, for example, 'post-Popperians' or 'post-Feyerabendians'. They are defining their work in opposition to Kuhn's, which is by itself an admission of Kuhn's centrality and influence.

In short, Kuhn did not establish a distinctive Kuhnian school of thought, yet the Kuhnians keep publishing works on Kuhn, to the point that they are growing as a separate sub-specialty in philosophy of science, while the post-Kuhnians react and aim at overthrowing a dominant Kuhnian image of science that maybe they have misunderstood or that probably is not even that dominant to begin with.

In my view, what (most) Kuhnians and post-Kuhnians have in common is a *Structure*-centred approach. Philosophers like Friedman, Scerri, and Ankeny and Leonelli clearly react against some of the central concepts of *SSR*, such as that of paradigm or scientific revolutions. Mizhrai rejects Kuhn's mature and post-*SSR* version of the incommensurability thesis, which he developed to respond to some of the criticisms to the *SSR* version. In other words, for Mizhrai, the central and most important element of the 'Kuhnian image of science' is still something that Kuhn introduced in *SSR* and lately reformulated. Kuhnian scholars, in turn, often responds to the criticisms levelled against Kuhn's model by stressing once again (and not without frustration) that: scientific revolutions are not sudden and irrational shifts; incommensurability does not imply incommunicability or incomparability; Kuhn was not interested only in scientific theories but also (and especially) in the practical problem-solving activity carried out by the scientific community; paradigms are flexible entities that normal scientists develop, adjust, and apply to a growing number of problems; and so on and so forth. By doing so, however, they restate the centrality of the model presented in *SSR*, thus following the same *Structure*-centred approach of the post-Kuhnians.

These are, of course, coarse generalisations. But the aim of this section was not to make justice to every single philosopher who has written about Thomas Kuhn. Nor was it my intention to claim that, because they define themselves as post-Kuhnians, then there is no value in the works of Friedman, Ankeny and Leonelli, or others. (Indeed, there is a *great* value in the works of some of these philosophers, so much so that, in my view, they would have not needed the label of 'post-Kuhnian' to begin with.) My aim, rather, was to diagnose the problem with the *Structure*-centred approach, that this book aims at overcoming.

1.3 Towards a *Structure*-Decentred Approach

The *opportunity* to develop a *Structure*-decentred approach to the philosophy of Thomas Kuhn is given by how accessible many of his previously unpublished works have become in recent years.

In 2021, Kuhn's *Lowell Lectures*, originally delivered in 1951, were edited by George Reisch and published with the title *The Quest for Physical Theory* (*QPT*). The book contains the seminal version of the model of *SSR*. The argumentative structure of the two books is very similar. In both *QPT* and *SSR*, Kuhn begins by talking about what he aims at rejecting—'textbook science' and the dominant 'image of science' respectively. Then, he examines some historical case-studies that, in his view, indeed challenge the traditional image of science. He then concludes with some general remarks on the nature of scientific progress. In *QPT*, Kuhn has not developed yet his famous philosophical vocabulary: the *Lowell Lectures* makes no mention of paradigms or incommensurability. Yet, the central ideas of what eleven years later will become the history-based model of *SSR* are already present in *QPT*.

The publication of *QPT* is important for several reasons. First, if Kuhn had already presented the central ideas of his model of science back in 1951—which means that he had already worked on such ideas even before that—than his relationship with other philosophers must be reassessed. For example, some interpret Kuhn as being heavily influenced by Wittgenstein, whom he cited in *SSR*. This interpretation rests on the hypothesis that Kuhn read *Philosophical Investigations*, first published in 1953, sometimes in 1959 (see Cederbaum 1983; Wittgensteinian interpretations of Kuhn have also been developed by Kindi 1995, and Sharrock and Read 2002). This interpretation, however, seem to conflict with some textual evidence. In fact, the penultimate draft of *SSR*, also known as 'proto-*SSR*', which was probably finished early in 1961 and then sent around to colleagues, does not contain any mention of Wittgenstein. Hoyningen-Huene (2015), then, openly challenges the idea that Kuhn was familiar with Wittgenstein since 1959 and speculates that he decided to cite him in the final version of *SSR* based on some of the colleagues' feedback to proto-*SSR*. With the publication of *QPT*, Hoyningen-Huene's hypothesis receives further support. If Kuhn's model of science had already began to take a distinctive shape back in 1951, then a direct influence of Wittgenstein's philosophy can be excluded even if one accepts that Kuhn read *Philosophical Investigations* in 1959. One can even concede that Kuhn found some usefulness in Wittgenstein's philosophy for *refining* his concept of paradigm. This is however different from claiming that Wittgenstein had a profound impact for the *formulation* of the concept.

Other philosophers have drawn parallels between *SSR* and Polanyi's *Personal Knowledge* (1958). Based on archival research, Moleski (2007) claims that Polanyi himself suspected Kuhn of plagiarism. Jacobs (2007) goes as far as claiming that, in writing *SSR*, Kuhn borrowed many of Polanyi's central theses without acknowledging him. More cautiously, Timmins (2013) poses the question of why Kuhn's *SSR* superseded Polanyi's *Personal Knowledge*. It must be noticed that Polanyi's book was based on his *Gifford Lectures* delivered at the University of Aberdeen (UK) in 1951–1952 as well as in several articles published between 1951 and 1958. Of course, we know that Kuhn read *Personal Knowledge* while he was writing *SSR*, and he even cites it (see *SSR*:44, ft.1). However, *QPT* demonstrates how he had been working on many of his views from at least 1951, years before the publication of Polanyi's book. While Timmins' question, which is about the reception and legacy of the two philosophy books, maintains its legitimacy, in the light of the publication of *QPT*, Kuhn's accusations of plagiarism or intellectual theft against Polanyi should simply fall away.

On the one hand, then, *QPT* dispels many doubts about the paternity of Kuhn's ideas, which sometimes become real accusations of intellectual dishonesty. On the other hand, it enables a consideration of Kuhn's philosophy on its own merits. Sometimes, Kuhn is portrayed as a physicist and a historian of science who entered the philosophical arena with the same enthusiasm and provincial naivete of someone who has just discovered a new field and believes that their ideas are going to be genuinely groundbreaking and original. So, for example, Bird (2000, 2002) maintains that Kuhn was simply too ignorant about philosophy to realize that many of his central theses had already been developed within the same logical-empiricist tradition he aimed at rejecting. It is true that Kuhn was a physicist and that he began his academic career as a history of science lecturer. It is also true, however, that after writing his PhD thesis in physics and while working as Assistant Professor of General Education and History of Science at Harvard University, Kuhn was *also* part of the Harvard Society Fellow, from 1948 to 1951. As the historian Joel Isaac

(2012) explains, the Society Fellow was an 'interstitial academy', with an interdepartmental and quasi-institutionalized status, that encouraged interdisciplinary interactions and convergence between the sciences and the humanities. Natural scientists, economists, sociologists, psychologists and philosophers found in the Society a *milieu* of mutual exchange and reciprocal enrichment. For Isaac, the Harvard Society Fellow allowed the emergence of American *scientific philosophy*, which looks at scientific practice and attempts to interpret it using concepts and methods from the social sciences and the humanities. In his view, this milieu allowed the emergence of key figures such as W.V.O. Quine and Thomas Kuhn. Indeed, it was at the Harvard Society Fellow that Kuhn met Quine, who back then was a senior fellow and who made analytic philosophy circulate at Harvard. It is very likely that Kuhn read a draft of Quine's 'Two dogma of Empiricism" (1952) at the Society and, also, that he encountered the philosophy of logical empiricism through the filter of Quine's criticisms. Available evidence indicates that during his time at Harvard, particularly as a junior fellow at the Harvard Society Fellow, Kuhn also encountered Percy W. Bridgman's operationalism, the works of A.J. Ayer, and a range of philosophical and scientific literature. It was during this time that he began to build his intellectual identity based on a unique blend of philosophy, sociology, psychology and history (see Mayoral 2017; Basafa 2025). Kuhn's unique approach to philosophy of science is already on full display in the *Lowell Lectures*.

Finally, even though, in 1951, many of them were still embryonal, some of Kuhn's ideas are *better* expressed in *QPT* than in *SSR*. For example, *QPT* makes it clearer what is the 'image of science' he was aiming to discard. And, contrary to the claim that he did not know what he was talking about when he was talking about logical empiricism, in Lecture VII of *QPT* Kuhn develops an argument that explains how his own philosophical outlook is not only different from, but also incompatible with that of logical empiricists (more on this in Chap. 2).

In 2023, a year after the publication of *QPT*, Kuhn's unfinished *PW* was finally edited by Bojana Mladenović and published together with some of his other last writings (see Kuhn 2022). *PW* structure is different from that of *QPT* and *SSR*. Instead of introducing the image of science he wants to challenge and then proceeding to do so via historical case studies, Kuhn begins by juxtaposing the image of science he rejects and the new image of science he proposes in its place. In other words, *PW* clarifies from the beginning what Kuhn's positive proposal amounts to: not a specific history-based model of science, but a new philosophical perspective on science. Philosophers who defend different history-based models of science may still look at science from the same perspective. Indeed, at the beginning of *PW*, Kuhn claims that philosophers like Feyerabend and Lakatos—who certainly do not agree with him about normal science, the necessity of paradigms, or revolutions—share with him the new philosophical perspective. Post-empiricist philosophers decided to look at the history of science thanks to such a perspective. The study of the history of science, however, poses several problems. For instance, historically oriented philosophers of science may stumble on passages in old scientific books that make little sense to the contemporary reader. They will conclude, then, that past scientific belief systems are radically different from the present one, and that such a difference is expressed, among other things, by their attribution of what appears as non-customary meanings to some scientific terms. This issue requires a philosophical theory of how meaning changes, which in turn requires a general theory of meaning and of meaning acquisition, that Kuhn begins to develop based on experiments in developmental psychology. This is, in a nutshell, the argument developed in the published parts of *PW*.

As I said earlier, the publication of *QPT* and *PW* provides me with an opportunity for writing about Kuhn's philosophy from the vantage point of more available primary sources to discuss. My *philosophical reason* for doing so, however, is the need to address, make explicit, and analyze Kuhn's philosophical perspective on science. As Kuhn hints at the beginning of *PW*, such a perspective plays a fundamental role in his whole philosophy. Without it, he would not have looked at the history of science and, therefore, he would not have developed his history-based model of scientific progress. Moreover, I contend that Kuhn's philosophical perspective clarifies and sheds a new light on the model he presented in *QPT*, *SSR*, and other writings.

By looking at his infamous history-based model first only to consider some of its philosophical implications later, the *Structure*-centered approach simply puts the cart before the horses. Kuhn himself may have contributed to this way of reading his work: as already said, in both *QPT* and *SSR* he mainly focusses on discussing how some historical case studies may challenge the traditional view on science and then, only towards the end, he engages briefly with some general (or even generic) philosophical arguments about progress, truth, and objectivity. At the same time, he lamented that what he regarded as his main contribution to philosophy of science, namely his new philosophical perspective, had remained underdiscussed if not unnoticed.

The times are now rip for a re-evaluation of Kuhn's philosophy through a *Structure*-decentered approach.

1.4 The Central Concepts of *Structure*: A Quick Outline

As it has become clear by now, this book is about the philosophy of Thomas Kuhn, *not* about *SSR*. At the same time, avoiding *any* reference to his most famous book could be an impossible task, especially considering that, in my view, the Kuhn's philosophical perspective clarifies some aspects of the model of science presented in *SSR*. Rather than devoting a whole chapter (or a whole book) to *SSR*, this section briefly outlines some of its central concepts. The rest of the book will assume familiarity with them.

1.4.1 The Model

In *SSR*, Kuhn distinguishes two phases of scientific progress. During periods of so-called *normal science*, scientists work on the resolutions of problems in the light of a dominant and unchallenged *paradigm*, which pre-selects scientific problems, tells scientists how to solve them and how to evaluate the proposed solutions. During these periods, the progress of science is cumulative. By articulating the dominant paradigm and by trying to solve as many problems as possible, the scientific community may stumble across some unexplained phenomena, or even recalcitrant problems. To solve these *anomalies*, scientists keep on articulating the paradigm, making it increasingly sophisticated, until a growing sense of dissatisfaction spreads within the scientific community. This is when normal research enters a period of *crisis*, during which scientists begin to question the validity of the paradigm. When the scientific community begins to look for a new paradigm it enters a period of *extraordinary science*, which ignites a *scientific revolution*, at the end of which the old paradigm is discarded, a new paradigm is accepted by the scientific community and, consequently, a new period of normal science begins. Science makes progress not only during normal periods, but through revolutions too: it is through a revolution that scientists find a new paradigm capable of solving recalcitrant problems. The resolution of recalcitrant problems is indeed a mark of progress, yet the type of progress brought about by a revolution is not cumulative. A scientific revolution represents a rupture with the dominant research tradition and the pre- and post-revolution paradigm are *incommensurable*.

This is, in a nutshell, the model of science contained in *SSR*. Such a model attracts two orders of criticism. To begin with, many scholars question its *descriptive* validity. History of science does not neatly display long periods of cumulative progress punctuated by sudden changes. It may also turn out that what Kuhn indicates as periods of normal science were not characterised by a unanimous consensus upon a paradigm but, rather, by a considerable degree of pluralism. Based on these and similar criticisms, the model of *SSR* may not be the best tool to analyse and explain several episodes in the history of science. On the other hand, philosophers criticised the *normative* side of *SSR*. Kuhn's model describes how mature sciences developed through history but, in a sense, it also tells how science in general ought to behave. For any science to make progress, it ought first to establish a paradigm and enter a phase of normal science. The cycle that goes from normal science to revolution to a new period of normal science will then begin. Many philosophers, however, were repulsed by the idea of dull and mindless normal scientists engaged in the resolution of esoteric puzzles, as well as by the idea of incommensurability that, in their view, threatened the rationality of science.

The challenges to the descriptive elements of *SSR* ought to be addressed at the descriptive level, that is through the study of the history of science (but without forgetting that different histories of science are produced by different historiographies of science and that, therefore, 'pure historical evidence' capable to arbitrate between conflicting historiographies may often be unavailable). Nevertheless, it is worth reminding that the concepts of normal science, paradigm, or revolution are *abstractions* Kuhn uses to provide a general explanation of how the mature sciences have behaved throughout their history. As such, they do not need to be always perfectly instantiated in the history of science (more on this in Sect. 3.3).

In the remainder of this section, I will briefly address some of philosophers' criticisms of the normative side of Kuhn's model. Even though such criticisms often stem from a superficial reading of *SSR*, or from blatant misinterpretations, many philosophers find some validity in them and end up restating views that Kuhn himself first and several Kuhnian scholars later have debunked time and time again.

1.4.2 Scientific Revolutions

Kuhn did not invent the expression 'scientific revolutions'. The Chemical Revolution, for example, was called so long before he used it as an example in *SSR*.[2] It must be specified, however, that Kuhn did not use such an expression in a colloquial or generic sense. Rather, he employed it within a model of scientific progress and to describe the process that leads to the abandonment of an old paradigm in favour of a new one. Kuhn explains that, after a revolution, some of the successes of the old paradigm are retained and reinterpreted in the light of the new paradigm, while some of the old problems are dismissed as uninteresting or even as unscientific. Once the new paradigm is adopted, scientific textbooks are rewritten and the old mode of conducting science goes into oblivion.

One of the most persistent misunderstandings of Kuhn's view is the idea that scientific revolutions are 'sudden and abrupt breaks' with a period of normal science. Critics of the concept, such as Laudan (1984), observe that scientific research is characterised by gradual and continuous changes, rather than by dramatic ruptures. A scientific revolution, however, is a process that unfolds through time and, as Kuhn explicitly says, it may take one or more generations of scientists to be concluded. In *The Copernican Revolution* (*CR*), Kuhn reconstructs the shift from the geocentric to the heliocentric cosmology as a four-century long process. In *Black-Body Theory and the Quantum Discontinuity 1894–1912* (*BB*), he analyses how Max Plank's revolutionary ideas were fully understood as such only by the posterior generation of scientists. Wray (2011) rightly observes that what characterises a revolution is "the thoroughness of the changes, rather than the brevity of the event" (Wray 2011:42). To explain this point, he returns to the parallel between political and scientific revolutions. Although there can be 'revolutionary acts', a political revolution needs time to be accomplished: the French Revolution may have begun with the storming of the Bastille, but it was nevertheless a complex process that required a considerable amount of time to be completed (and that, for some historians, it has not fully come to completion yet). Similarly, in science there can be

[2] For a historical analysis of the concept of scientific revolutions, see Shan (2023).

relevant discoveries or new theories that are usually regarded as marking the beginning of a revolution, which however is not accomplished overnight.[3]

Toulmin (1970) advances a more pernicious criticism. If normal science, engaged in the articulation of the paradigm, is a period in which there are continuous and incremental changes, and if scientific revolutions are also processes extended through time in which there can often be continuous and incremental changes, how is it possible to distinguish between the two?

To answer this question, it is important to consider how what may look like small continuous changes may have far reaching implications. As an example, it is worth to briefly look at the history of the formulation of the so-called Planck Law, which Kuhn recounted in *BB*. The Planck Law, or Planck relation, states that the energy of a photon is proportional to its frequency, and it is considered as one of the fundamental laws in quantum mechanics. Contrary to the traditional account, which describes Planck as a genius who single-handedly laid the conceptual foundations of quantum mechanics, *BB* tells a much more complicated (and perhaps more fascinating) history.

Following Kuhn's reconstruction, in its earlier formulation, dating back to 1901, the equation expressing the Planck Law was written as

$$E = hf$$

where E is the energy of an oscillator, h is the Planck constant (which Planck postulated in 1900), and f is the frequency of a radiation.

Following Kuhn, Planck reformulated his law between 1906 and 1914, and the early equation became

$$E = nhf$$

where n is a variable which can take *only* the value of integer numbers (i.e., 1, 2, 3, etc.). This indicates that the total energy values of an oscillator are not continuous; instead, they occur in discrete quantities known as quanta.

What shall we make of the story of the two formulations of Planck Law? Surely, the introduction of n in the equation could count as a 'small change', motivated by

---

[3] The parallel between scientific revolutions as changes of paradigm and political revolutions as changes of political regime makes sense intuitively. Interestingly, one of the sources of disagreement between Kuhn and Feyerabend was precisely on the political metaphor for scientific revolutions. On commenting on an early draft of *SSR*, Feyerabend wrote to Kuhn: "Remember my reservations concerning your comparing political revolutions with scientific revolutions. The most fundamental revolution, to me, in the domain of knowledge, would be the transition from a stage of dogmatism to a stage where replacement of any paradigm is possible (although, as you have pointed out, one will be very careful not to give up the old paradigm too quickly and too lightheartedly). Now it seems to me that political revolutions are more closely related to this fundamental revolution than to changes of paradigms about nature. That is they are more closely related to attitudes concerning any possible paradigm (methodological attitudes, as I would call them) than to the change of attitude with respect to one particular paradigm" (Feyerabend to Kuhn, in Hoyningen-Huene 1994:373).

the need of adjusting the law to the available data. The same Planck interpreted it in this way, and he did not think that he had overcome the paradigm of classical mechanics. It was only with the following generation of scientists, which included Einstein and Lorentz, that the revolutionary significance of Planck's 'small' change was fully appreciated.

Therefore, a scientific revolution may be initiated by changes that appear like the continuous and cumulative changes going on during normal science. However, some small changes may have an 'avalanche effect' the revolutionary effects of which may require some time to be noticed. Often, the extent with which relatively small changes reverberate throughout the whole conceptual system, to the point of overthrowing some of its central assumptions, may elude the scientists that operated them and can be fully evaluated only with hindsight.

1.4.3 Incommensurability

In its mathematical form, incommensurability is a relation between two magnitudes lacking a common measure for their comparison—e.g., the radius and the circumference of a circle. Kuhn uses the term metaphorically to indicate the lack of a neutral and absolute ground for inter-paradigm comparison. But what does he exactly mean?

Even though 'incommensurability' is one of the less frequently used terms in *SSR*, it is employed with several different meanings. Excluding the discussion found in the Postscript of the second edition, the terms 'incommensurability' and 'incommensurable' appear eight times throughout *SSR*, each instance reflecting a slightly distinct interpretation. The term 'incommensurability' appears four times, with regards to: normal scientific traditions standing at the opposite sides of a scientific revolution (*SSR*:148); standards (*SSR*:149); competing paradigms (*SSR*:150); and paradigms (*SSR*:157). The adjective 'incommensurable', used three times, qualifies: ways of seeing the world and practising in it (*SSR*:4); pre- and post-revolutionary scientific traditions (*SSR*:103); the worlds (or parts of the worlds) which scientists inhabit (*SSR*:112). Finally, the noun 'incommensurable' describes the transition between paradigms, which is indeed a transition between incommensurables (*SSR*:150). To sum up, in *SSR* 'incommensurability' and 'incommensurable' are used to describe the relation between scientific traditions, standards, paradigms, competing paradigms, ways of seeing the world, ways of practising in the world, worlds, sections of the world.

In the light of its many meanings, some philosophers have distinguished different incommensurability theses (or different variations of the incommensurability thesis). Sankey and Hoyningen-Huene (2001), for example, distinguish between *semantic incommensurability*, involving the lack of a common language for inter-theoretic comparison, and *methodological incommensurability*, involving differences in interpreting and applying evaluative standards of scientific research (Hoyningen-Huene and Sankey 2001:vii–xxxiv).

The semantic version of incommensurability attracted considerable attention. As Kuhn explained, scientists supporting competing paradigms speak incommensurable scientific languages. They may even attach different meanings to the same terms, ending up talking at cross-purposes. Similarly, the historically oriented philosopher who reads old science textbooks, will experience incommensurability when encountering apparently senseless passages. Some have argued that the incommensurability experienced by scientists is different from that experienced by historians and philosophers (Demir 2008). In any case, in so far as it is a thesis about meaning variation across paradigms, the semantic incommensurability thesis states that the mutual translatability between incommensurable scientific languages is impossible.

Critics argued that such a thesis is self-defeating. Kuhn, it was said, claims that the scientific theories of the past are expressed in a language which is 'incommensurable' with respect to the language of our current theories; yet he himself does exactly what his notion of incommensurability seems to forbid when, as a historian, he renders past scientific theories into the contemporary language that we can all understand (Shapere 1966; Scheffler 1967). Furthermore, Putnam (1975) developed a theory of reference stability in the face of meaning changes, which undermines the very idea of incommensurability.

Kuhn (1983) replies to his critics by maintaining that incommensurability does *not* imply incomparability: in fact, it only involves a relatively small set of interdefined terms, grouped within the same 'cluster' which, if altered, may lead to the phenomenon of semantic incommensurability. If scientists use the same terms—e.g., 'mass' or 'force'—but in different clusters, they will experience the communication breakdown symptomatic of incommensurability. At the same time, however, considerable portions of scientific language may remain unaltered even after episodes of revolutionary change, not to mention those very large chunks of common everyday language that scientists rely on during their day-by-day communication.

Because it involves only a small set of interrelated terms, therefore, Kuhn's mature version is best described as *local incommensurability,* and it does not impede inter-theoretical comparison.

> Insofar as incommensurability was a claim about language, about meaning change, its local form is my original version. If it can be consistently maintained, then the [...] criticism directed at incommensurability must fail. The terms that preserve their meanings across a theory change provide a sufficient basis for the discussion of differences and for comparisons relevant to theory choice. (Kuhn 1983, in *RSS*:36)

Furthermore, (local) semantic incommensurability does not impede cross-paradigm communication. Scientists from incommensurable research traditions can become 'bilingual' by learning each other's scientific language. Yet, such a process does not rely on the possibility of a *translation*, that is, of the one-to-one substitution of the words of the theoretical vocabulary of a scientific language into the theoretical vocabulary of the other. Instead, for Kuhn, scientists acquire a 'foreign' scientific language through *interpretation*, a process through which they learn a 'constellation of meanings', that is the way the other language groups its words in different clusters (more on this on Chap. 6).

Incommensurability, however, is not only semantic. The long debates about Kuhn's theory of meaning notwithstanding, it looks like the interest for the methodological aspects of incommensurability is regaining momentum among philosophers. Some argue, for instance, that some episodes of revolutionary scientific change can be better understood as characterised by methodological rather than by semantic incommensurability (Chang 2012). Okasha (2011), who has re-read Kuhn's view on methodological incommensurability through the lenses of Social Choice Theory, has also generated considerable debate (see Bradley 2017; Morreau 2015; Okasha 2015; Stegenga 2015).

The cognitive/perceptual aspects of the early formulation of the incommensurability thesis have been expanded upon by Bird (2005, 2007). Kuhn, it should be remembered, was inspired by the experiments on perception and recognition conducted by Bruner and Postman (1949). From these empirical studies, he conjectured that scientists from different paradigms see the world differently. For example, when looking at a pendulum, an Aristotelian sees a grave striving to get back to its natural place, whereas a Newtonian sees a problem to be solved by applying the laws of motion of classical mechanics. In this sense, incommensurability is a thesis about how scientists belonging to incommensurable paradigms see the world (of scientific research) in irreconcilable ways.

The literature on the incommensurability thesis is vast. While I will return to some aspects of in some of the following chapter, throughout the book, following Kuhn's own clarifications on the matter, I will presuppose that incommensurability does not imply incomparability and incommunicability.

1.4.4 Reality

SSR makes occasional remarks about how scientists from the opposite sides of a revolution live in 'different worlds'. Many commentators have found the idea that a change of paradigm corresponds to a world change rather puzzling, if not discomforting.

It does not help that, in other works, Kuhn claims that

> the world is not invented or constructed. The creatures to whom this responsibility is imputed, in fact, find the world already in place, its rudiments at their birth [...]. That world, furthermore, has been experientially given, in part to the new inhabitants directly, and in part indirectly, by inheritance, embodying the experience of their forebears. As such, it is entirely solid: not in the least respectful of an observer's wishes and desires; quite capable of providing decisive evidence against invented hypotheses which fail to match its behaviour. Creatures born into it must take it as they find it. (Kuhn 1991, in *RSS*:101)

It appears that, for Kuhn, the world is not invented, and it is solid, yet it can change following the adoption of a new paradigm. These seemingly contradictory statements could be explained by the Kantian interpretation developed by Hoyningen-Huene (1993). From such Kantian perspective, Kuhn can be read as saying that what is neither invented nor constructed is the impenetrable and unknowable *world-in-itself*, while what changes after a revolution is scientists' *phenomenological world*. Kuhn's proposal would amount to a sort of 'collective transcendentalism', rather than an individualistic or even solipsistic idealism. In other words, the phenomenal world is constituted and shared by the members of a community, not constructed by a solitary epistemic subject. After all, "the world [in this interpretation, the phenomenological world] is *our* representation of *our* niche" (Kuhn 1991, *RSS*:103, *my emphases*).[4]

Other remarks cannot be accommodated within a Kantian framework so easily. For example, while expressing his distaste for the most extreme derives of the Strong Program in the sociological studies of science, Kuhn (1992) claims a role of 'nature' in a process of *negotiation*.

> These [sociological] studies have dealt, in microscopic detail, with the processes within a scientific community or group from which an authoritative consensus finally emerges, a process this literature often refers to as *negotiation*. […] What negotiation, so-called, seeks to establish is the facts from which scientific conclusions should be drawn, together with the conclusions—the new laws or theories—which should be based upon them. These two aspects of the negotiation—the factual and the interpretive—are carried on concurrently, the conclusions shaping the description of facts just as the facts shape the conclusions drawn from them. Such a process is clearly circular, and it becomes very difficult to see what role experiment could have in determining its outcome. […] The question therefore arose, how can a process so nearly circular and so largely dependent on individual contingencies be said to result in either true or probable conclusions about the nature of reality? […] [The] form taken by studies of 'negotiation' has […] made it hard to see what else may play a role as well. […] These newer formulations freely acknowledge that observations of nature do play a role in scientific development. But they remain almost totally uninformative about that role—*about the way, that is, in which nature enters the negotiations that produce beliefs about it.* (Kuhn 1992, in *RSS*:109–110, *my emphases*)

It is not entirely obvious whether 'nature' should be identified with the 'phenomenal world' or with the 'world in itself'. But Kuhn is not entirely clear even regarding what he means by 'negotiation'. Sometimes, he says that "the term 'negotiation' seems to me just right, except that, when I say 'letting nature in', it's clear that that's an aspect of it to which the term 'negotiation' applies only *metaphorically*, whereas it's fairly literal in the other cases. *But you are not talking about anything worth calling science if you leave out the role of [nature].* (Kuhn 1995, in *RSS*:317, *my emphases*). Some would perhaps say that the metaphorical character of negotiation

[4] Kuhn's last remark seems to paraphrase "the world is my representation", the opening sentence of Arthur Schopenhauer's *The World as Will and Representation*, first published in German in 1818. Of course, such a sentence has a remarkable Kantian flavour, Schopenhauer being one of the most devoted and knowledgeable commentators of Kant—so much, in fact, that his whole philosophical system is based on the critical re-development of Kant's original insights. It would be too much to expect the 'Darwinian neo-Kantian' Kuhn was aware of Kant's scholarship in late German Romanticism, unless the summer course in History of Philosophy he took as an undergraduate was much more thorough than he describes it in Kuhn (1995). However, it must be noticed that "the world is my representation" is also the closing sentence of Erwin Schrödinger's *What is Life?*, published for the first time in 1944. It is more plausible to think that, perhaps, Kuhn paraphrases Schrödinger, who in turn had read, and perhaps had been somehow influenced by, Schopenhauer's philosophy.

is explainable by the obvious fact that 'nature', not being a moral or rational agent, cannot really 'negotiate' anything with anyone. The 'negotiation with nature' is therefore metaphorical in the sense that such an expression gives a sort of anthropomorphic description of nature, describing it as being able to interact with us, to communicate meanings, to have intentions and desires. But then, what would be the cases in which the meaning of the negotiation with nature would be 'fairly literal'?

The Kantian reading of Kuhn, and the Kantian metaphysics in general, does not address a further, difficult question. If we cannot know anything about the world-in-itself, but only about the phenomenal world, how can we know that the world-in-itself exists? The problem, in other words, is the following: we cannot know any attribute of the world-in-itself; existence is an attribute; therefore, we cannot attribute existence to the world-in-itself. One could legitimately conclude, therefore, that the existence of the world-in-itself is a rather unjustified presupposition we can, or we should, get rid of. In this way, we would be left with a world of phenomena with no noumenal foundation, which would make the Kantian metaphysics indistinguishable from fully fledged idealism.

It has been suggested that Kuhn arrived at the notion of world-in-itself through a process of 'subtraction'; that is, by trying to subtract all the subject-sided elements from the perception of the phenomenal world—which led him to conclude that it is impossible to have an absolute object-sided access to the substratum of the perceived phenomena. However, the Kantian interpreter of Kuhn concedes that "the theoretically efficacy of this postulated world is, at least within the bounds of a theory like Kuhn's, highly *questionable*" (Hoyningen-Huene 1993:268, *original emphasis*). It becomes even more questionable once it is considered that, sometimes, Kuhn seems not to postulate any world-in-itself whatsoever, such as when he says that "[the] view which I grope would also be Kantian, but without 'things in themselves' and with categories of the mind which could change through time" (Kuhn 1979, in *RSS*:207).

There is, however, another possible interpretation. Joseph Rouse (2003a, b, 2013, 2024) interprets Thomas Kuhn as a *philosopher of scientific practice*. In his view, Kuhn's greatest contribution to the philosophy of science is his focus on the practical problem-solving and paradigm-articulating activity carried out by the scientific community, rather than in the logical reconstruction of scientific theories. This represented a shift from a 'theory-centred' to a 'practice-centred' approach in philosophy of science and, under this respect, Kuhn was ahead of his times. So much so that, however, most of his commentators and critics failed to appreciate, or even to notice, such a shift and continued to interpret Kuhn from their theory-centred perspective. For example, they kept reducing paradigms to theories, or to talk about scientific revolutions as changes of scientific theories.

What does this have to do with the world change problem? In both his pre- and post-*SSR* writings, Kuhn is clearer about, on the one hand, his identity as a philosopher of scientific practice and, on the other hand, on the kind of change he meant in *SSR*. As already hinted in some of the quotations above, in some of his mature writings, Kuhn does not speak about 'the world'. Rather, he speaks about scientists practicing within their 'niche'. The animals' activity within their niche may modify the niche itself, which in turn will require the species to adjust to such environmental changes. Similarly, scientists work within their field may lead to unexpected discoveries or anomalies that may reshape the field itself, in turn forcing the scientific community to re-adapt. In *QPT*, Kuhn speaks about how scientists inhabit a 'behavioural world'. A transition between different 'behavioural worlds', then, would involve a change in the actions, responses, and patterns of behaviours at the basis of the scientific activity. That niches or behavioural worlds change does not imply that the world itself changes at the whims of scientists adopting a new paradigm.

Although interesting, these subtle metaphysical and ontological issues will not be further discussed in this book, since I will presuppose that Kuhn's talk of world changes (or changes of niches, or of behavioural worlds) does not pose an obstacle for understanding all the other aspects of his work.[5]

To sum up, in the rest of the book I will presuppose that: scientific revolutions are not sudden ruptures, but *processes* that may take a long time to be concluded; incommensurability does *not* imply incommunicability or incomparability; semantic incommensurability is *not* the only kind of incommensurability; the phenomenon of 'world changes' is neither mysterious nor far-fetched, but requires a deeper understanding of Kuhn's philosophy of scientific practice as well as of his metaphysics (the latter being a project that I do not pursue here).

1.5 Moving Forward

Writing about the philosophy of Thomas Kuhn requires not only a deep knowledge of the works *of* Kuhn, but also acquaintance and engagement with the works *about* Kuhn. Such an engagement is not risk free. If I discuss in some rather common interpretations of Kuhn in very broad terms, I could be accused of making bold and unwarranted generalisations. If, to justify some of my general claims, I refer to some philosophers leaving out others, then I could be accused of cherry-picking. These are risks that are worth taking. The reader who wants to challenge me by showing the inadequacy of my hasty generalisations and the misleadingness of my cherry-picking can certainly do but, I hope, not *before* considering the central elements of my interpretation.

By no means am I the first to suggest moving beyond *SSR* to give a more complete and articulated assessment of Kuhn's views. Many of the things I will say in this book have already been anticipated by other and far more talented philosophers. For example, I agree with Joseph Rouse's interpretation of Kuhn as a *philosopher of*

[5] In recent years, Hoyningen-Huene has returned to the issue of Kuhn's metaphysics (see Hoyningen-Huene 2021, 2022, 2023). Others, like Massimi (2015), have dispelled the worry that Kuhn's position corresponds to a form of constructionism and to the negation of the existence of a mind-independent world. These works contribute to make Kuhn's doctrine of world changes less obscure.

scientific practice, mentioned in the previous section. Like Wray (2011), I take the *collective dimension* of Kuhn's theory of scientific rationality very seriously. I also sympathise with Kuukkanen (2007), who interprets Kuhn as a *coherentist*. For reasons of space, often it will not be possible for me to examine in more depth the views of the philosophers with whom I disagree and, sadly enough, of those with whom I agree. At the same time, I do think that my interpretation possesses a 'unifying' quality. By explicitly addressing his philosophical perspective, I will also explain how the interpretations of Kuhn as a philosopher of scientific practice, as a philosopher interested in collective rationality, and as a coherentist epistemology can be harmonised within one interpretive framework.

In developing my interpretation on Kuhn's philosophy, I follow Bertrand Russell's maxim: "In studying a philosopher, the right attitude is neither reverence nor contempt, but first a kind of hypothetical sympathy, until it is possible to know what it feels like to believe in his theories" (Russell 1945:39). Kuhn himself had this attitude when he studied past scientific texts. But he also claimed that he was somehow able to "get inside the heads of people who wrote them" (Kuhn 1995, in *RSS*:276). I will not try to get inside the head of Thomas Kuhn. I am not even interested in doing such a thing. My own approach is not biographical or psychological, but hermeneutical. In my view, the meaning of Kuhn's writings is given by his whole oeuvre, and the meaning of his whole oeuvre is constituted by his writings. This is the 'hermeneutic circle' which I attempt to break in. The circle, however, involves only written texts, not what was inside the head of his author. In line with hermeneutical tradition, I think that sometimes texts may say something more, or perhaps even something different, then what their authors had originally in mind. Therefore, whenever in this book I use expressions such as "Kuhn thought that", "Kuhn believed", "Kuhn wanted", and so on, I am not talking about Kuhn's actual thoughts, beliefs, wishes, and desires, which of course I cannot know. The Kuhn of this book is the 'philosophical Kuhn', not Kuhn as a person.

Even though this book is not an intellectual biography of Kuhn, I am not against such a kind of project in principle. As a matter of fact, I have enormously benefited from reading several works on Kuhn's interactions with other colleagues and personalities of the academic world, or on the preparatory works that paved the path towards *SSR* (Pinto de Oliveira 2007, 2017; Mayoral 2017; Reisch 2019; Wray 2021). Sometimes, the examination of archival material even contributed to change my mind on some issues. For example, in a previous article of mine (Politi 2020) I claimed that up until his "Commensurability, comparability, communicability" (1983) Kuhn only spoke about networks of inter-related kind terms, whereas the concept of a conceptual taxonomy, which possess a more rigid hierarchical structure, emerged only later in "What are scientific revolutions?" (1987). Melogno (2023) and Melogno and Giri (2023) have proved me wrong, since they demonstrate that Kuhn began to use the concept of a taxonomy *at least* from 1980. The *logical* point however remains: a network is different from a taxonomy, and perhaps Kuhn did not need to adopt the latter instead of the previous (see Chap. 6). Indeed, these works cement the idea that Kuhn's views evolved in a rather continuous way, instead of undergoing sudden twists and turns. In this book, however, I simply follow a different approach, which I take to be complementary and not opposed to the more exegetical approaches.

The book is articulated in three parts.

The three chapters of the first part are about Kuhn's *philosophical perspective*. This is the perspective that led Kuhn to look at the history of science, which in turn allowed him to develop the famous model of *SSR*. In other words, Kuhn's philosophical perspective is prior to his history-based model of science and, therefore, it must be discussed and analysed explicitly. Since he developed his new perspective in opposition to a dominant image of science, in Chap. 2 I explain what the target of Kuhn's criticisms was. Contrary to a rather traditional interpretation, I will explain that Kuhn's aim was not to confront logical empiricism directly. Rather, he aimed at discarding what he called the *static image* of science, of which logical empiricism provides only one possible version. In Chap. 3, I make explicit Kuhn's new philosophical perspective, which he calls the *dynamic perspective*. Kuhn developed such a perspective by rejecting one of the fundamental assumptions of the static image and then considering the philosophical implications of such a rejection. In the same chapter, I clarify the relation between Kuhn's dynamic perspective, his works in the history of science, and his history-based model of science. Chapter 4 focuses on one implication of the dynamic perspective, namely the need to take a *collective approach* to scientific rationality. For Kuhn, there is not such a thing as a set of universal rules that every rational individual scientist can use. Scientific rationality, therefore, must be studied by describing how actual groups of scientists make their decisions. While this chapter dispels the accusations of irrationality, it also opens the problem of how to delimit and describe the agent of science, that is the scientific community.

Chapters 5–8, which constitute the second part of the book, explain how Kuhn attempted to solve such a problem. In fact, they reconstruct the philosophical reasons behind Kuhn's adoption of various concepts, such as paradigm, paradigm, disciplinary matrix, exemplars, taxonomies, and lexicons, which Kuhn used to explain what differentiate a scientific community from a group of random individual scientists. In my view, in fact, Kuhn struggled with two important problems that have not been appreciated enough so far. The first problem is methodological. Historians and philosophers of science, he believes, should not study complete and well-formed scientific theories, as if they inhabited a sort of Platonic and a-temporal dimension. Rather, they should be like ethnographers who observe a scientific community as if it were an 'alien society', so that they can reconstruct its collective culture and practice. The problem, well known to ethnographers, anthropologists and social scientists, is then: *how* can this shared culture and practice be described? Kuhn's changing concepts were an attempt to answer this question. But there is also another, perhaps more philosophical problem, lurking in the background: namely, the relationship between *individuals* and *communities*. Like the individuals belonging to the same culture do not think and behave in the same way, so scientists belonging to the same scientific community do not need to agree on everything, make the same choices, and interpret data, theories, or models in the same way. Actually, individual variation within the scientific community is necessary to collective rationality. Given the existence of individual differences, does it even make sense to talk about and focus on communities? Kuhn attempted to solve these problems by considering the scientific community as analogous to a linguistic community. The study of the linguistic features of science allowed him to individuate the scientific communities of the past. In his view, furthermore, a scientific lexicon is flexible enough to guarantee individual variations within the scientific community and yet rigid enough to guarantee communication and the permanence of the scientific/linguistic community through time.

The third part brings Kuhn's dynamic perspective 'beyond Kuhn' himself, in the sense that I will apply it to issues that Kuhn neglected, or even to contemporary debates. In Chap. 7, I discuss the phenomenon of scientific specialisation, which Kuhn began to tackle in some of his mature writings, but in a way that I do not find satisfying and that it does not cohere with his own dynamic perspective. I also discuss interdisciplinarity, which Kuhn surprisingly overlooked. In Chap. 8, I correct some misreading of Kuhn in relation to the so-called Value-Free Ideal and I show how his view may help see some of current debates on the roles of non-epistemic values in science and on the science-society relationship under a different light. In Chap. 9, I put Kuhn's ideas on scientific progress 'in conversation' with the contemporary debate on the topic. Chapter 10 concludes the book with some reflections on Kuhn's philosophical legacy.

Chapters 4 and 5 contains parts of "Taxonomies, Networks, and Lexicons: a study of Kuhn's post-'Linguistic Turn' philosophy" (Politi 2020), here reprinted with permission from Francis & Taylor. Chapter 7 contains parts of "Specialisation and the incommensurability among scientific specialties" (Politi 2019), here reprinted with permission from Routledge, and of "Specialisation, interdisciplinarity, and incommensurability" (Politi 2017), here reprinted with permission from Francis & Taylor. It also contains parts of "Scientific revolutions, specialisations, and the discovery of the structure of DNA: toward a new picture of the development of science" (Politi 2018), here reprinted under the terms of the Creative Commons Attribution 4.0 International License.

References

Andersen, Hanne. 2001. *On Kuhn*. Belmont: Wadsworth/Thomson Learning, Inc.

Andersen, Hanne, Peter Barker, and Xian Chen. 2006. *The Cognitive Structure of Scientific Revolutions*. Cambridge: Cambridge University Press.

Ankeny, Rachel, and Sabina Leonelli. 2019. Repertoires: A post-Kuhnian perspective on scientific change and collaborative research. *Studies in History and Philosophy of Science* 60:18–28.

Basafa, Mark Nader. 2025. Thomas Kuhn's "pre-structure" affinities with Arthur Lovejoy's critical realism. *HoST - Journal of History of Science and Technology* 19:158–172. https://doi.org/10.2478/host-2025-0007.

Bird, Alexander. 2000. *Thomas Kuhn*. Chesham: Acumen and Princeton: Princeton University Press.

Bird, Alexander. 2002. Kuhn's wrong turning. *Studies in History and Philosophy of Science* 33:443–463.

Bird, Alexander. 2005. Naturalising Kuhn. *Proceedings of the Aristotelian Society* 105:109–127.

Bird, Alexander. 2007. Incommensurability naturalised. In *Rethinking Scientific Change and Theory Comparison*, ed. Lena Soler, Howard Sankey, and Paul Hoyningen-Huene, 21–39. Dordrecht: Springer.

Bradley, Seamus. 2017. Constraints on rational theory choice. *British Journal for the Philosophy of Science* 68:639–661.

Bruner, Jerome, and Leo Postman. 1949. On the perception of incongruity: A Paradigm. *Journal of Personality* 18:206–223.

Cederbaum, Daniel. 1983. Paradigms. *Studies in History and Philosophy of Science* 14:173–213.

Chang, Hasok. 2012. Incommensurability: Revisiting the chemical revolution. In *Kuhn's The Structure of Scientific Revolutions Revisited*, ed. Vasso Kindi and Theodore Arabatzis, 153–177. New York: Routledge.

Demir, Ipek. 2008. Incommensurabilities in the work of Thomas Kuhn. *Studies in History and Philosophy of Science* 39:133–142.

Devlin, William, and Alissa Bokulich. 2015. *Kuhn's Structure of Scientific Revolutions - 50 Years On*. London: Springer.

Friedman, Michael. 2001. *Dynamics of Reason*. Stanford: CSLI Publications.

Giri, Leandro, Pablo Melogno, and Miguel Hernán, eds. 2023. *Perspectives on Kuhn*. Cham: Springer.

Hoyningen-Huene, Paul. 1993. *Reconstructing Scientific Revolutions*. Chicago: Chicago University Press.

Hoyningen-Huene, Paul. 1994. Two letters of Paul Feyerabend to Thomas Kuhn on a draft of 'the structure of scientific revolutions'. *Studies in History and Philosophy of Science* 26:353–387.

Hoyningen-Huene, Paul. 2015. Kuhn's development before and after structure. In *Kuhn's Structure of Scientific Revolutions—50 Years On*, ed. William Devlin and Alisa Bokulich, 185–195. London: Springer.

Hoyningen-Huene, Paul. 2021. The genealogy of Thomas Kuhn's metaphysics. In *Interpreting Kuhn: Critical Essays*, ed. K. Brad Wray, 2–26. Cambridge: Cambridge University Press.

Hoyningen-Huene, Paul. 2022. Is Kuhn's "world change through revolutions" comprehensible? *Epistemology and Philosophy of Science* 4:55–72.

Hoyningen-Huene, Paul. 2023. The plausibility of Thomas Kuhn metaphysics. In *Perspectives on Kuhn*, ed. Leandro Giri, Pablo Melogno, and Miguel Hernán, 139–154. Cham: Springer.

Isaac, Joel. 2012. *Working Knowledge: Making the Human Sciences from Parsons to Kuhn*. Cambridge: Harvard University Press.

Jacobs, Struan. 2007. Michael Polanyi and Thomas Kuhn: Priority and credit. *Tradition and Discovery* 33:25–36.

Kindi, Vasso. 1995. Kuhn's the structure of scientific revolutions revisited. *Journal for General Philosophy of Science* 26:75–92.

Kindi, Vasso, and Theodore Arabatzis, eds. 2013. *Kuhn's Structure of Scientific Revolutions Revisited*. New York: Routledge.

Kuhn, Thomas S. 1979. Metaphor in science. In *Metaphor and Thought*, ed. Andrew Ortony, 409–419. Cambridge: Cambridge University Press. Reprinted in *RSS*:197–215.

Kuhn, Thomas. 1983. Commensurability, comparability, communicability. In *PSA: Proceedings of the Biennial Meeting of the Philosophy of Science Association 1982*, ed. P. Asquith, T. Nickels, and A. Arbor, 669–688. Reprinted in *RSS*:33–57.

Kuhn, Thomas. 1987. What are scientific revolutions? In *The Probabilistic Revolution*, ed. Lorenz Krüger, Lorraine Daston, and Michael Heidelberger, 13–32. Cambridge: MIT Press. Reprinted in *RSS*:13–32.

Kuhn, Thomas S. 1991. The road since *Structure*. *PSA: Proceedings of the Biennial Meeting of the Philosophy of Science Association 1990*: 2–13. Reprinted in *RSS*:90–104.

Kuhn, Thomas S. 1992. The trouble with the historical philosophy of science. *Occasional Publication of the Department of History of Science*. Cambridge: Harvard University. Reprinted in *RSS*: 105-120.

Kuhn, Thomas S. 1995. A physicist who became a historian for philosophical purposes: A discussion between Thomas Kuhn and Aristides Baltas, Kostas Gavroglu, and Vassiliki Kindi. In *RSS:* 255–323. MIT Press.
Kuhn, Thomas S. 2022. *The Last Writings of Thomas S. Kuhn*, ed. Bojana Mladenović. Chicago: University of Chicago Press.
Kuukkanen, Jouni-Matti. 2007. Kuhn, the correspondence theory of truth and coherentist epistemology. *Studies in History and Philosophy of Science* 38:555–566.
Kuukkanen, Jouni-Matti. 2008. *Meaning Changes: A Study of Thomas Kuhn's Philosophy.* Saarbrucken: VDM - Verlag Dr. Mueller e.K.
Laudan, Larry. 1984. *Science and Values*. Los Angeles: California University Press.
Marcum, James. 2005. *Thomas Kuhn's Revolution: An Historical Philosophy of Science*. London: Continuum.
Mayoral, Juan V. 2017. *Thomas S. Kuhn: La búsqueda de la estructura.* Zaragoza: Prensas de la Universidad de Zaragoza.
Melogno, Pablo. 2023. Kuhn's 'The natures of conceptual change': The search for a theory of meaning and the birth of taxonomies (1980–1994). *International Studies in the Philosophy of Science* 36:87–103.
Melogno, Pablo, and Leandro Giri. 2023. Towards a genealogy of Thomas Kuhn's semantics. *Perspectives on Science* 31:385–404.
Massimi, Michela. 2015. 'Working in a new world': Kuhn, constructionism, and mind-dependence. *Studies in History and Philosophy of Science* 50:83–89.
Mizrahi, Moti. 2015. Kuhn's incommensurability thesis: What's the argument? *Social Epistemology* 29:361–378. Reprinted in *The Kuhnian Image of Science: Time for a Decisive Transformation?*, ed. Moti Mizrahi, 5–44, 2018. London: Rowman and Littlefield.
Mladenović, Bojana. 2017. *Kuhn's Legacy. Epistemology, Metaphilosophy, and Pragmatism.* New York: Columbia University Press.
Moleski, Martin. 2007. Polanyi vs. Kuhn: Worldviews apart. *Tradition and discovery* 33:8–24.
Morreau, Michael. 2015. Theory choice and social choice: Kuhn vindicated. *Mind* 124:239–262.
Okasha, Samir. 2011. Theory choice and social choice: Kuhn versus arrow. *Mind* 477:83–115.
Okasha, Samir. 2015. On arrow's theorem and scientific rationality: A reply to Morreau and Stegenga. *Mind* 124:279–294.
Pinto de Oliveira, J.C. 2007. Carnap, Kuhn, and revisionism: On the publication of structure in the encyclopedia. *Journal for General Philosophy of Science* 38:147–157.
Pinto de Oliveira, J. C. 2017. Thomas Kuhn, the image of science and the image of art: The first manuscript of *Structure*. *Perspectives on Science* 25:746–765.
Politi, Vincenzo. 2017. Specialisation, interdisciplinarity, and incommensurability. *International Studies in the Philosophy of Science* 31:301–317.
Politi, Vincenzo. 2018. Scientific revolutions, specialisation and the discovery of the structure of DNA: Toward a new picture of the development of the sciences. *Synthese* 195:2267–2293.
Politi, Vincenzo. 2019. Specialisation and the incommensurability among scientific specialties. *Journal for General Philosophy of Science* 50:129–144.
Politi, Vincenzo. 2020. Taxonomies, conceptual networks, and lexicons: A study on Kuhn's post-'linguistic turn' philosophy. *International Studies in the Philosophy of Science* 33:87–103.
Putnam, Hilary. 1975. *Mind, Language and Reality: Philosophical Papers, Vol. 2*. Cambridge: Cambridge University Press.
Reisch, George. 2019. *The Politics of Paradigms*. New York: SUNY Press.
Richards, Robert, and Lorraine Daston, eds. 2016. *Kuhn's 'Structure of Scientific Revolutions' at Fifty*. Chicago: Chicago University Press.
Rouse, Joseph. 2003a. *How Scientific Practices Matter: Reclaiming Philosophical Naturalism.* Chicago: University of Chicago Press.
Rouse, Joseph. 2003b. Kuhn's philosophy of scientific practice. In *Thomas Kuhn*, ed. Thomas Nickles, 101–121. Cambridge: Cambridge University Press.
Rouse, Joseph. 2013. Recovering Thomas Kuhn. *Topoi* 32:59–64.

Rouse, Joseph. 2024. Has there been a second Kuhnian revolution? In *Rethinking Thomas Kuhn's Legacy*, ed. Yafeng Shan, 11–30. Cham: Springer.
Russell, Bertrand. 1945. *A History of Western Philosophy*. New York: Simon and Schuster.
Sankey, Howard, and Paul Hoyningen-Huene. 2001. Introduction. In *Incommensurability and Related Matters*, ed. Paul Hoyningen-Huene and H. Sankey, vii–xxx. Dordrecht: Springer.
Scerri, Eric. 2016. *A Tale of Seven Scientists and a New Philosophy of Science.* New York: Oxford University Press.
Scheffler, Israel. 1967. *Science and Subjectivity*. Indianapolis: Bobbs-Merrill.
Shan, Yafeng. 2023. The historiography of scientific revolutions: A philosophical reflection. In *Handbook of the Historiography of Science*, ed. Mauro Condé and Miriam Salomon, 257–273. London: Springer.
Shan, Yafeng, ed. 2024. *Rethinking Thomas Kuhn's Legacy*. London: Springer.
Shapere, Dudley. 1966. Meaning and scientific change. In *Mind and Cosmos: Essays in Contemporary Science and Philosophy*, ed. Robert Colodny, 41–85. Pittsburgh: Pittsburgh University Press.
Sharrock, Wes, and Rupert Read. 2002. *Kuhn: Philosopher of Scientific Revolutions*. Oxford: Polity.
Stegenga, Jacob. 2015. Theory choice and social choice: Okasha versus Sen. *Mind* 124:263–278.
Timmins, Adam. 2013. Why was Kuhn's *Structure* more successful than Polanyi's *Personal Knowledge*? *HOPOS* 3:306–317.
Toulmin, Stephen. 1970. Does the distinction between normal and revolutionary science hold water? In *Criticism and the Growth of Knowledge*, ed. Imre Lakatos and Alan Musgrave, 39–47. Cambridge: Cambridge University Press.
Wray, Brad K. 2011. *Kuhn's Evolutionary Social Epistemology*. Cambridge: Cambridge University Press.
Wray, Brad K. 2021. *Kuhn's Intellectual Path: Charting the Structure of Scientific Revolutions*. Cambridge University Press.
Wray, Brad K., ed. 2024. *Kuhn's The Structure of Scientific Revolutions at 60*. Cambridge: Cambridge University Press.

Part I
Thomas Kuhn's Philosophical Perspective

Chapter 2
The Static Perspective

Abstract Commonly held interpretations suggest that Kuhn's primary aim was to reject the image of science advanced by logical empiricism. This chapter challenges that assumption. I argue that Kuhn's target was in fact broader—what he later described as the 'static perspective' on science. This is not to say that he lacked specific criticisms of logical empiricism, which are already evident in some of his earliest writings. I also examine the differences between Kuhn's, Carnap's, and Quine's conceptions of the role of philosophy. This will help clarifying the distinctive features of Kuhn's philosophical project.

Keywords Thomas Kuhn · Logical empiricism · Rudolf Carnap · W.V. Quine · Logical analysis of science · Holism

2.1 Introduction

In the very first page of *SSR*, Kuhn famously claimed that "History, if viewed as a repository for more than anecdote or chronology, could produce a decisive transformation in the image of science by which we are now possessed" (*SSR*:1). Leaving aside the issue of the relationship between history and philosophy, about which much more will be said in the next chapter, here I want to focus on the *image of science* that Kuhn wanted to transform and liberate us from. Since it appears that he developed his own views in opposition to such an image, it makes sense to try to understand the target of Kuhn's criticisms before looking into the more positive aspects of his proposal.

In Sect. 2.2, I examine the widespread association between the image of science described at the beginning of *SSR* and the philosophy of logical empiricism. According to the orthodox interpretation, one of Kuhn's primary goals was to overcome logical empiricism and initiate a 'post-empiricist' turn in philosophy of science. More recently, revisionist interpretations have suggested that Kuhn's views were not only compatible with, but in some respects continuous with, logical

V. Politi, *Thomas Kuhn's Philosophical Perspective*, Boston Studies in the Philosophy and History of Science 348,
https://doi.org/10.1007/978-3-032-13540-7_2

empiricism. Yet both interpretations presuppose that Kuhn's main target was logical empiricism itself. As I argue in Sect. 2.3, however, Kuhn's actual target was what he later described as the 'static perspective'—a broader and historically deeper philosophical perspective on science, of which logical empiricism is only a recent version. This is not to say that Kuhn had no specific criticisms of logical empiricism, as I discuss in Sect. 2.4. Already in his early work, in fact, Kuhn argued that attempts to formalise scientific language fail to capture the dynamic and open-ended nature of scientific practice. Finally, in Sect. 2.5, I briefly compare the respective views of Carnap, Quine, and Kuhn on the role and aims of philosophy. While the 'Carnap–Kuhn' and 'Carnap–Quine' relationships have received significant scholarly attention, Kuhn's relationship with Quine has been comparatively overlooked. As I will show in this and following chapters, although Kuhn appreciated his critique of the analytic/synthetic distinction, he also expressed reservations about key aspects of Quine's project. Understanding Kuhn's philosophy as a critique of the static perspective—developed in dialogue with, but also in opposition to, both Carnap and Quine—sheds light on its distinctive character.

Before I begin, I must specify that Kuhn often speaks both of a 'static perspective on science' and of a 'static image of science'. The two are different but clearly related concepts, in so far as the static image results from the static perspective. However, the differences between the two being minimal, in what follows I may often use 'static perspective' and 'static image' interchangeably. This will also be the case in the next chapter, where I will talk about the 'dynamic perspective on science' and the 'dynamic image of science'.

2.2 What Was Kuhn Arguing Against?

2.2.1 The Orthodox Interpretation

According to the most popular interpretation of Kuhn, *SSR* aimed at discarding the 'received view' in philosophy of science attributed to logical empiricism. Emerged as the more a sophisticated and 'liberalised' development of the earlier European logical positivism, logical empiricism established itself as the dominant school in philosophy of science around the 1940s, when many of its leaders flew the Nazi regime and settled abroad, mainly in the US.[1] This philosophical school descends from British empiricism, exemplified by the works of philosophers such as Francis Bacon, John Locke, David Hume, and George Berkeley. For these philosophers, all

[1] Despite their differences, it has become somehow customary for several philosophers, including some of those mentioned in this chapter, to use 'logical positivism' and 'logical empiricism' interchangeably (Richardson and Uebel 2007:1, *ft.* 1). Since, as I will shortly explain, Kuhn's view is often compared to Carnap's, and since the more mature works of Carnap exemplify some of the central tenets of logical empiricism, in this chapter I will use the previous expression but not the latter.

knowledge comes from sensory experience.[2] Around the end of the XIX century, empiricism was revamped by Ernest Mach, who regarded the laws of physics not as representations of the world as it is but, rather, as economical idealisations of sensations (that is, of the world as it is perceived).

To the empiricist tradition, and as their name suggests, logical empiricists added a 'logical' twist. In fact, they sought to provide a logical explication of the conceptual language of science. In their view, there are two classes of sentences: *synthetic sentences*, whose truth-value is established a posteriori, based on some state of affairs we have evidence of, and *analytic sentences*, which are true or false a priori, based on definitions alone and that, as such, do not add anything to our knowledge. Importantly, logical empiricists reject the Kantian idea of *synthetic a priori sentences*, whose truth value does not depend on experience but that, at the same time, add something new to our knowledge.

At the same time, they do recognize the possibility of different scientific 'languages'—or to use the terminology of Carnap (1937), 'linguistic frameworks'. Roughly speaking, for Carnap a linguistic framework is a formal meta-language, that is a language about an 'object language'. In the context of science, a linguistic framework encompasses the logical rules of the language in which a scientific theory is formulated. But while scientific theories are always accepted or rejected on the face of the available evidence, in choosing between linguistic frameworks one cannot appeal to observations or experiments. Strictly speaking, in fact, linguistic frameworks are neither correct nor incorrect. Their choice, therefore, is based on stipulated conventions.

Ultimately, for the logical empiricists, the progress of science consists in the accumulation of empirical knowledge, expressed by synthetic sentences. The philosophers' work consists in clarifying the logical structure of linguistic frameworks, to clarify and explicate vague or ambiguous theoretical concepts, and to analyse the relationship between observational and theoretical language. Even though, as rightly pointed out by Creath (2023), logical empiricism is best described not as a rigid doctrine, but as a variegated movement, with its different members not always

[2] British empiricists were not the first to argue that all knowledge comes from experience. Since ancient times, several philosophers have argued in favour of similar views. What distinguishes the works of Locke or Hume from those of the ancient and medieval empiricists is the level of analysis of the psychological mechanisms that lead to progressive levels of abstraction and theorisation. While for some Ancient Greeks humans are born with a mind that is *tabula rasa*, ready to receive and be modelled by external stimuli, for the British empiricists things are not as clear-cut. In their view, although the content of knowledge is given by sensory experience, human intellect possesses some innate functions (memory, ability to make associations, ability to make abstractions, and so on) which bring order to the flux of sensations and allow us to get from stimuli to theoretical knowledge. In the XX century, the ancient idea of a completely unadulterated mind waiting to receive external stimuli was not very common. Logical empiricists themselves investigated the relationship between stimuli (expressed by the 'observational language') and more abstract knowledge (expressed by the 'theoretical language'). But were they also interested in cognitive and psychological mechanisms? In other words, was their brand of empiricism closer to ancient or modern empiricism? For the purposes of this book, I will have to leave these questions unanswered.

sharing the same views, what I have said so far serves as a very rough and general synthesis of some of its central and most known tenets.

The decline of logical empiricism was accelerated by the landmark works of, among the others, Karl Popper (1959), Michael Polanyi (1958), and Norwood Hanson (1958). Even though he was familiar with all of them, perhaps the philosopher who influenced Kuhn the most was W.V.O. Quine, whom he met at the Harvard Society of Fellows (see Sect. 1.2). Quine famously criticised logical empiricism in his influential "Two dogmas of empiricism" (1951). Specifically, he critically examined and ultimately rejected two foundational assumptions of logical positivism, namely: the *analytic–synthetic distinction* (the idea that there are sentences that are true purely by virtue of meaning, and others that are true due to how they relate to the world); and *reductionism* (the idea that each meaningful statement can be translated into a logical construct based on immediate sensory experience). Quine argues that the distinction between analytic and synthetic statements is ill-defined and circular, relying on vague notions such as synonymy or necessity, which themselves require clarification. He also contends that the reductionist project of grounding all knowledge in sensory experience has failed. Instead, Quine proposes a *holistic view of knowledge*, according to which our beliefs face the 'tribunal of experience' not individually but as a system. It is our whole web of beliefs, then, that must adjust in response to empirical inputs.

> The totality of our so-called knowledge or beliefs, from the most casual matters of geography and history to the profoundest laws of atomic physics or even of pure mathematics and logic, is a man-made fabric which impinges on experience only along the edges. Or, to change the figure, total science is like a field of force whose boundary conditions are experience. A conflict with experience at the periphery occasions readjustments in the interior of the field. Truth values have to be redistributed over some of our statements. [...] Any statement can be held true come what may, if we make drastic enough adjustments elsewhere in the system. Even a statement very close to the periphery can be held true in the face of recalcitrant experience by pleading hallucination or by amending certain statements of the kind called logical laws. Conversely, by the same token, no statement is immune to revision. Revision even of the logical law of the excluded middle has been proposed as a means of simplifying quantum mechanics; and what difference is there in principle between such a shift and the shift whereby Kepler superseded Ptolemy, or Einstein Newton, or Darwin Aristotle? (Quine 1951:39–40)

Right after the publication of its first edition, *SSR* was read as an attack to logical empiricism by both its supporters (like Hesse 1963) and its detractors (like Shapere 1964). The relationship between Kuhn's and Quine's views, instead, is interpreted in different ways. For example, some claim that Kuhn provides historical 'support' to Quine's arguments against the analytic/synthetic distinction. Or, in other words, "Quine rejects as philosophically untenable a conception of truth that Kuhn was to show historically untenable" (Roth 2012:113). Stephen Schwarz, in his book about the history of analytic philosophy, also claims that "the view of science that emerges from Kuhn's historical work is close to Quine's and supports his criticisms of the logical positivism" (Schwarz 2012:91). In his work, by contrast, Kitcher (2012) highlights both the similarities and some of the differences between Quine and Kuhn and poses major emphasis on the positive aspects of Kuhn's philosophy.

In very general terms, then, the orthodox interpretation claims that:

(a) *SSR* overthrows the image of science held by logical positivism, and
(b) *SSR* provides historical examples to Quine's arguments against the analytic/synthetic distinction, which Kuhn accepted.

More recent readings challenge the orthodox interpretation, especially when it comes to (a).

2.2.2 The Revisionist Interpretation

Some philosophers have examined how effective Kuhn's critique of logical empiricism was, particularly given certain noted similarities between his perspectives and those of Rudolf Carnap, a prominent figure within the logical empiricist movement.

Irzik and Grünberg (1995), for example, argue that Carnap's linguistic frameworks and Kuhn's paradigms, or his more mature concept of a lexicon (see Chap. 7), serve equivalent functions. Both provide the condition of possibility for the formulation of theories that are evaluated using empirical data, even though neither can be empirically tested themselves. Kuhn's normal science would then correspond to the activity within a Carnapian linguistic framework, while a scientific revolution would correspond to changing that framework. Furthermore, Carnap, like Kuhn, held a principle of semantic holism, for which the meaning of theoretical terms depends on the larger context of the language their part of. In virtue of semantic holism, Carnap's view strikingly resembles Kuhn's incommensurability thesis, as evidenced by passages like the following:

> In translating one language into another the factual content of an empirical statement cannot always be preserved unchanged. Such changes are inevitable if the structures of the two languages differ in essential points. For example: while many statements of modern physics are completely translatable into statements of classical physics, this is not so or only incompletely so with other statements. The latter situation arises when the statement in question contains concepts (like, e.g. 'wave function' or 'quantization') which simply do not occur in classical physics; the essential point being that these concepts cannot be subsequently included since they presuppose a different form of language. This becomes still more obvious if we contemplate the possibility of a language with a discontinuous spatio-temporal order which might be adopted in a future physics. Then, obviously, some statements of classical physics could not be translated into the new language, and others only incompletely. (This means not only that previously accepted statements would have to be rejected; but also that to certain statements—regardless of whether they were held true or false—there is no corresponding statement at all in the new language.) (Carnap 1936:126)[3]

[3] If this is the case—that is, if Carnap anticipated Kuhn's incommensurability thesis or, at least, its semantic version—then one of the tasks for the historian of the philosophy of science would be to understand why Kuhn's thesis is heavily criticised as incoherent and threatening the rationality of science, while Carnap's view is not. In fact, if *both* Carnap *and* Kuhn believed in the incommensurability/untranslatability between incompatible scientific languages, then why did Kuhn's incom-

Irzik and Grünberg conclude that both Carnap and Kuhn describe science through a "pragmatically oriented semantic conventionalist framework", where the selection between different linguistic frameworks or paradigms is not based on their truth or falsity, but is instead guided by conventions. In their view, therefore, the orthodox interpretation of Kuhn as the philosopher who ignited a post-empiricist revolution in philosophy of science is wrong. Apart from some terminological differences, in fact, the similarities between Kuhn and Carnap run deep, and their approaches are best seen as complementary, rather than anti-thetical.

For other philosophers, the very fact that *SSR* was published as a volume of the *Encyclopedia of Unified Science*, the flagship project of logical empiricism, is evidence of the deep affinities between Kuhn and the philosophical movement he aimed at discarding. Reisch (1991) examines the two letters that Carnap sent to Kuhn after reading drafts of *SSR*. Reisch believes Carnap welcomed Kuhn's monograph because it echoed many of his own ideas. As argued by Uebel (2011), although Carnap knew very little about the history of science, he supported the publication of *SSR* in the *Encyclopedia* because he thought that the historical approach nicely complemented his own logical-analytical approach.

Many proponents of the revisionist interpretation agree that, although Carnap promptly recognized the similarities between his philosophy and *SSR*, Kuhn did not initially perceive these connections. For Bird (2000, 2002), this happened because of a lack of a formal philosophical training: Kuhn was simply too ignorant about philosophy to realise that many of his most central thesis, such as semantic holism and incommensurability, belong to the empiricist tradition he was criticising. For Richardson (2007), Kuhn did not grasp the subtleties and complexities of Carnap and other philosophers. Rather, he reacted against a sort of 'straw version' of logical empiricism, which had become popular in the 1950s and 1960s, and which many logical empiricists themselves may have had unwittingly contributed to creating.

Not every philosopher considers the relationship between Kuhn and Carnap either in terms of stark opposition or of straightforward compatibility. For example, Friedman (2003) considers Kuhn and Carnap to share a post-Kantian root, which they decline and expand in different ways. His erudite historical-philosophical reconstruction allows us to consider *both* the similarities *and* the important differences between the two philosophers. For example, Friedman rightly stresses how for Kuhn, but not for Carnap, scientific change is cognitively and philosophically relevant.[4]

mensurability thesis cause such a fuss while Carnap was not charged with irrationalism and relativism?

[4] "The point, I take it, is that change of language involves an external question for Carnap and is therefore merely pragmatic, and not cognitive or epistemic in the only sense of 'epistemology' Carnap recognizes. For, although Carnap [...] does connect his notion of change of language with scientific revolutions, he never discusses such revolutions in any serious way. Such a historical investigation could never be a part of what Carnap himself preserves of epistemology, namely, *Wissenschaftslogik* [the logic of science]—the formulation and examination of a variety of possible linguistic frameworks within which the results of the special sciences may be represented. What is crucial, for Carnap, is that the only remaining properly philosophical problems are purely

It is true that, later in his life, Kuhn regretted not engaging more with Carnap. Responding to Earman (1993), he admits that, when he wrote *SSR*, he was not familiar with Carnap's mature philosophy. Of the two letters the latter had written to him, Kuhn says that he believed they contained nothing more than formal compliments, not that they expressed genuine interest in his ideas.[5] *Immediately* after that, however, he also remarks that, apart from some similarities, there are also some crucial differences between Carnap's and his own philosophical approach.

> Nevertheless, the passages which John [Earman] quotes to show the deep parallels between Carnap's position and mine also show [...] a correspondingly deep difference. Carnap emphasized untranslatability as I do. But, if I understand Carnap's position correctly, the cognitive importance of language change was for him merely pragmatic. [...] Concerned from the start with the *development* of knowledge, I have seen each stage in the evolution of a given field as built—not quite squarely—upon its predecessor, the earlier stage providing the problems, the data, and most of the concepts prerequisite to the emergence of the stage that followed. In addition, I have insisted that some changes in conceptual vocabulary are required for the assimilation and development of the observations, laws, and theories deployed in the later stage (whence the phrase "not quite squarely" above). Given those beliefs, the process of transition from old state to new becomes an integral part of science, a process that must be understood by the methodologist concerned to analyze the cognitive basis for scientific beliefs. Language change is *cognitively* significant for me as it was not for Carnap. (Kuhn 1993, in *RSS*:227–228, *original emphasis*)[6]

While maybe he did not have a thorough knowledge of the many facets of logical empiricism *when he wrote SSR*, it looks like *after SSR* Kuhn did catch up with some of the relevant philosophical literature. Whether the revisionist interpreters like it or not, the post-*SSR* Kuhn was not so philosophically naive anymore (and, anyway, it is not even clear that he was that philosophically naïve even *before* then; see Chap. 1).

Interestingly, while revisionist philosophers often emphasize those passages where Kuhn admits the little knowledge he had of the mature Carnap when he wrote *SSR*, they tend to ignore or dismiss those where he explains the differences between his philosophical project and Carnap's. For example, Irzik and Grünberg (1995) downplay Kuhn's claims about the 'cognitive significance' of scientific change,

---

formal—belonging to the application of logic to the language of the special sciences. Although many interesting empirical questions may arise in analysing the historical transitions from one theory to another during a scientific revolution [...], the only properly philosophical questions here concern the (purely timeless) articulation of the logical structures of the two different languages under consideration. For Kuhn, by contrast, as the very first chapter of *Structure* makes clear, the point is precisely that historical examination of scientific change can, above all, be genuinely philosophical" (Friedman 2003:21–22).

[5] "[When I wrote *SSR*] I was almost totally innocent of the post-*Aufbau* Carnap, and discovering him has distressed me acutely. Part of my embarrassment results from my sense that responsibility required that I know my target better, but there is more. When I received the kind letter in which Carnap told me of his pleasure in my manuscript, I interpreted it as mere politeness, not as an indication that he and I might usefully talk. That reaction I repeated to my loss on a later occasion" (Kuhn 1993, in *RSS*:227).

[6] Compare this with Friedman (2003), see ft. 4 above.

again arguing that such claims result from a superficial knowledge of logical empiricism. Yet, in the quote above, Kuhn explicitly rejects Carnap's conventionalist conception of language choice, as well as the idea that such a choice involves two complete and fully formed languages, which have nothing in common (for more on this, see Chap. 3).

Some philosophers have reacted against the revisionist interpretation. Tsou (2015) argues that such an interpretation is the result of a misleading 'Kuhnification' of Carnap. What revisionists do, in fact, is to establish a parallel between linguistic frameworks and paradigms, based on some shallow similarities between the two. As Tsou rightly points out, however, Carnapian linguistic frameworks are logical reconstructions of the conceptual foundations of scientific theories, whereas Kuhnian paradigms attempt to capture the constellation of commitments binding a scientific community (methods, values, theoretical language, and so on). I would even go further and examine a deeper level of the problem: the 'Kuhnification' of Carnap's linguistic frameworks Tsou speaks about became possible only after that Kuhn's paradigm has been 'Carnapified', that is interpreted as a containing only scientific theories. So, in the end, the revisionist interpretation does not do justice to either Carnap or Kuhn. More importantly, as Tsou stresses, such an interpretation misses the relevant differences between their philosophical outlooks: logical analysis, on the one hand, and the development of a general model of science grounded on the descriptions of actual scientific activity, on the other.

As for the publication of *SSR* in the *Encyclopedia*, revisionists may read too much into it. As Pinto de Oliveira (2007, 2015) argues, Carnap appreciated *SSR* as a work of history of science, not of philosophy of science. He maintained a strict separation between 'internal' and 'external' questions about science, and believed that philosophy, in the form of logical analysis, answered the previous, while empirical disciplines, such as history, sociology, or psychology, answered the latter. Despite his two enthusiastic letters, Carnap never referred to Kuhn's ideas in his own philosophical works. And why didn't he do so? Revisionist philosophers do not comment on this point.[7]

Summing up: for the orthodox interpretation, Kuhn aimed at rejecting the 'received view' of science developed by the logical empiricists, while for the revisionist interpretations many of Kuhn's views are compatible with, if not an extension of, those of Rudolf Carnap. All these interpretations (the orthodox, the

[7] Following Pinto de Oliveira's argument, I would like to point out that from the fact that a monograph was published in the *Encyclopedia* does not imply that it aligned with logical empiricism. One of the earliest monographs published as part of the project was authored by John Dewey (1939), whose approach could hardly be considered as compatible or complementary to that of logical empiricism. Another volume contained two separate essays: the first, by Giorgio de Santillana, provides a historical reconstruction of the doctrine of rationalism and lamented its demise in contemporary philosophy and science in favor of empiricism, while the other, by Edgar Zilsel, elaborates a sociological and 'externalist' history of the emergence of the mechanistic conception of nature as being caused by the prevalence of mechanical technology (de Santillana and Zilsel 1941). These two essays did not align at all with logical positivism, as evidenced by the rather spirited review by Ernest Nagel (1941).

revisionist, and even the 'anti-revisionist') share the same presuppositions, namely: that what Kuhn wanted to reject was logical empiricism. From there, they proceed assessing Kuhn's philosophy based on whether he managed to hit or miss such a target. Such a presupposition, however, can be challenged.

2.3 The Static Perspective

Does "the image of science we are possessed" Kuhn wanted to discard truly correspond to that of logical empiricism? In *SSR*, the answer to this question is not clear, since the book does not discuss explicitly the view of any philosopher contemporary to Kuhn. Interestingly, if one looks at his *earliest* and *last* works, one may get a better understanding of what, exactly, Kuhn was reacting against.

In the first of his *Lowell Lectures*, Kuhn speaks about 'textbook image of science'. From scientific textbook, science emerges as a systematised and complete body of objective knowledge. Such a knowledge describes reality without the corrupting filters of personal biases and irrational beliefs. This is possible because science has its own method, which is universal. The growth of science corresponds to the addition of new pieces of truth to the existing body of knowledge. In most cases, the textbook image of science describes current science as providing complete knowledge of nature.

The textbook image of science relies on a philosophical perspective that has a long history, and that begins with the views held by Francis Bacon, John Locke and John Stuart Mill, who were among the first philosophers engaged in the project of establishing the foundations for a rational, universal, and a-historical scientific method. However, philosophers are not the only ones holding the conception that objective knowledge is obtained through such a universal scientific method. In fact, Kuhn also quotes the British statistician Karl Pearson:

> The scientific man [*sic*] has above all things to strive at self-elimination in his judgements, to provide an argument which is as true for each individual mind as it is for his own. The classification of facts, the recognition of their sequence and relative significance is the function of science, and the habit of forming a judgment upon these facts, unbiased by personal feelings, is characteristic of what may be termed the scientific frame of mind. (Pearson 1900, quoted in *QPT*:1–2)

For Kuhn, the validity of this philosophical image of science—which is held by philosophers and scientists alike, and which informs the way scientific textbooks are written—is implicitly taken for granted, to the point that it is not even subjected to scrutiny.

> Twentieth-century methodology has achieved a subtlety of analysis neither dreamed of by Bacon nor realized by Pearson. But the basic premise of empiricism, that the scientist, or at least the ideal scientist, proceeds from objective experimental facts or meter readings to the unique laws which govern them, still underlies the most sophisticated methodological discussion. And this same conception of the godlike objectivity of science infects our very language. To be scientific in an argument or an analysis is to be something worthy of praise;

> nothing annoys us more to be told that our pet theory is "unscientific", for this we take to mean biased or based on personal prejudice. Even though our contemporary disillusionment has deprived us or our image of the scientist as a messiah, we preserve our symbol of the scientist as the man [sic] in the highly starched, gleaming white coat who, in the laboratory as in the dentifrice ad, abandons all prejudice so that he may proceed firs to a dispassionate analysis of all the facts and then to the formulation of the immutable law which governs them. (*QPT*:2–3)

Such an image of science, however, is wrong. On the one hand, the observations of 'pure facts' is impossible. On the other hand, as shown by the historical cases Kuhn examines in the following of his lectures, actual science is much more flexible and 'creative' than how it is portrayed by the textbook image.

In some of his last writings, Kuhn returns to the characterisation of the image of science he wants to discard (see, especially, *SKHP*, in Kuhn 2022:2–7). In those writings, the image of that scientific knowledge as a corpus of fixed beliefs that gets expanded through the simple addition of new items is regarded as the result of something Kuhn calls the *static perspective on science*. As in *QPT*, Kuhn explains that this perspective originated as early as the seventeenth century. Alongside Francis Bacon, however, this time he also attributes to Renée Descartes part of the paternity of the static perspective.

As already mentioned, Bacon is one of the earliest modern empiricist philosophers, of whom the logical empiricists are XX century descendants. He maintained that observations and experiments provide the basis for scientific knowledge and, in his view, the there is something like a universal scientific method, which proceed by inductive observation. Such a method could free humanity from the 'idols of the mind' and allow us to get to objective knowledge. It is not too difficult to see how, in the hands of logical empiricists, Bacon's ideas were modified in the belief that all knowledge is grounded on observable phenomena and that scientific theories should be reformulated by linking abstract or theoretical statements into the statements expressed in the observational language. Moreover, Bacon's attack against the 'idols of the mind' appears not too dissimilar from the logical empiricists' rejection of metaphysics, that is speculative and ultimately meaningless, and that must be replaced by scientific research. So far, therefore, in his last writings Kuhn moves in a rather familiar territory.

The fact that he credits Descartes as one of the earliest representatives of the static perspective, however, may come as a surprise—especially if one equates the static perspective with empiricism. Descartes, in fact, was anything but an empiricist. He maintained that the basis of objective knowledge is not empirical experience, which is pray to errors as well as to pernicious sceptical objections about the very existence of the world. Rather, for Descartes, certitude is grounded on a set of clear, distinct, and self-evident ideas, from which the rest of knowledge can be derived through deductive reasoning.

Even though they disagreed on its nature, however, both Bacon and Descartes attempted to establish an absolute foundation upon which the whole of scientific knowledge could be secured. More than empiricism per se, then, one of the main presuppositions of the static image of science is *foundationalism*. As Kuhn explains, foundationalism is 'constructive': once the foundations are provided, special rules and methods allow to 'ascend' from them and produce new knowledge. Here is where both the Baconian and the Cartesian project ran into problems. On the one hand, getting at general truths from the observation of particulars is prey to the well-known difficulties associated with induction. The deductive method, on the other hand, proceeds from the general to the particular in a logical way, but it appears incapable of discoveries and ampliative inferences.

Despite these difficulties, the two foundationalist projects were not abandoned. The idea that all scientific knowledge comes from the examination of facts that are neutral with respect to individual bias or cultural influences, either through observation or through experimentation, remains a central tenet of the contemporary versions of the static image of science. The kind of 'pure' evidence postulated by the static perspective must be.

> purely sensory, dependent only on the biological makeup of the sense organs shared by all normally equipped human observers. Evidence able to satisfy [this] requirement must, it was thought, be compounded from pure sensations—a colour seen, a warmth felt, a sound heard. These sensations provided the simple ideas of sense from which, in Locke's pioneering version of the viewpoint, the complex ideas of bodies in the physical world were to be constructed (*PW*, in Kuhn 2022:124)

However, as Kuhn points out, "Discovering neutral atomic sensations from which the molecular sensation of a chair or a billiard ball could be compounded presented great difficulties" (*PW*, in Kuhn 2022:124).

In the more recent versions of the static perspective, the idea of the neutrality of observation has been modified into a thesis about a neutral observational language.

> The conclusions deduced from laws and theories take the form of statements about what should be observed if those laws and theories are true. Statements can, however—it was then realised—be compared only with other statements, not with sensory observations directly. What objective testing therefore seemed to require was not pure sensations alone but also a neutral observation vocabulary able to express them. It was statements cast in the observation vocabulary that must command assent, independent of culture or prior to belief, if the objectivity of science was to be preserved. (*PW*, in Kuhn 2022:124)

Yet, the pursuit of such a neutral observational language has never been achieved.

Even though contemporary supporters of the static perspective believe that all scientific knowledge begins with empirical experience, they do not dispense with deductionism completely. Rather, they reduce its scope to the evaluation and justification of the conclusions derived from the set of generalisations already at hand. This led, especially in the philosophy of science of the first half of the XX century, to the introduction of an increasingly categorical distinction between the *context of discovery* and the *context of justification*. The distinction, introduced by Reichenbach (1938), quickly became one of the most debated topics in the philosophy of science of the past century. What exactly goes under the general label of 'context of discovery/context of justification distinction', however, may consist of several different distinctions. Hoyningen-Huene (1987, 2006) shows how the discovery/justification distinction may be read as, among other things: a chronological distinction (first discovery, then justification), the distinction of two different processes in science, the distinction between descriptive and normative aspects of given episodes in the history of science, or the distinction between the object of study of philosophy of science (the logic of science) and the object of the empirical studies in history, sociology, and psychology of science.

For his part, in his last writings Kuhn characterises the discovery/justification distinctions in the following way:

> Once hope for a constructive method had been given up, discovery was relegated by the [static perspective] to psychologists and sociologists. Only justification, the evaluation of proposed laws and theories, remained the proper concern of philosophy of science. As the logical empiricists, in particular, have emphasised, laws and theories may come into being in many ways: accidents or personal idiosyncrasy often play a role; the special concerns and training of the investigator always do. But it is not their manner of generation that renders the resulting innovations contributions to knowledge. The same processes could as well have led to egregious error, and whether they have in fact done so can be determined only by one or another form of testing, validation, confirmation. These processes lie within the context of justification; it is when they are deployed that deductive methodology functions; and it is they alone which the tradition has taken to be of philosophical concern (*SKHP*, in Kuhn 2022:5)

Kuhn also specifies that, in the logical empiricist version of the distinction, the context of justification, governed by deductive reasoning, necessarily deals with *propositions*. If deductive methodology was to apply, the knowledge claims to be evaluated had necessarily to be embodied in a set of timeless statements. Scientific knowledge thus came to be viewed as a collection of propositions—that is, as a collection of statements whose truth or falsity is independent of the time, the circumstances, and the language of their utterance. Correspondingly, the philosopher's job becomes the specification of rational techniques for determining which of the propositions in the collection—which of the generalisations in a scientific text, for example—were true and which one were false. Proposed solutions took the form of logical relationships that were to provide criteria of acceptability (*SKHP*, in Kuhn 2022:5).

Kuhn maintains that 'justificationalism' and 'propositionalism', grounded on the distinction between context of discovery and context of justification, have at least two important implications. The first is what he defines as "*methodological solipsism*" (*SKHP*, in Kuhn 2022:7, 11). Like mathematical proofs, the justification of the logical consequences of established generalisations does not admit individual variations in the application of deductive methodology. The presence of individual variations and eventual disagreements would indicate that at least one of the disagreeing parties is departing from the rules of rationality and tainting the evaluative process with subjectivity; that is, that at least someone is being irrational or wrong. For the static image of science, then, the individual rational scientist stands for all rational scientists. Or, to recall Pearson, "The scientific man [...] provide[s] an argument which is *as true for each individual mind as it is for his own*". This leads to the conclusion that the justification of the evaluation made by the scientific group is nothing more than the scaled-up version of the evaluation made by the rational individual.

The second implication is a kind of *confirmational holism* that runs into the problems posed by the so-called Duhem-Quine thesis. Since evaluation involves several logically connected propositions, failure to grant a conclusion will be reflected on them all. In other words, if what is tested is not an individual claim, but always a collection of claims, then it is impossible to confirm or disconfirm a hypothesis in isolation with respect to a bundle of auxiliary hypotheses. In case of empirical disconfirmation, it becomes impossible to ascertain what exactly has been disconfirmed, whether the tested hypothesis or the bundle of auxiliary hypothesis and background assumptions it is part of. For Kuhn, holism derives from justificationalism, and, at the same time, it hampers it.

In short, Kuhn characterises the static perspective as consisting of:

- f*oundationalism*, that is idea that there exists a neutral and absolute foundation of scientific knowledge; often such a foundation is identified with a neutral empirical basis, either as it is directly experienced or as it expressed in an observational language
- the idea that scientific knowledge consists of *timeless truths*; its contemporary variation amounts to *propositionalism*, that is the idea that scientific knowledge consists of true propositions
- *methodological solipsism*, following from the idea that all rational individuals must comply with the same universal rules of scientific rationality
- the *context of discovery/context of justification distinction*, associated to the idea that justification is concerned only with the analysis of the logical connections between propositions
- *confirmational holism*, following from the idea that propositions are all logically connected

Many philosophers interpret *SSR* as attacking logical empiricism, although implicitly. *QPT* and his last writings offer support for a different interpretation. Kuhn challenged the static perspective on science promoted by philosophers, scientists, and textbooks, which assumes a neutral, timeless foundation for scientific knowledge. This presupposition entails several implications, such as the view that scientific knowledge consists of enduring truths established upon an absolute and objective foundation, and that any rational person may acquire this knowledge by following the universal principles of the scientific method. Logical empiricism in only one the most recent versions of the static perspective. At the same time, in his last writings Kuhn seems to discuss more directly the views held by logical empiricists. For example, he talks about the implications of propositionalism, a variation of the idea that scientific knowledge consists of universal truths, which take the form of the discovery/justification distinction and confirmational holism. What, then, is the value of distinguishing between a general static perspective and the philosophy of logical empiricism, which in any case partakes in such a perspective?

To begin with, while logical empiricists endorse a version of the static image, not all the people who endorse the static image are logical empiricists. Indeed, Kuhn aims at discarding an image of science the validity of which is presupposed by philosophers, scientists and laypeople. It is the view that informs how scientific textbook are written, how junior scientists are trained, and how we think about the nature of the scientific enterprise. This is a general view and goes beyond what logical empiricists claim. Furthermore, following Kuhn, it appears that at the heart of the static perspective there is not empiricism but, rather, foundationalism. It is true that, throughout history, most philosophers have identified the foundations of science with empirical knowledge. However, rationalist philosophies *à la* Descartes are also foundationalist and, therefore, they support an image of science as built upon secure, universal and ultimately static foundations.

Finally, the discussion about Kuhn's view on the static image of science allows to bring into attention another interesting issue. Kuhn is often read as a supporter of Quine's holism. In this reading, the whole paradigm, like a 'web of beliefs', faces the tribunal of experience. Anomalies experienced at its periphery may lead to internal readjustments or even to a paradigm change. However, because of propositionalism, Kuhn attributes a 'strong holism' to the contemporary version of the static image as well. Propositional holism faces the well-known Duhem-Quine problem of confirmation. But if holism is a source of problems for the logical empiricists, then it should be a source of problems for Quine too. And if Kuhn follows Quine, then his view should also be prey to the problem of confirmational holism. As I will show in the next chapter, Kuhn's new perspective aims at solving the problems faced by the static perspective, including those associated with confirmational holism. This means that he does not endorse Quine's strong holism, and that some elements of Quine's philosophy may be a target of his criticisms as much as some elements of the philosophy of the logical empiricism.

2.4 Against Formalisation

As discussed in the previous section, in his last writings Kuhn critiques the contemporary version of the static perspective, which reinterprets the idea that scientific knowledge is a collection of timeless truths in terms of *propositionalism*. According to Kuhn, this stance generates conceptual difficulties that the static perspective is ill-equipped to resolve. It is not hard to discern the philosophy of logical empiricism within the contemporary formulation of the static perspective. Already *QPT*, however, Kuhn articulates arguments that challenge core tenets of logical empiricism more explicitly. Specifically, such arguments target the *methodology* of logical empiricists, which consists in the logical analysis of the language of science.

Kuhn begins Lecture VII of his *Lowell Lectures* by specifying that he is not against logic per se. In fact, he looks with admiration at the logicist project in mathematics, that aims at demonstrating how mathematical theorems are deduced from logical principles. As he explains, every deductive system possesses two main characteristics, namely: the relation of *necessity between premises and conclusions*, and the *dependence of the validity of the conclusions on the formal structure of the argument*, rather than on knowledge of matters of fact. With these two characteristics in mind, Kuhn proceeds to discuss how some mathematical postulates can be deduced from logic.

For example, logicians may introduce two sets of entities, one containing lower-case letters (a, b, c...) and the other containing upper-case letters (A, B, C...), as well as some symbols, such as ∊. Then, assuming that there are only two entities in the first set, a and b, and that they are distinct (that is, $a \neq b$), and assuming that there are only two entities in the second set, A and B, it is possible to develop some logically valid arguments. If we accept as premises that $a \in A$, $b \in A$, $a \in B$, and $b \in B$, then we get the conclusion that $A = B$. The conclusion *necessarily* follows from the premises, but what does the whole argument *mean*? As Kuhn observes, we may interpret the lower-case entities as 'points', the upper-case entities as 'straight lines', and the symbol ∊ as indicating a relation of belonging. Then the argument would state that if two points, a and b, belong to two straight lines, A and B, then the two straight lines are one and the same. Or, in other words, that only one straight line can connect two different points: this is a fundamental axiom of Euclidean geometry and, as the example shows, it is deduced by logic.

For Kuhn, the effect of the logicist project has been "to expand and to generalise the subject matter of logical and mathematical studies", which in turn resulted in "the emancipation of logic and mathematics from the study of the real or apparently real world" (*QPT*:117). As his examples shows,

> Instead of points, lines, and planes we might speak of ziggles, zaggles, and daggles, without any reference to the sorts of things which we denote by these words. And this lack of reference or meaning then applies not only to the entities of our mathematical systems—the points, lines, or planes—but also to the relationships between them, relationships like points lying on a line. (*QPT*:120–121)

In short, logic and mathematics are purely formal languages. They are 'about themselves'; in other words, they do not have a 'content'. This means that the rules of these formal languages are not drawn from some characteristics of the world; rather, they are *conventions*. In a sense, Kuhn says, such conventions are 'linguistics' because they are the pre-requisites for talking about the world, communicating, and even thinking. Conventions can be changed, and a change of conventions leads to the creation of new formal languages, which allow new ways of thinking and talking about the world.

Some philosophers believe that the edifice of the empirical science, like that of mathematics, can be secured upon the foundations of logic. The project of these philosophers, therefore, begins with the formalisation of scientific language. As Kuhn explains, these philosophers need, first of all, a much richer and more complex formal language than the one used for the formalisation of geometry. Apart from the a's and A's interpreted as dots and lines, in fact, one needs to introduce *new symbols* that would stand for theoretical concepts such as force, mass, and so on. Furthermore, one also needs to introduce *new rules* for connecting the formal and 'contentless' symbols with the language expressing our experience of the world. These rules are necessary to give a meaning to the purely formal language.

For the philosophers embarking in the project of the formalisation of scientific language, then,

> all scientific language, all science is reduced to convention and its consequences plus meaning-relations. And the ultimate outcome of this program would be to separate entirely the empirical portion of our knowledge from the portion due to the conventions we employ in gaining and formalising it. (*QPT*:131)

Kuhn does not deny the usefulness of the formalisation of some portions of the scientific language. Such an operation could lead, for example, to the refinement of 'vague' concepts. The explication of formal categories and the refinement of a logical-theoretical language, in turn, may direct observations and experiments towards new aspects of the world. Under this respect, then, the formalisation of scientific language is a fruitful enterprise, since the analysis of formal categories may lead to an expansion of the empirical basis of science. At the same time, however, some formal categories may conflict with empirical observations, especially when they have become so formally rigid—that is, when every vagueness, ambiguity and open-endedness has been removed from their definitions—that they cannot capture unexpected experiences coming from the complex reality. Between 'theoretical' and 'observational language', in short, there is a two-way relationship. While an enrichment of the previous may expands the latter, empirical experience too can influence the formal language, especially when the latter is unable to deal with new observations. Such a situation will require an adjustment in the categories of the 'core' of the formal system, or even a substitution of one formal system with another. In this case, however, the choice between different formal systems will not be based on conventions, but it will be based on how each system is able to deal with the new observations. For Kuhn, then, "the formalisation can be a source of scientific crisis or it can be an aid in the resolution of such crisis" (*QPT*:132). In any case, even at its most fruitful, formalisation is a tool for improving the relationship between theoretical and observational language, not an aim in itself.

Moreover, Kuhn observes that formalisation is successfully applied only on an *existing body of knowledge*. Some philosophers, however, pursue a much bolder project.

> Basically it is a proposal that not only the product of research but the research itself be designed and expressed in terms of a formal language plus meaning system. Fundamentally, it calls for the construction of a language adequate to deal unambiguously with infinite variety presented by the totality of our perceptual experience. And this I think is clearly impossible. […] [In fact,] such a language would either be infinitely complex and therefore totally unmanageable, or alternatively […] it would be incapable of dealing with *new* experience. (*QPT*:133–134, *original emphasis*)

In short, for Kuhn, the formalisation of scientific language is either useful but limited to existing organised bodies of knowledge, or it is an impossible project. Claiming that formalising existing knowledge is all we need, furthermore, would mean treating the current state of science as the end point of research. Or, in other words, "supposing the project completed, its results would be to *freeze* scientific attention upon just those aspects of nature which are embraced by contemporary science" (*QPT*:134, *my emphasis*). The project of a complete formalisation of scientific language, therefore, presupposes the validity of the static image of science and, in a sense, it reinforces it.

The *Lowell Lectures*, it is worth stressing, were delivered to a non-specialist audience. In the lectures, Kuhn popularises complex scientific and philosophical theories with an accessible expository style. For this reason, one should not be surprised by the lack of references to other philosopher or of more detailed and technical discussions. Even so, it is not difficult to recognize that Lecture VII summarizes and then criticizes some central elements of the philosophy of logical empiricism. Not only does Kuhn explains the distinction between 'theoretical' and 'observational language' and the introduction of 'rules' for connecting the two. He also summarises what Carnap maintained, namely: that the truth of logical languages are conventions, that such languages are neither true nor false, and that it is possible to have different logical languages. Kuhn may have even been aware that, for Carnap and the other logical empiricists, the meaning of theoretical statements depends on the language they are stated in, and that this view resembles what in *SSR* will become the incommensurability thesis. The point, however, is not whether *some* of the logical empiricists' ideas may be similar to *some* of Kuhn's ideas. The point is, rather, that Kuhn criticises the very philosophical method of formalisation of scientific language.

As discussed in Sect. 2.2.2, some philosophers speculate that Carnap appreciated *SSR* because he considered Kuhn's historical approach to philosophy of science as complementary to his own logical approach. No matter what Carnap thought, Kuhn clearly did not consider the logical and the historical approach as complementary, but as conflicting. The logical approach, in his view, makes sense only from a static perspective, which he rejects.

2.5 On Carnap's, Quine's, and Kuhn's Philosophical Projects

From what I have said so far, it should appear clear that Kuhn's philosophy cannot be regarded as a complement or an extension of Carnap's, just because the two may have hold some superficially similar ideas about how theoretical terms acquire their meanings. And while I have already begun to explain how Kuhn, in criticising the static perspective, also stressed the need to find a solution to the problem of confirmational holism, posed by Quine, the next chapters will clarify how he also endorsed a weaker form of holism and how he developed a theory of meaning and language acquisition in response to Quine's theory. For now, however, I want to compare the three philosophers' views on what philosophy of science should be and should aim for. Of course, the reader should bear in mind that Carnap, Quine, and Kuhn are towering figures in philosophy and have generated copious debates. Therefore, this section cannot help but being rather generic, and perhaps even a bit superficial. Despite that, it may still be helpful in questioning the comparisons of Kuhn's philosophy with Carnap's, or the assumption that Kuhn took for granted the validity of Quine's views.

Carnap, alongside other logical empiricists, maintained the distinction between the context of discovery and the context of justification. As already mentioned in Sect. 2.3, this is a distinction of the different 'moments' of scientific research: first there is discovery, which may be fortuitous, serendipitous, and so on, then there is justification, that relies on rational rules and epistemic reasons. At the same time, it is a disciplinary distinction: while empirical disciplines, like history, sociology, or psychology, look at the details and contingencies of the context of discovery, philosophy's domain is the context of justification. In Carnap's view, there is no such a thing as 'philosophical knowledge'. The only possible knowledge is scientific knowledge. Philosophy is neither a doctrine nor a source of knowledge. Rather, it is an activity that consists in the logical analysis of scientific language, to be carried out within the limits of the context of justification.

Like Carnap, Quine believes that all possible knowledge is exhausted by scientific knowledge. In his view, "it is within science itself, and not in some prior philosophy, that reality is to be identified and described" (Quine 1981:21). However, his 'naturalistic stance' is so strong that, contrary to Carnap, he denies that philosophy could be an independent discipline, with its own methods and domain of investigation. If knowledge is only scientific knowledge, then the role of philosophy would be to explain how we acquire such knowledge; however, philosophy must do so by adopting scientific methods. Epistemology, that is the theory of knowledge, must be itself a branch of science. It must be, in other words, "contained in natural science, as a chapter of psychology" (Quine 1969:83). The psychology Quine talks about is the branch of empirical psychology that is concerned with the study of perception and cognition. The long-term naturalistic project of Quine consists in explaining how we can ascend from the stimuli coming from the external world to theoretical scientific knowledge (see Quine 1995). At the time when he launched the project for a 'naturalised epistemology', experimental psychology was still in its infancy. The development of science, however, would eventually make it possible for psychology to fully encapsulate philosophy.

For Quine, science has no other foundation other than science itself. He is therefore against the philosophers' normative pretentions. In what he calls 'regimented theory'—which corresponds to the ideal sum of all our best scientific theories or, in other words, to our 'total knowledge'—philosophy can play nothing more than a descriptive or clarificatory role.

There are clear differences between Carnap's and Quine's views. While Carnap defends philosophy as a separate discipline, Quine predicts the absorption of philosophical theories into scientific knowledge. At the same time, they both attribute to philosophy a clarificatory purpose, although each in their own way. For Carnap, logical analysis may bring clarity and precision to vague theoretical concepts and, in this sense, may be useful to the progress of science. For Quine, philosophy does not logically explicate scientific concepts for the benefit of scientists; rather, it clarifies science to non-scientists.

Especially in the light of the previous section, it appears clear that Kuhn rejects the method of logical analysis as appropriate to a philosophy that wants to capture the creative and expansive features of science. In his works, furthermore, he relies on historical case studies as well as on empirical studies on perception to demonstrate the impossibility of accessing 'pure facts'. This may lead to think that, in a sense, he practices what Quine preaches, namely a form of naturalised epistemology. Kuhn's naturalism, however, is different from Quine's. As I will discuss in the next chapter, he uses empirical data and theories not to elucidate how what current science amounts to, but to develop a *philosophical perspective* on science itself, capable of overcoming the static perspective. Such a philosophical perspective does not presuppose the correctness of current scientific theories, because it interprets science as a creative, open-ended and ever-developing attempt to capture an infinitely complex world. From this perspective, it makes no sense to postulate an ideal regimented theory as the end point for scientific research, because it makes no sense to postulate *any* ideal end point for science. At the same time, philosophy still have its own normative function, which does not derive from logical analysis or abstract theories of scientific rationality, but, rather, from the description of scientific activity. Finally, and contrary to Quine, Kuhn believes that while we ascend from stimuli to theoretical knowledge, our theoretical knowledge *also* influences how we perceive stimuli: every observation, after all, is theory laden.

To sum up, Carnap holds that philosophy is an independent discipline, and its subject matter is the rational justification of science. In his view, philosophy may provide some norms to science through logical analysis and formalisation. Quine is a naturalist, who regards philosophy as continuous to scientific theories and methods, to the point that it will be eventually absorbed by science. In his view, the only function left to philosophy is to clarify and explain scientific knowledge to non-scientists. In as much as he relies on empirical evidence and methods, Kuhn is also a naturalist. Contrary to Quine, however, he does not believe that philosophy is absorbed by science. In his view, in fact, philosophy must provide a perspective on science. Furthermore, philosophy may still play a normative role but, contrary to Carnap, such norms do not emerge from the logical analysis carried within the context of justification but, instead, from the descriptions of the scientific activity as it unfolds in the context of discovery.

2.6 Conclusions

Distinguishing between the static perspective and the philosophy of logical empiricism clarifies Kuhn's original intents. Kuhn aimed at rejecting a general perspective on science which is shared by philosophers, scientists, and lay-people, informs how science textbooks are written and, therefore, how science is taught, and influence how we look at the history of science. Such a perspective is presupposed; its validity is unquestioned. Logical empiricists presuppose the static perspective on science, just like everyone else. Kuhn's philosophical project consists in questioning the validity of such a presupposed perspective in order to develop a new and better one.

While many would agree that, at least at the descriptive level, the static perspective provides a rather unrealistic image of science, it must be stressed, as Kuhn does, that it is also riddled by several philosophical difficulties. To begin with, there are the difficulties surrounding the concept of an absolute and neutral foundation of scientific knowledge. Such a foundation can be interpreted as a 'pure' empirical basis, but not necessarily. The idea of a pure empirical basis may become, in some of the recent versions of the static perspective, a pure 'observational language', but not necessarily. Kuhn may criticise some of the implications of the logical empiricism (i.e., the problem of confirmational holism) in so far as he addresses one of the contemporary versions of the static perspective, but the main target of his criticisms is more general and more fundamental. At the same time, it is also possible to see how, for Kuhn, one of main problems with logical empiricism was the attempt to formalise and logically reconstruct the language of science. By 'freezing' scientific knowledge to the contemporary scientific theories, such a methodology reinforces the static image of science.

An alternative philosophical perspective ought to provide not only a less idealised and more descriptively adequate account of actual scientific research. It ought also to be able to overcome the philosophical problems plaguing the static image of science. The positive aspect of Kuhn's project, therefore, consists in the description of the actual scientific activity but also, and perhaps more importantly, in the establishment of an overall better philosophical perspective.

References

Bird, Alexander. 2000. *Thomas Kuhn*. Chesham: Acumen and Princeton: Princeton University Press.

Bird, Alexander. 2002. Kuhn's Wrong Turning. *Studies in History and Philosophy of Science* 33:443–463.

Carnap, Rudolf. 1936. Truth and confirmation. In *Readings in Philosophical Analysis*, ed. Herbert Feigl, 119–127. New York: Appleton Century Crofts.

Carnap, Rudolf. 1937. *The Logical Syntax of Language*, English translation by Amethe Smeaton, London: Routledge. Original German: *Logische Syntax der Sprache*. Vienna: Springer, 1934.

Creath, Richard. 2023. Logical Empiricism. In *The Stanford Encyclopedia of Philosophy* (Winter 2023 Edition), ed. Edward N. Zalta and Uri Nodelman. https://plato.stanford.edu/archives/win2023/entries/logical-empiricism/

de Santillana, George, and Edgar Zilsel. 1941. *The Development of Rationalism and Empiricism* (International Encyclopedia of Unified Science: Foundations of the Unity of Science, vol. II-8) Chicago: Chicago University Press.

Dewey, John. 1939. *Theory of Valuation*. (International Encyclopedia of Unified Science: Foundations of the Unity of Science, Vol. II-4.) Chicago: Chicago University Press.

Earman, John. 1993. Carnap, Kuhn, and the philosophy of scientific methodology. In *World Changes: Thomas Kuhn and the Nature of Science*, ed. Paul Horwich, 9–36. Cambridge: MIT Press.

Friedman, Michael. 2003. Kuhn and logical empiricism. In *Thomas Kuhn*, ed. Thomas Nickles, 19–44. Cambridge: Cambridge University Press.

Hanson, Norwood Russell. 1958. *Patterns of Discovery*. Cambridge: Cambridge University Press.

Hesse, Mary. 1963. Untitled review of Kuhn's *The Structure of Scientific Revolutions*. *Isis* 54:286–287.

Hoyningen-Huene, Paul. 1987. Context of discovery and context of justification. *Studies in History and Philosophy of Science* 18:501–514.

Hoyningen-Huene, Paul. 2006. Context of discovery versus context of justification and Thomas Kuhn. In *Revisiting Discovery and Justification*, ed. Jutta Schickore and Friedrich Steinle, 119–131. Dordrecht: Springer.
Irzik, Gürol, and Teo Grünberg. 1995. Carnap and Kuhn: Arch enemies or close allies? *British Journal for the Philosophy of Science* 46:286–307.
Kitcher, Philip. 2012. *Preludes to Pragmatism*. Oxford: Oxford University Press.
Kuhn, Thomas S. 1993. Afterwords. In *World Changes: Thomas Kuhn and the Nature of Science*, ed. Paul Horwich, 311–341. Cambridge: MIT Press. Reprinted in *RSS*:224–252.
Kuhn, Thomas S. 2022. *The Last Writings of Thomas Kuhn*, ed. Bojana Mladenović. Chicago: Chicago University Press.
Nagel, Ernest. 1941. Review of *The Development of Rationalism and Empiricism. The Journal of Philosophy* 38:795–696.
Pearson, Karl. 1900. *The Grammar of Science*. 2nd ed. London: Adam and Charles Black.
Pinto de Oliveira, J. C. 2007. Carnap, Kuhn, and revisionism: On the publication of structure in the *Encyclopedia. Journal for General Philosophy of Science* 38:147–157.
Pinto de Oliveira, J. C. 2015. Carnap, Kuhn, and the history of science: A reply to Thomas Uebel. *Journal for General Philosophy of Science* 46:215–223.
Polanyi, Michael. 1958. *Personal Knowledge: Towards a Post-Critical Philosophy*. London: Routledge.
Popper, Karl. 1959. *The Logic of Scientific Discovery*, translation by the author of *Logik der Forschung* (1935). London: Hutchinson.
Quine, William V. O. 1951. Two Dogmas of Empiricism. *The Philosophical Review* 60:20–43.
Quine, William V. O. 1969. *Ontological Relativity and Other Essays*. New York: Columbia University Press.
Quine, William V. O. 1981. *Theories and Things*. Cambridge: Harvard University Press.
Quine, William V. O. 1995. *From Stimulus to Science*. Cambridge: Harvard University Press.
Reisch, George. 1991. Did Kuhn kill logical positivism? *Philosophy of Science* 58:264–277.
Reichenbach, Hans. 1938. *Experience and Prediction*. Chicago: University of Chicago Press.
Richardson, Alan. 2007. "That sort of everyday image of logical empiricism": Thomas Kuhn and the decline of logical empiricist philosophy of science. In *The Cambridge Companion to Logical Empiricism*, ed. Alan Richardson and Thomas Uebel, 346–370. Cambridge: Cambridge University Press.
Richardson, Alan, and Thomas Uebel. 2007. Introduction. In *The Cambridge Companion to Logical Empiricism*, ed. Alan Richardson and Thomas Uebel, 1–10. Cambridge: Cambridge University Press.
Roth, Paul. 2012. The philosophy of the social sciences in the twentieth century. In *The Sage Handbook of the Philosophy of the Social Sciences*, ed. Ian Jarvie and Jesús Zamora Bonilla, 103–118. London: Sage.
Schwarz, Stephen. 2012. *A Brief History of Analytic Philosophy*. London: Wiley-Blackwell.
Shapere, Dudley. 1964. Untitled review of Kuhn's *The Structure of Scientific Revolutions. Philosophical Review* 73:383–394.
Tsou, Jonathan Y. 2015. Reconsidering the carnap-kuhn connection. In *Kuhn's Structure of Scientific Revolutions—50 Years On*, ed. William Devlin and Alisa Bokulich, 51–69. Cham: Springer.
Uebel, Thomas. 2011. Carnap and Kuhn: On the relation between the logic of science and the history of science. *Journal for General Philosophy of Science* 42:129–140.

Chapter 3
The Dynamic Perspective

Abstract This chapter is about the dynamic perspective on science, which Kuhn developed in opposition to the static perspective. First, I will spell out how Kuhn built his new perspective by drawing the logical and philosophical implications of the rejection of foundationalism. Then, I will explain why the dynamic perspective is different and more fundamental than the model of science of *SSR*. The analysis of the relationship between his philosophical perspective, his work on the history of science, and his history-based model of science will make it clearer which parts of Kuhn's thought are more susceptible to empirical testing than others. The chapter concludes with a discussion of Kuhn's (unexpected) defence of so-called Whig history of science in some of his last writings.

Keywords Thomas Kuhn · Historiography of science · History and philosophy of science · Naturalised philosophy · First principles · Presentism

3.1 Introduction

A philosophical image of science can be replaced only by another philosophical image of science.

As explained in the previous chapter, Kuhn criticised the static image because, in his view, it does not provide a realistic description of the scientific activity and, more importantly, it is plagued by several problems. Already in *QPT*, he described how the 'textbook science' is unable to account for 'creative science' (*QPT*: Lecture I) or 'constructive research' (*QPT*: Lecture VIII). Recognising the limitations of the static perspective, however, is not be enough for discarding it once and for all. What is needed is a whole new philosophical perspective on science. Throughout his whole career, Kuhn attempted to develop such a new perspective that, in some of his last writings, he defined as the *dynamic perspective*.

In Sect. 3.2, I will show how Kuhn develops his new perspective from the rejection of the idea of an absolute foundation of science, that is one of the main tenets

V. Politi, *Thomas Kuhn's Philosophical Perspective*, Boston Studies in the Philosophy and History of Science 348,
https://doi.org/10.1007/978-3-032-13540-7_3

of the static perspective. One could reject the idea of an absolute foundation of science by just glimpsing at some empirical data—for example, from the history of science or psychology. However, the dynamic perspective per se is built on the examination of the philosophical implications of such a rejection. In other words, Kuhn did not reject the static perspective mainly by testing it empirically but, rather, through philosophical counterarguments. In Sect. 3.3, I discuss how some philosophers interpret this method as a relapse into armchair philosophizing, while others argue that Kuhn's philosophy has always been a priori anyway. Following on that, in Sect. 3.4 I explain how distinguishing Kuhn's dynamic perspective from his works in the history of science and from his *SSR* model of science brings more clarity to many aspects of his philosophical project. Finally, in Sect. 3.5, I discuss some aspects of Kuhn's rather surprising late defence of presentist and 'Whig' history of science.

3.2 The Dynamic Perspective

The specification of the main features of the static perspective is the foil Kuhn uses to explain his own *dynamic perspective* (or historical, or developmental perspective). This perspective rejects the idea of an absolute foundation of scientific knowledge. As explained in the previous chapter, if interpreted as a neutral empirical ground, such an absolute foundation would require either a well-supported theory about the neutrality of observation (or of sensory experience in general), or the articulation of a neutral observational language. As Kuhn points out, both these projects have never been satisfactorily completed, so that one of the main assumptions of the static perspective remains unwarranted. Moreover, findings from experimental psychology support the view for which the act of perception is never 'passive' and its content never 'pure', that is unadulterated by subjective assumptions, expectations, biases, and so on. These findings can be used to reinforce the rejection of the idea of a neutral empirical basis of science. Nevertheless, it is important to stress that, *philosophically*, such a rejection is only the first step towards the resolution of the problems with the static image of science. In other words, challenging or rejecting one of the central ideas of the static perspective is not enough: either the old perspective is amended, or a new perspective is developed. Kuhn takes the second route and, from the rejection of the belief in the neutral basis for science, he draws several consequences.

To begin with, without the assumption of the existence of a permanent and absolute basis as the foundation for scientific knowledge, the idea of an "Archimedean platform outside of history, outside of time and space is gone beyond recall" (Kuhn 1992, in *RSS*:115). In its place and considering that scientific knowledge never unfolds into a historical, social, and epistemic vacuum, the dynamic perspective focusses on something much more context-dependent, namely the knowledge held by the scientific community at a particular time.

> A scientist *must* accept a great many of the current knowledge claims of his community, for they are constitutive of the community's practice, of the form of life, that is, of an enduring tribe. To refuse to accept them would be to decline membership in that tribe and thus to refuse the practice of science. Though there is a great deal about which members of an individual tribe—physicists or chemists, for example—may disagree, those very disagreements are made possible, recognisable, and discussable only by the far larger body of beliefs that members share, beliefs that unite them as members of a single tribe. That much of their body of beliefs is simply among the givens of a given time. (*SKHP*, in Kuhn 2022:10, *original emphasis*)

Acceptance of current scientific knowledge, to the point of presupposing its validity, is imperative for individuals who are members of a scientific community. Rejecting such a knowledge means giving up the membership to the community. At the same time, sharing some presuppositions does not imply that everybody will always interpret things in the same way, or that everyone will reach the same conclusions. Disagreement is possible within the same scientific community, as long as it is against the background of some shared assumptions.

New pieces of information can be added to the corpus of established scientific knowledge. Sometimes, however, the acquisition of new knowledge may require the adjustment of some or even many parts of the currently held system. In the most extreme cases, it may require a change of system of knowledge (or 'belief system'). Lacking an Archimedean platform, scientists do not test the single statements of each competing belief system directly with pure facts. Rather, they compare two different systems with empirical observations *and* with each other. In the dynamic perspective, therefore, evaluation is always *comparative*.

For Kuhn, this is one of the most illuminating contributions of his dynamic perspective. From this perspective, the task for philosophers of science is not to assess the rationality of one particular belief system (*via* logical analysis or by other means), but to explain why the change from one system to another was indeed rational, despite the absence of a neutral basis for their comparison. Since science is necessarily bound to a specific context (scientists, after all, do not have a God's eye view on reality and on themselves), a theory of scientific rationality cannot rely on universal and timeless principles, but must take such a context in consideration, a task that could be supported by historical reconstructions. Consequently, the distinction between the context of discovery and the context of justification must be rethought: what goes on in the context of discovery is indeed part of scientific justification.

Kuhn criticised the discovery/justification distinction since (at least) *SSR*. In the introduction of the book, he writes

> History [...] is a purely descriptive discipline. The theses of [*SSR*] are, however, often interpretive and sometimes normative. Again, many of my generalisations are about the sociology [...] of scientists; yet at least a few of my conclusions belong traditionally to logic or epistemology. [...] I may even seem to have violated the very influential contemporary distinction between the 'context of discovery' and the 'context of justification'. (*SSR*:8–9)

Starting from at least his *TL* (see Melogno 2019), Kuhn returns on the discovery/justification distinction, to make explicit his dynamic perspective. Expanding on his earlier rejection of the distinction, he comments

> As evaluation becomes comparative, considerations once thought relevant only to the context of discovery become critically important to the context of justification as well. To understand a scientific discovery or the invention of a new scientific theory one must first find out what the members of the relevant scientific specialty knew, or thought they knew, before the discovery or invention was made. In addition, one must determine what part, if any, of that earlier body of knowledge claims needed to be set aside and replaced as the innovation was accepted. […] Justification thus requires knowledge of the body of beliefs accepted by scientists immediately before the innovation to be evaluated was made. (*SKHP*, in Kuhn 2022:9)

For Kuhn, questioning of the discovery/justification distinction leads to the dissolution of the problems arising from the Duhem-Quine thesis. A new piece of knowledge may require a change of beliefs, leading to the competition of belief systems, which must be therefore compared. During the evaluation process, however, scientists do not compare the two competing systems as wholes. To begin with, in most cases they found themselves in the situation of comparing an articulated and established belief system with a newly emerging and underdeveloped system which, since it has not been accepted yet, did not have the opportunity to be fully articulated yet. The comparison, therefore, involves only *parts* of the two systems—namely, the parts they compete about.

As an example, Kuhn mentions Torricelli's experiment with the barometer (*PW*, in Kuhn 2022:126–127). Galileo maintained that vacuum exists in nature. He demonstrated this by showing how the water contained in a pipe put upside down in a vat, which is also filled with water, does not flow completely into the vat, leaving an empty space at the top of the pipe. Galileo held the idea that the vacuum dispersed into materials gives them cohesive force. Therefore, he interpreted the empty space left by water on the top of the upside-down pipe as an indication of the quantity of vacuum required to sustain water in such a situation. Torricelli thought that, if this was the right explanation, then liquid with different densities should leave empty spaces of different lengths at the top of the pipe. This is so because heavier liquids would put more stress to the sustaining force of vacuum. Torricelli repeated the experiment by using mercury, a much heavier liquid than water, which should have left a bigger empty space on the top of the pipe. However, the result of the experiment did not confirm Galileo's hypothesis. Torricelli's alternative hypothesis was that the pressure of air over the liquid contained in the vat was the cause of the raising of the liquid inside the vertical pump.

From a static perspective, the results of Torricelli's experiment are preys to the problems associated with the holistic character of confirmation. The results violate the expectations, but what is violated exactly? Galileo's hypothesis, the conjunction of Galileo's hypothesis with some auxiliary hypothesis, Galileo's whole theoretical system? In the dynamic perspective, by contrast, two different belief systems are compared with one another. Despite their differences, the two systems share several presuppositions. In the case of Torricelli's experiment, the laws of hydrostatics were the *undisputed accepted common ground* between a Galilean and a Torricellian interpretation. In other words, hydrostatics was premised and taken for granted by both parties of the dispute: no experimental data was taken to either confirm or disconfirm it. The dynamic perspective escapes the Duhem-Quine problems because considering what two competing systems share and what they compete about provides a principled way to understand what, *in a specific comparative context*, is being confirmed or disconfirmed.

One may question whether the Galileo-Torricelli dispute is a good example. After all, unlike the disputes between Galileo and the Aristotelians, Torricelli was not overthrowing the Galilean paradigm, he was only reacting against one of his hypotheses. In this scenario, it is easy to see how Galileo's and Torricelli's hypotheses, although different, shared several assumptions. In the end, however, even the difference between the Galileo-Torricelli dispute and the conflict between incommensurable paradigms may be of degree, rather than of kinds. A new paradigm is always born as a response to the current paradigm. Two competing paradigms are always part of the same historical and scientific context and, apart from their incompatible parts, they may share several presuppositions, for example about the functioning of experimental devices. But it is important to stress that this does *not* mean that experiments are 'theory free'. Experiments and observations are theory laden, but the comparison between two competing paradigms does not test the theories informing them, nor does it question their validity.

Two more implications of the new perspective are the abandonment of methodological individualism in favour of a theory of *collective rationality* and the *abandonment of the correspondence theory of truth.* These two implications of the dynamic perspective will be discussed separately, in Chaps. 3 and 9 respectively.

Finally, though Kuhn does not make them explicit, there are at least two more implications of the dynamic perspective. The first is a new object of study for the philosophy of science. Philosophers holding a static perspective do not look at the historical dimension of science, which do not pertain to the context of justification. As such, the objects of their analyses are *well-formed* and *complete* scientific theories. By contrast, Kuhn and the other philosophers who endorsed a dynamic perspective aimed at analysing what allows scientists to develop theories to begin with. For Dudley Shapere (1984), the historically oriented philosophers of science substituted scientific theories with the 'presuppositions' of science as the object of their analyses. In Shapere's view, Kuhn's 'paradigms', Toulmin's 'ideals of natural order', or Feyerabend's non-customary conception of 'theories' are all instances of the kind of presuppositions these philosophers were after. As we shall see in following chapters, even after he gave up the concept of paradigm, Kuhn kept using concepts that referred to something more fundamental than scientific theories. 'Lexicons' and 'systems of beliefs', in fact, are what allows the formulation and development of scientific theories; they are, in a sense, their conditions of possibility (see Chap. 7).

Strictly associated with this point is the inadequacy of logical analysis and the need for a different kind of method for the philosophy of science. As discussed in the previous chapter, logical analysis can be surely conducted on well-formed and complete scientific theories. It is, however, a kind of analysis that can be only done *post hoc*, that is, *after* the theories have been fully developed. If one is interested in the dynamics of science and, therefore, to theories and beliefs *formation* as well as to what allows it, then logical analysis will be of no use. Kuhn seems to have developed an aversion for the project of formalising scientific language very early on in his career and his late writing about the dynamic perspective only reinforce such a stance.

To sum up, in opposition to the static perspective, Kuhn's dynamic perspective can take the form of the following argument:

1. *Premise*: the assumption that there is an absolute neutral foundation of science is unjustified and must be rejected
2. from (1), scientific evaluation is not about the comparison of a system of beliefs with a fixed reality, meant to be the absolute neutral foundation of science; rather, it is about the comparison of different systems of beliefs held by a given community at a given time
3. from (2), it becomes clear that contextual elements are necessary to scientific justification, so that the distinction between context of discovery and context of justification is blurred
4. from (2) and (3), the problems associated with the Duhem-Quine thesis are dissolved, because the context of the comparison dictates what parts of two competing paradigms are tested and what parts are presupposed by both of them
5. from (2), since systems of beliefs are held by communities, science requires a theory of collective rationality
6. from (1) and (2), the correspondence theory of truth is abandoned

Furthermore, considering what has been discussed in this section and in the previous chapter, I would add:

7. the dynamic perspective does not look at well-formed complete scientific theories, but their 'presuppositions'
8. from (7), as well as from the criticism to formalisation as the chief method in philosophy of science (see Sect. 2.4), scientific presuppositions cannot be studied *via* logical analysis

Two things need to be stressed. First, my claim is not that Kuhn had always had clear in mind what the dynamic perspective amounted to since the times of *SSR*, or even before that. Very likely, he could see all the details and implications of his own perspective only in his last writings, where in fact he attempts to explain it in more details. By contrast, some philosophers interpret Kuhn's philosophy as having exclusively negative character; or, rather, a 'therapeutic' aim *à la* Wittgenstein (Sharrock and Read 2002; Kindi 2005, 2013). In this interpretation, Kuhn only wanted to liberate us from the image of science we are possessed from, without proposing any alternative, aware of the fact that there are many sciences and many ways of doing scientific research, and that they are all grouped together through 'family resemblance'. Therefore, offering *one* alternative to the traditional image of science would be impossible and Kuhn never attempted to do so. However, the publication of *QPT* demonstrates that Kuhn always had in mind the rejection of the 'textbook science', or 'static perspective', *and* that, to do so, he needed to develop another philosophical perspective. The publication of his last writings confirms this hypothesis and, contrary to the Wittgensteinian interpretation, should lead to a better appreciation of the positive parts of Kuhn's philosophical project.

Second, it is crucial to understand that Kuhn decided to look at the history of science and developed his model of scientific progress *from—and because of*—the dynamic perspective. The explication of such a perspective may reverberate through Kuhn's historical studies and history-based model of science, bringing a new light on some old issues, as I will explain in the next section.

3.3 The History of Science and the Philosophy of Science

All it takes for the dynamic perspective to be developed as a full-fledged philosophical position is the rejection of one of the central tenets of the static perspective, namely the unjustified assumption about the existence of a neutral foundation for science. All the other tenets of the dynamic perspective follow from such a rejection. Put it in this way, it appears that Kuhn's philosophical perspective does not need much recourse to empirical studies to be developed. It remains to be seen, therefore, whether Kuhn's increasing focus on a philosophical perspective, rather than on concrete historical case studies, also represents a rupture within his own intellectual development.

Notoriously, in 'The Trouble with the Historical Philosophy of Science' (Kuhn 1992), after providing a general overview of the philosophical perspective that he and other 'historically oriented' philosophers developed in opposition to the received image of science, he concluded that "one can reach many of the central conclusions we drew with scarcely a glance at the historical record itself [...] [In fact] many of the most central conclusions we drew from the historical record can be derived instead from first principles (Kuhn 1992, in RSS:111–112). This and similar claims come as a surprise when considering that, in *SSR* and in other works, Kuhn seemed to have given much more than a scarce glance to the historical records. As pointed out by Andersen, "what the nature of these 'first principles' would be remained an unsolved problem in Kuhn's philosophy of science" (Andersen 2001:76). Even though, then, it will never be possible to get at what Kuhn *exactly* had in mind when he talked about first principles, it is still possible to discuss whether, following Bird (2000, 2002), his mature philosophy made a 'methodological turn' from his early naturalism towards a kind of armchair speculative style.

Some philosophers, for example, resist this kind of reading and interpret Kuhn's philosophical development as continuous. In their view, he was never much concerned with historical details to begin with and, even in its 'early' phase, his philosophy contained several a priori elements. Kindi (2005) argues for the primacy of the philosophical elements over the historical and empirical ones throughout Kuhn's whole intellectual trajectory. In her view, Kuhn's project is "a transcendental one, offering the conditions of possibility of science" (Kindi 2005:513). As Kindi points out, the *SSR* concepts of normal science and scientific revolutions are in a dialectical relationship: normal science is the condition of possibility for a revolution to occur, and the condition of possibility of normal science is the consensus bestowed upon a paradigm. These concepts, in other words, entertain logical relations of necessity, which can be explored without any empirical input. In Kindi's view, therefore, the history of science only played a 'genetic' role for Kuhn's philosophy of science: it allowed him to develop his concepts, but the analysis of their logical connections and the philosophical image of science which emerges from them do not need the recourse to the history of science. Furthermore, the fact that the Kuhnian project is 'philosophical', rather than 'empirical', would save it from empirical refutation. Criticising its limited empirical basis or finding counterexamples cannot falsify Kuhn's model, which is not arrived at through inductive generalisations but, rather, it is, and it has always been, a philosophical argument.

Although suggestive, Kindi's interpretation has some issues. First, qualifying a project as 'philosophical' is not enough to disqualify it as naturalistic: exception made for the most extreme versions of eliminativism, naturalised philosophy, grounded on and continuous with the empirical sciences, *is* a way of philosophising. Such a style of doing philosophy, furthermore, did not emerge with Quine's invitation to naturalise epistemology (Quine 1969), but has its roots in the history of philosophy itself. As argued by Kitcher (1992), what we now call 'naturalised epistemology' is not even a novelty emerged in the second half of the XX century but, rather, the revival of the philosophical approach followed by modern thinkers such as Bacon, Descartes, Locke, Hume, or Kant. These philosophers grounded their normative theories of knowledge and rationality on the psychological and anthropological studies of their times, and embedded epistemology within a general view of knowing subjects and of the social and cultural world in which they interact. For Kitcher, returning to this approach is the remedy to the 'Fregean' methodology, consisting in reducing epistemological problems to questions about conceptual and logical analysis. Kuhn, at least in *SSR*, seems to have done precisely that. Kitcher's account of the 'pre-Fregean' approach to philosophy, therefore, allows us to understand why the Kuhn of *SSR* can be considered, without conflict or contradiction, as *both* someone who relies on the empirical findings of the sciences of his time *and* as a philosopher close in spirit to the likes of Kant.

Second, the idea that 'philosophical' (as opposed to 'empirical') images of science cannot be in principle challenged by empirical evidence is rather unwarranted. To make just few examples: Kant's conception of absolute Space and Time as the pure intuitions making everybody's experience possible has been challenged by the revolutions in physics and geometry occurred in the XX century; thinkers like Hegel or Marx are criticised on the ground that their philosophical systems are contradicted by the facts; and so on. Of course, once a philosophical system is empirically challenged, it does not need to be quickly condemned to oblivion. In fact, Kant's transcendental philosophy was reinterpreted by the neo-Kantians; some of Hegel's and Marx's ideas could still be fertile if revisited from within more realistic frameworks; and so on. Therefore, to say that Kuhn always had in mind a philosophical project does not bear the conclusion that such a project, being philosophical, is immune from empirical disconfirmation.

Finally, one should also wonder about the usefulness of an a priori philosophy of science that in principle cannot be refuted by the empirical studies of actual scientific activity. For Kindi, such a philosophy would be 'therapeutic' in a Wittgensteinian sense: it would free us from the received conception of science as progressively getting at the truth by observing pure facts. Undoubtedly, Kuhn's philosophy is *also* therapeutic in this sense. However, I find it difficult to regard it *only* as such. Part of Kuhn's project is also *normative*. To make progress in a cumulative way, scientists *ought to* bestow their consensus upon a paradigm and put on 'standby' the more foundational questions, so that they can focus on solving problems. To resolve a crisis, scientists *ought to* consider alternatives and make a revolution. These are all normative claims that tell what scientists ought to do for the sciences to make progress. Such claims, furthermore, are drawn from the description of how the mature sciences have behaved through history to make progress. Both the normative and the descriptive claims are therefore related to the history of science, and necessarily so, and they could have been hardly derived by first principles alone (granted that we do not even know what these first principles that Kuhn speaks of are, nor do we know how he could have derived conclusions from them).

Recently, Mladenović (2007, 2017) has proposed to overcome an unnecessary dichotomy presupposed by some of Kuhn's commentators, namely the idea that a model of science can only be *either* 'historical/naturalistic' *or* 'philosophical' (Mladenović 2017:60–61). In her view, normal science, revolution, incommensurability, and all the other concepts of *SSR* are neither a priori categories nor concepts directly drawn from the historical records through inductive generalisation. Like the explanatory categories used in social science, such as Max Weber's (1949) *ideal types*, the concepts Kuhn introduced in *SSR* are 'grounded' on the empirical evidence provided by the historical records but, being abstractions (or, indeed, 'idealisations'), they are not perfectly instantiated by every single historical case study. This interpretation saves the historical/naturalistic elements of Kuhn's philosophy, appreciated by Bird, but, at the same time, shows how an eventual empirical disconfirmation would require much more than a handful of counterexamples, thus avoiding the problems Kindi worries about. At the same time, I think that Mladenović's interpretation inherits some of the limits of the previous two interpretations.

It is true that ideal categories are difficult to disconfirm empirically. However, it is also true that their eventual confrontation with the empirical data is a necessary step to test their validity and usefulness. Using history as a ground for the development of explanatory ideal types and then not testing them back with new historical records would correspond to developing, at best, circular explanations, perpetually confirmed by the very things they seek to explain; and, at worst, 'empty explanations', that is explanans without any real explanandum. Mladenović's 'third way' between naturalism and apriorism needs to decide whether empirical data have the power to eventually challenge and disconfirm Kuhn's model or whether, by contrast, his ideal types cannot be empirically disconfirmed in principle. In the first case, talking about ideal types only postpone the problem of deciding how much evidence and of which kind is needed to empirically challenge Kuhn's philosophical system. In the second case, it would be not too dissimilar to Kindi's view for which such a system cannot be empirically challenged in principle.

Although for different reasons, furthermore, Mladenović, like Bird, seems to support some version of the 'turn' thesis. In her view, the 'early' and 'late' Kuhn had different targets. When he wrote SSR, he wanted to dislodge the traditional image of science and, to this aim, he relied on something that logical empiricism either ignored or dismissed as 'external': namely, the history of science. By the 1990s, the rejection of the traditional image of science through micro-histories had become mainstream, leading to constructionist and relativistic positions, the most extreme versions of which Kuhn found repugnant. His new challenge, then, was to show how, contrary to the constructionists and relativists, the history of science contributes to the rejection of the logical empiricist conception of rationality and progress, but *not* to the rejection of the very ideas of rationality, progress, and cognitive authority of science. It is to respond to *this* challenge that, for Mladenović, Kuhn began to talk about 'first principles'. Such an expression "was intended to mark the status of his response to this threat as a philosophical response rather than a historiographical or sociological one" (Mladenović 2017:61). In short, Mladenović argues that the 'early' Kuhn used history for philosophical aims, whereas the 'late' Kuhn began to shift his meta-philosophical position about the nature of the relation between history and philosophy of science in order not to reduce the latter to the previous, and to avoid demolishing, or 'deconstructing', the cognitive authority of science.

I find it rather puzzling to regard Kuhn's reference to first principles as just aiming at setting the 'philosophical tone' of his reply to scholars of constructionist persuasions. A philosophical argument does not need to be developed from first principles to be, indeed, 'philosophical' rather than historiographical or sociological—and this is something that Mladenović herself maybe would agree with. More importantly, like many other 'turn' interpretations, it is not entirely clear *when* Kuhn underwent it. Mladenović, in fact, overlooks the fact that Kuhn began his meta-philosophical reflections about the relation between history and philosophy of science much earlier than in the 1990s.

At the second Biennial Meeting of the Philosophy of Science Association, held in 1970 (the same year the second edition of SSR came out), Kuhn reacted against the peculiar characterization of the history of science defended by Imre Lakatos (1970), who had introduced a novel way for considering the internal-external dichotomy in the history of science. Traditionally, *internal history* of science is taken to be focused exclusively on the 'epistemic factors' driving science (i.e., the content of the scientific theories held by a given scientific community of the past, the epistemic reasons behind episodes of scientific change, and so on). *External history*, instead, considers the wider social, political, and institutional context in which science unfolds. Lakatos interpreted internal history of science as producing *rational reconstructions*, which are however much narrower than traditional internalist historical narratives, since they carefully omit many elements. Lakatos was aware of the limits of his rational reconstructions and suggested a certainly original method to overcome them: "[one] way to indicate discrepancies between history and its rational reconstruction is to relate the internal history in the text and indicate in the footnotes how actual history 'misbehaved' in the light of its rational reconstruction" (Lakatos 1970:107).

This way of doing history of science—or, rather, rational reconstructions—risks to remain biassed in favour of preconceptions about what it counts and what it does not count as rational. In Lakatos' view this is not even a problem and maintains that "[history] without some theoretical bias is impossible" (Lakatos 1970:107). The internal history of science, intended as rational reconstruction, is "a history of events which are selected and interpreted in a normative way" (Lakatos 1970:108). At the same time, he claims that "philosophy of science provides normative methodologies in terms of which the historian reconstructs 'internal history' and thereby provides a rational explanation of the growth of objective knowledge" (Lakatos 1970:91). In other words, a particular internal history of science is always produced in the light of a particular philosophy of science. This 'philosophy-derived history' is then used as an empirical tribunal for evaluating different philosophical theories about science: in other words, "two competing methodologies can be evaluated with the help of (normatively interpreted) history" (Lakatos 1970:91).

Kuhn points out that "because different methodological schools provide different criteria of rationality, internal accounts of scientific development vary from school to school" (Kuhn 1970:181). If every philosophical school always writes its own history of science, it follows that two competing philosophical theories will be underdetermined by the same pieces of historical evidence, which they both reinterpret from their own perspective. In short, for Kuhn, "what Lakatos conceives as history is not history at all but philosophy fabricating examples", which in principle cannot have "the slightest effect on the prior philosophical position which exclusively shaped it" (Kuhn 1970:143). For this reason, he also argues that "[history] done for the sake of philosophy is often scarcely history at all" (Kuhn 1970:183).

Kuhn then argues in favour of the *mutual independence* of history and philosophy. The two disciplines are distinct and separate, although discussing the historical dimension of science can sometimes be helpful to both historians and philosophers. Kuhn could not be more explicit about this point: he welcomes the dialogue between the two fields of study, but he also stresses that such a dialogue "must be inter- not intra-disciplinary" (Kuhn 1970:4).

Throughout his career, Kuhn remained consistent on this point. In reviewing Howson (1976) a collection of historical studies influenced by Lakatos' methodology, he even stressed why the independence of history from philosophy is beneficial to both disciplines. Historical narratives can be improved by gathering and interpreting new historical data. Contra Lakatos, however, even though they may follow historiographical approaches, historians do not need a philosophical theory of what counts as 'internally rational' to improve their narratives. Furthermore,

> Once the historian has provided such a narrative, the philosopher (sometimes the same person, but wearing a different hat) may examine it, asking about its significance for current doctrines concerning scientific method. If history is to have a chance of influencing those

> doctrines, however, such questions should be withheld, insofar as possible, until the task of the historian is complete. (Kuhn 1980:185)

Kuhn himself often operated like someone capable of wearing the 'historical' and the 'philosophical hat', but not at the same time. On the one hand, the way he discusses historical case studies in his philosophical writings is far from the richness of details expected in historical narratives proper. On the other hand, his historical work does not make use of his famous philosophical jargon. For example, in his *BB*, which tells the detailed history of the reformulations of Planck Law, leading the quantum revolutions, there is no mention of 'paradigm', 'incommensurability', and 'revolutions'. This even led to puzzlement about the relationship between *BB* and the philosophical model of science of *SSR*, to the point that Kuhn had to write an afterword explaining how the history of the quantum revolution does indeed follow the model of *SSR* even though BB, that is a work in the history of science, not in the philosophy of science, had to be written without presupposing its validity (Kuhn 1984; see also Timmins 2019; Potters 2022).

Philosophy and history are two separate disciplines and Kuhn kept them that way in his own work. That the same person can use the methods of the two disciplines does not mean that the two disciplines share the same aims, methods, and perspectives. While historians look at the historical details to construct a narrative as coherent as possible, philosophers are concerned with more general issues. As a result, "[the] Galileo or Descartes who [appears] in the philosophers' papers [is] a better scientist or philosopher but a less plausible seventeenth-century figure than the figure presented by the historians" (Kuhn 1970:7). Moreover, while historians need "a vast amount of research [...] before [beginning] to write" (Kuhn 1970:8), philosophers do not conduct research of this sort. However, they are engaged in something unknown to historians and scientists alike, namely a "Socratic [...] juxtaposition of views forged from each other through critical confrontation and analysis" (Kuhn 1970:9). These methodological differences derive from a fundamental difference between philosophers and historians, which is a difference of goals: the aim of the historian is the construction of a fine-grained narrative based on the details and particulars of the past, whereas the philosopher aims at drawing general conclusions. History and philosophy of science, then, are and must be kept separate: everything a philosopher needs from history "is less likely to be produced by marriage than by active discourse" (Kuhn 1970:20).

It is not too difficult to see how the 'Socratic juxtapositions of views through criticism and analysis' Kuhn talked about in 1970 anticipate his reference to a philosophy from first principles he hinted at in 1992. It is also not too difficult to see how the argument supporting his dynamic perspective, and clarified in the previous section, is developed through a 'Socratic juxtaposition' with respect to some central assumptions of the static view. Finally, it appears clear that Kuhn used the argument about the separation between history and philosophy not only to react against the irrationalist drift of the social constructionists, in the 1990s, but also, as early as the 1970s, to respond to a 'hyper-rationalist' philosopher like Lakatos. Such an argument is not a mere rhetorical device to mark his own position with respect to a particular target of criticism, as Mladenović argues. Rather, it is an argument that will allow us to disentangle the many elements present in Kuhn's work, and to look at how they are related.

3.4 Philosophical Perspectives, History of Science, and History-Based Models of Science

Kuhn wanted to develop a new philosophical perspective on science, and he also looked at the history of science. At the same time, he maintained that history and philosophy of science must be kept separate. How to make sense, therefore, of the co-presence of historical and philosophical elements in Kuhn's work?

For Hoyningen-Huene (1992), these elements are all inter-related, in the sense that "the history of science already determines, among other things, the realm of questions that can, in a sociological or philosophical perspective, be sensibly asked with respect to science" (Hoyningen-Huene 1992:490). In other words, the historical records pose questions about the nature of scientific progress and the community-based character of scientific research, which must be answered through philosophical arguments and the social study of the scientific community. I am very sympathetic with Hoyningen-Huene's reading, which has the merit of not dismissing the *many* elements of Kuhn's work. This reading, however, seems to presuppose a sort of unidirectionality: history asks, philosophy answers. By contrast, I propose to look at Kuhn's work as including (at least) three elements interacting in the following ways.

First and foremost, there is Kuhn's *dynamic perspective*, which challenges the assumptions of the static perspective and has several philosophical implications. Such a perspective is developed through 'Socratic juxtaposition' with the traditional, 'static' image of science. As already said, it is not clear whether so-called 'historical philosophers of science', who share the dynamic perspective, could have reached all their conclusions from 'first principles'. As a matter of fact, Kuhn's references to first principles are rare: for instance, there is no mention of them in Kuhn (2022). What Kuhn does mention, however, are two 'oddities' with the very title of 'historical methodology' attributed to the members of the new philosophy of science (*PW*, in Kuhn 2022:115–123). To begin with, none of them was a trained historian when they started to look at history. Moreover, the recourse to historical case studies in philosophy of science was not even a novelty. What was new, rather, was the *way of looking at* historical case studies. For instance, while the empiricist tradition relied on oversimplified and sketchy historical examples for illustrative purposes, the supporters of the dynamic perspective believed that the history of science could reveal something important about the nature of scientific development.

Being a perspective that determines how to look at and interpret empirical data about science (especially the records from the history of science), the dynamic perspective is less prone to empirical disconfirmation. This does not mean that new empirical data cannot challenge the dynamic perspective. Just as empirical evidence from the history of science and psychology challenges some of the presuppositions of the static perspective but it is not sufficient to discard it, the dynamic perspective can only be replaced by another philosophical perspective.

Then there is the *history of science*, which Kuhn arrived at thanks to the dynamic perspective. Kuhn was a philosopher *and* a historian of science, even though he was not both at the same time. He published two historical monographs—namely *CR* and *BB*—as well as several articles in international history of science journals. He was one of the leaders of the project for the creation of the *Sources for the History of Quantum Physics* archive. Any interpretation of Kuhn cannot elude the fact that he knew and practised the history of science. To construct his historical narratives, Kuhn used the framework developed by the so-called 'new historiography of science'. The 'old historiography' looks at the past as an imperfect version of the present and describes scientific progress as the gradual accumulation of successes which led to the present (more on this in the next section). The novel approach, by contrast, seeks to understand the past in the past's own terms, and accounts for both the successes and failures in science. Kuhn always recognized his intellectual debt towards notable French historians, such as Émile Meyerson, Hélène Metzger, and Alexandre Koyré, to whose works he referred often (see *SSR*: xi, 3, 48 f. 3, 49 f. 6, 67 f. 2). In his historical works, he follows their approach.

Through generalisation from some episodes of the history of science, interpreted with the tools of the new historiography of science, Kuhn develops a general *history-based model of science*, initially presented in *SSR*, which is one of the most influential books in philosophy of science even though philosophers themselves do not seem to agree on how to define it. Alexander Bird qualifies it as "not primarily a philosophy text [but] a work in what I call 'theoretical history'" (Bird 2000:viii). Michael Friedman maintains that Kuhn's account of scientific progress through revolutions is "our best current historiography of science" (Friedman 2001:47). For some, *SSR* developed "a very large-scale narrative of the development of science" (Golinski 2012:21), while for others it presented a "philosophical history in the Hegelian sense" (Graham 1997:32–33, 127). I am not sure whether these definitions are appropriate. *SSR* does not examine any case study with the level of detail typically found in scholarly works in the history of science, nor is Kuhn's contribution to the historiography of science unprecedented, as the new historiographical approaches were already established and he chose to draw upon them.

I take *SSR* to be a book in philosophy of science because, by relying on existing historiographical approaches, Kuhn describe several historical case studies that also play a normative function. In other words, SSR does not only describe what happened during the Chemical Revolution, or what happened during the transition from classical mechanics to quantum mechanics and Einstein's relativity. It also develops a general philosophical theory for *why* scientific progress requires periods of normal science guided by a paradigm; how the development and articulation lead to the discovery of anomalies and, ultimately, to its demise; and therefore, why scientific revolutions are the natural consequence of normal science, and why during periods of extraordinary science rational scientists may disagree. This is the model discussed and criticised by philosophers. As Kindi suggests, one should consider the relations of necessity between the central concepts of *SSR* (i.e., there would be no revolutions without normal science, the two are in a 'dialectical relationship'). However, it is undeniable that these concepts, which a philosophical analysis may reveal as being related by logical necessities, are drawn from the study of the history of science. As Mladenović would say, they are 'grounded' on the historical data, meaning that they are never perfectly instantiated in reality but, rather, they are ideal types that help us explain the case studies they are grounded on and, possibly, new case studies.

The history of science and the history-based model of science mutually interact. History provides the data that the model explains and generalises from. The model, in turn, may be used to explain new historical data. New and deeper looks at the historical data, however, may challenge the model, leading to its modification or even to its rejection. *Pace* Kindi, even philosophical models can be challenged by new empirical data, especially if they are *both* normative *and* descriptive. And what Mladenović does not stress enough is that even ideal types can be modified or rejected should they be found to be inadequate to satisfactorily explain the empirical data they are grounded on.

In *SSR*, Kuhn devotes a large chunk of the book to the history-based model of science, relegating to the last few pages the critique to the teleological account of the progress of science and the discussion of how scientific development is pushed from behind, rather than towards the ultimate truth. In *RSS* and in *PW*, by contrast, he focuses and expands on his philosophical perspective that, he laments, is yet to be fully appreciated, or even understood. Kuhn's mature writings, then, invite us to consider the elements of his philosophical system in a reverse order with respect to how he himself presented them in *SSR* and are normally discussed. Instead of considering the history of science and the history-based model of science first, and then their eventual epistemological and philosophical implications, one should examine Kuhn's philosophical perspective first and then consider how this led him to look at the history of science and to develop a history-based model of the development of science. Of course, this reverse order should not be intended as having only an expository function. It is, in fact, a reversal of the causal relations among the three elements of Kuhn's thought. As Kuhn says numerous times, the so-called 'historical philosophers of science' were not primarily concerned with the history of science per se. Rather, they were driven by their dissatisfaction with the empiricist static perspective on science and, to overcome its difficulties, aimed at developing a new perspective. Motivated by such a philosophical project, they decided to look at the history of science in a new way and, eventually, to develop history-based models of the progress of science.

In short, Kuhn's philosophical perspective is logically prior to his work in the history of science and to his history-based model of science. This does not mean that Kuhn himself always understood all the implications of his own perspective. He managed to explain of what he was arguing against (the static image) and of what he was proposing instead (the dynamic perspective) only in his last writings. And even such last insights need to be carefully interpreted and systematised.

Nevertheless, and to repeat, if he had not pursued the philosophical project of developing a new perspective on science, he would have not looked at the history of science to begin with. At the same time, beginning the analysis of Kuhn's philosophy from his dynamic perspective does not allow to uncover just his motivations to look at history. In fact, it also allows to clarify some elements of his history-based model of science. For example, arguing against the incommensurability thesis by showing that two theories are not logically contradictory completely misses the point. From a dynamic perspective, incommensurability is never about the comparison of two fully complete and fully formalised scientific theories, since scientists during periods of extraordinary science compare an established paradigm with an emerging one, which is far from being complete. Therefore, from a dynamic perspective, their choices cannot be justified by the retrospective reconstruction of the logical relationship of two complete theories.

Some final considerations are in order. First, the historically oriented philosophers of science all looked at history motivated by the same philosophical project and, in so doing, they may have ended up developing different or even conflicting history-based models of science. For example, Kuhn and Lakatos developed two different models of the historical development of science, but they did agree on the fact that evaluation in science is always comparative and on many other central tenets of the dynamic perspective. Despite the evident differences in their models of science, Kuhn and the other 'historically oriented' philosophers shared a philosophical perspective, which deserves to be discussed in all its implications.

Second, unlike the history-based models, this philosophical perspective can hardly be challenged by empirical observations. A philosophical perspective can be challenged only by philosophical arguments. The dynamic perspective is, in a sense, Kuhn's 'philosophical core' that, as Kindi would agree, cannot be empirically refuted.

Third, the recourse to the history of science is not even necessary to the dynamic perspective. As I will explain in the next chapter, Kuhn's philosophical perspective urged him to look at the actual scientific activity. He and other philosophers looked at the history of science, but they could have decided to look at other disciplines, such as sociology or the ethnography of science. Whether they would have developed sociology-based or ethnography-based models of science, and whether such models would have been consistent with their history-based models, is an open question.

3.5 Presentism

Having clarified the three elements present in his work, what could still perplex the reader of Kuhn's mature writings is not much his talk about philosophical perspectives and first principles but, rather, his defence of the value and usefulness of so-called Whig history. This may come as a shocking surprise, considering that Kuhn was one of its fiercest critics.

Towards the end of his third *SML*, Kuhn distinguishes two kinds of history. On the one hand, there is *hermeneutic history*, which attempts to understand the past on its own terms. This is the kind of history Kuhn practised throughout his career. He compares the complex interpretive approach of hermeneutic history to the ethnographic method followed by anthropologists when they observe and study an alien culture. To do hermeneutic history, the historian-ethnographer of science must learn to see the world and to think like the members of the culture under study (that is, the members of a scientific community of the past). On the other hand, there is the much-maligned *Whig history*, which considers the past as an imperfect version of the present. Whig history of science characterises past scientific knowledge as incomplete at best, and the scientists who produced it as naive or simply mistaken. The narratives produced by this kind of history conspire to create an image of the development of science as a process that gradually corrects mistakes and inevitably leads to the present state of the art.

> This sort of history does not attempt to acquire and deploy the older lexicons used by the communities under study; instead, it tells the story of the past in translation, abandoning the attempt to preserve truth values in exchange for preserving a constant truth. The resulting narrative does not proceed blindly but is constantly guided by what the translator—already possessed of and by the future—thinks is important to achieve. (*SML*, in Kuhn 2022:87)

As it is well known, Kuhn was one of the most strenuous critics of Whig historiography. Yet, towards the end of his career, he hinted at the possibility that such a historiography may still play some important functions.

> If one's concern is, like mine, primarily philosophical, directed to the nature of the historical process or the nature of human knowledge, there is no question which of these modes [the hermeneutical or the Whig mode] is to be preferred. I am well known for the consistency with which I have denounced and ridiculed Whig history. Given my concerns, I do not mean to quit. But it is correspondingly important for me to insist that Whig history has an indispensable human function that cannot be fulfilled by the sort of history in which I most believe. It supplies community members with a past which is not foreign but domestic, which can be assimilated directly, and which can serve as a platform from which to move ahead. (*SML*, in Kuhn 2022:87–88)

Whig history had been already criticised decades before Kuhn. For instance, Herbert Butterfield (the one who coined the very expression 'Whig history') wrote: "The study of the past with one eye, so to speak, upon the present is the source of all sins and sophistries in history, starting with the simplest of them, the anachronism" (Butterfield 1931:32). For a long time, avoiding Whiggisms has been the dominant norm among historians, in general, and historians of science, in particular. The work of Thomas Kuhn, as well as that of Meyerson, Metzger, and Koyré before him, surely contributed to its consolidation. More recently, however, some historians and philosophers of science have begun to challenge the anti-presentist norm.

Nick Jardine (2004), for instance, applies the concepts of *emic* and *etic standpoints*, developed by anthropologists, to the history of science. The emic standpoint is that of the observers who immerse themselves in a foreign culture, so that they can acquire and understand the point of view of the subjects under study. Anthropologists seeking to understand an alien culture in its own terms are therefore conducting emic interpretations. The etic standpoint is the observer's own point of view. Anthropologists must explain the culture they have studied in terms that can be understood by their own group of origin. To do so, they must rely on at least some of the categories of their native culture, thus engaging in etic interpretation. Similarly to the anthropologist, for Jardine, historians of science must interpret past science in its own terms, but the use of at least some of the contemporary categories is inescapable. To begin with, historians select historical periods and data that are deemed as interesting from the present perspective. Many studies in the history of science, in fact, attempt to uncover facts, events, and causal relations among them that may shed a new light on the present itself. For example, the history of modern medicine and epidemiology enriches our understanding of the causal mechanisms behind the spreading of some plagues of the past and this, in turn, may help us draw parallels with contemporary pandemics. Following Jardine, furthermore, and without denying the necessity of an emic component in the history of science, the reconstruction and interpretation of 'past science' presupposes a criterion of demarcation between 'science' and 'pseudo-science'. Even if, in an emic perspective, we may say that necromancy was considered a science in a specific context at a certain period, we also know from our present point of view that it is not. Failing to apply our current criteria could produce historical narratives that are incapable (or unwilling) to specify which parts of the past belief systems were wrong, so that they had to be modified or abandoned. In short, "a total ban on appeal to knowledge not possessed by the agents studied is a recipe for historiographical paralysis" (Jardine 2004:272). To make an example: a full history of witch hunts must also clarify that witchcraft does not exist, because failing to do so would impede us to condemn the unjust prosecution and killings of innocent women based on prejudices and false beliefs. To complement Jardine's use of the emic and etic categories, then, one could say that the history of science also possesses a moral dimension, which require the recourse to the contemporary viewpoint (Alvargonzález 2013).

A growing number of philosophers agree that there is more to presentism than Whig history, and that not all forms of presentism are bad. Some of them have distinguished and defined different 'presentisms', attempting to draw a line between legitimate and illegitimate forms of presentism (see, among the others, Loison 2016; Chang 2021; Barseghyan 2022). Those who defend the legitimacy of at least some forms of presentism stress the idea of the ineluctability of the present perspective to select salient features of the past, the 'genealogical' causal lines that connect the past with the present, the 'critical' function of history, or even, especially in Hasok Chang's proposal for an 'active pluralistic presentism', the recovery of abandoned past theories whose usefulness in solving some problems and making correct predictions may be re-assessed in the light of the present knowledge (see Chang 2009, 2010, 2013, 2021). Some philosophers even go as far as finding compatibilities between presentism and anti-presentism (Pulkkinen 2023).

So far, Kuhnian scholars have all the reasons to cheer. Not only Kuhn was one of the first philosophers to 'denounce and ridicule' Whig history at a time in which mainstream philosophers regarded the history of science as the immature prologue of contemporary science. It also turns out that, at a time when many philosophers, and in part thanks to his own work, openly endorsed and even took for granted anti-Whiggism, Kuhn became one of the very firsts to re-evaluate presentism, anticipating of various decades some recent debates. Of course, the temptation to 'mythologise' a philosopher ought to be resisted. Obviously, Kuhn did not anticipate everything that is discussed in recent philosophy of science. Besides, it appears that he did not have clear in mind the fact that while Whig history is a form of presentism, there are other forms of presentism that are not Whig history. Moreover, his remarks about the function of presentist history, although not as vague and sparse as those about first principles, are still very sketchy.

In the few passages where he speaks about the topic, Kuhn attributes mainly two functions to Whig history. The first is what I call an 'existential' function. By describing it in contemporary and, therefore, intelligible-for-us language, Whig history allows us to regard the alien past as *our own*, as something "which is not foreign but domestic" (*SML*, in Kuhn 2022:88).

Kuhn also speaks about a second function, which I call 'educational'.

> I was often asked whether science ought not to be taught historically, and I regularly responded that it should not, at least not to students who hoped later to practise science. These students need to acquire command of the current tools of the profession, both conceptual and instrumental; history is at best a slow and inefficient way to do that. Learning alternate possibilities—how Aristotle conceived motion or the void, how Volta conceived the battery, how Planck conceived his first derivation of the black-body law—these would be irrelevant distractions. Conceivably, it might even undermine the student's grasp of and faith in the tools required for current practice. [...] Like everyone else, their present identity requires an appropriate past, and that is provided by a sort of narrative which, at its not infrequent worst, simply attributes bits of current knowledge, items from current textbooks to historical figures who are supposed to have discovered them, displacing ignorance or superstition in the process. The very inadequacies, factual and conceptual, of that approach make it a more effective source of the identity required for the successful practice of science. If they have been formed by a lie—which I do quite intend to suggest—it is a noble one. (*SML*, in Kuhn 2022:88)

I find Kuhn's arguments rather unconvincing, if not a bit strange.

About the 'existential function' of presentism, he seems to think that to be grasped as our own, the past must be completely understandable by us in the present; presentism makes the past understandable; therefore, presentism is needed. The requirement of full understandability, however, may be too stringent. Much of childhood is opaque if not impenetrable to the adults we are today, yet we are able to recognise it as our own. On a larger scale, we do not understand what it is like to be a primate, yet we understand that humans and chimpanzees developed from a common ancestor. We simply do not need to recover and then translate in an adult language all the memories, experiences, and the full phenomenological dimension of childhood, nor do we need to look at the evolution of the living world as the progressive getting closer to humans, to understand the past, either individual or collective, as our own. In the same way, I see no difficulty in regarding past belief systems as part of our own heritage, despite their relatively 'alien' character.

As for the educational function, Kuhn is ambiguous. On the one hand, he claims that the 'noble lie' of presentism makes for an effective scientific training, and that hermeneutical history may be distractive. On the other hand, he also adds that "the sort of history I have practised should, I feel, be taught to a variety of audiences, *including scientists*, [though] it is not an appropriate route to professionalisation" (*SML*, in Kuhn 2022:88, *my emphasis*). Either hermeneutical history should be taught to a varied audience that includes scientists or, if there is a high risk of distraction, perhaps history should not be taught *at all* to scientists. Rather than with presentist history of science, scientists could just be trained with only present science. In this way, there would be no need to re-evaluate presentism.

Contemporary historians and philosophers of science are slowly challenging the largely unquestioned anti-presentist stance that has dominated the historical studies for several decades. Kuhn may have been one of the first philosophers to question such a stance, but not with the right arguments. On reading them, one has the impression that he was just beginning to think about these issues and to scratch the surface of the gulf between hermeneutical and presentist historiography. Unfortunately, whether his re-evaluation of presentism was supposed to play some role in his new articulation of the issue of scientific progress can never be known (see below, Chap. 10). At the same time, however, his short foray on these issues may shed some more light on his overall philosophical project.

Famously, at the very beginning of *SSR*, Kuhn stated that "History, if viewed as a repository for more than anecdote or chronology, could produce a decisive transformation in the image of science by which we are now possessed" (*SSR*:1). Clearly, viewing the past as more than a bunch of anecdotes require the hermeneutic approach to history. But *who*, exactly, is possessed by the current image of science and, more importantly, must be liberated? Apparently, not the scientists, who should not be taught hermeneutic history of science (or history of science at all). In his prescriptions, Kuhn is unequivocal: scientists must be trained in the light of the dominant paradigm, and their job is not to question it. A more historical and dynamic perspective could be not only distracting, but even counterproductive. But if this is the case, *for whom* did Kuhn develop such a perspective?

In Kuhn's view, scientists are part of a rather insulated community, which has its own norms, codes, and ethos. Guided by their paradigm, scientists decide which problems to solve and how. At the same time, and however insulated it may be, the scientific community is still part of society. For its part, society heavily relies on science to make many decisions. The problem is that, often, society relies on science without understanding how it works. Kuhn was aware of this problem since the publication of *CR*, where he observed that

> Contemporary Western civilization is more dependent, both for its everyday philosophy and for its bread and butter, upon scientific concepts than any past civilization has been. But the scientific theories that bulk so large in our daily lives are unlikely to be final [...]. The mutability of its fundamental concepts is not an argument for rejecting science [...]. But an age as dominated by science as our own does need a perspective from which to examine the scientific beliefs which it takes so much for granted. (*CR*:3–4)

The specific results of scientific research can be assessed by scientists only, who do not need to know the history of their own discipline to make it advance. However, science in itself can be looked at from a historical/dynamic perspective to free society from the static image, which depicts scientific progress as the infallible and inevitable piling up of new neutral facts over the old ones. This means that Kuhn develops his historical/dynamic perspective not for scientists, but rather for philosophers *as well as* for citizens, who live in a society that is strongly dependent on science even though it often misunderstands it. This perspective, in other words, is an instrument for the *social criticism* of science, where 'criticism' must be understood not as a way to deconstruct, demystify or reject science but, rather, as the activity of assessing its function and limits. (For a Kuhnian perspective on the science/society relationship, see below, Chap. 9).

3.6 Conclusions

What do we talk about when we talk about Thomas Kuhn?

Kuhn's philosophy is multifaceted and complex. Often Kuhn critics do not clarify which of its parts are discussing or criticizing, whether his history-based model of science, his way of doing history of science, or his more fundamental philosophical perspective. Other times, they misconstrue some of the parts of Kuhn's philosophy because they fail to understand properly how they relate to the others. For instance, rejecting Kuhn's philosophy and proclaiming the need of a 'post-Kuhnian' approach because empirical data challenge some parts of the *SSR* model does not consider that his dynamic perspective, being a philosophical perspective, would need another philosophical perspective to be fully replaced. (Arguments in favour of a 'post-Kuhnian' approach are even more puzzling when considering that often the historical data used to challenge the *SSR* are collected from a perspective that appears similar to Kuhn's.) Or, to make another example, rejecting the incommensurability thesis by showing that one complete scientific theory can be logically derived from another one means misunderstanding the dynamic perspective from which Kuhn developed his thesis (as well as the fact that actual scientists almost never have to choose between two well-formed and complete theories). And so on. Disentangling the dynamic perspective from the history of science and from the history-based model of science, and examining how they are related, is crucial to assess Kuhn's philosophy.

References

Alvargonzález, David. 2013. Is the history of science essentially Whiggish? *History of Science* 51:85–99.

Andersen, Hanne. 2001. *On Kuhn*. Belmont, CA: Wadsworth/Thomson Learning, Inc.

Barseghyan, Hakob. 2022. Selection, presentism, and pluralist history. *Studies in History and Philosophy of Science* 92:60–70.

Bird, Alexander. 2000. *Thomas Kuhn*. Princeton: Princeton University Press.

Bird, Alexander. 2002. Kuhn's wrong turning. *Studies in History and Philosophy of Science* 33:443–463.

Butterfield, Herbert. 1931. *The Whig Interpretation of History*. London: G. Bell and sons.
Chang, Hasok. 2009. We have never been whiggish (about phlogiston). *Centaurus* 51:239–264.
Chang, Hasok. 2010. The hidden history of phlogiston: How philosophical failure can generate historiographical refinement. *HYLE - International Journal for Philosophy of Chemistry* 16:47–79.
Chang, Hasok. 2013. *Is Water H2O? Evidence, Realism and Pluralism*. Dordrecht: Springer.
Chang, Hasok. 2021. Presentist history for pluralist science. *Journal for General Philosophy of Science* 52:97–114.
Friedman, Michael. 2001. *Dynamics of Reason*. Stanford: CSLI Publications.
Golinski, Jan. 2012. Thomas Kuhn and interdisciplinary conversation: Why historians and philosophers of science stopped talking to one another. In *Integrating History and Philosophy. Problems and Prospects*, ed. Seymor Mauskopf and Tad Schmaltz, 13–28. Dordrecht: Springer.
Graham, Gordon. 1997. *The Shape of the Past*. Oxford: Oxford University Press.
Howson, Colin. 1976. *Method and Appraisal in the Physical Sciences*. Cambridge: Cambridge University Press.
Hoyningen-Huene, Paul. 1992. The interrelations between the philosophy, history and sociology of science in Thomas Kuhn's theory of scientific development. *British Journal for the Philosophy of Science* 43:487–501.
Jardine, Nick. 2004. Etics and emics (not to mention anemics and emetics) in the history of the sciences. *History of Science* 42:261–278.
Kindi, Vasso. 2005. The relation of history of science to philosophy of science in *The Structure of Scientific Revolutions* and Kuhn's later philosophical work. *Perspectives on Science* 13:495–530.
Kindi, Vasso. 2013. The Structure's legacy: Not from philosophy to description. *Topoi* 32:81–89.
Kitcher, Philip. 1992. The naturalists return. *Philosophical Review* 101:53–114.
Kuhn, Thomas S. 1970. Notes on Lakatos. *PSA: Proceedings of the Biennial Meeting of the Philosophy of Science Association 1970*:137–146.
Kuhn, Thomas S. 1980. The halt and the blind: Philosophy and history of science [Review of *Method and Appraisal in the Physical Sciences: The Critical Background to Modern Science, 1800–1905*, by C. Howson]. *The British Journal for the Philosophy of Science* 31:181–192.
Kuhn, Thomas S. 1984. Revisiting Planck. *Historical Studies in the Physical Sciences* 14:231–252. Reprinted in the second edition of *BB*.
Kuhn, Thomas S. 1992. The Trouble with the Historical Philosophy of Science. *Occasional Publication of the Department of History of Science*. Cambridge: Harvard University. Reprinted in RSS, 105–120.
Kuhn, Thomas S. 2022. *The Last Writings of Thomas Kuhn*, ed. Bojana Mladenović, Chicago: Chicago University Press.
Lakatos, Imre. 1970. History of science and its rational reconstructions. *PSA: Proceedings of the Biennial Meeting of the Philosophy of Science Association 1970*:91–136.
Loison, Laurent. 2016. Forms of presentism in the history of science: Rethinking the project of historical epistemology. *Studies in History and Philosophy of Science* 60:29–37.
Melogno, Pablo. 2019. The discovery-justification distinction and the new historiography of science: On Thomas Kuhn's Thalheimer Lectures. *HOPOS* 9:152–178.
Mladenović, Bojana. 2007. "Muckraking in history": The role of the history of science in Kuhn's philosophy. *Perspectives on Science* 15:261–294.
Mladenović, Bojana. 2017. *Kuhn's Legacy. Epistemology, Metaphilosophy, and Pragmatism*. New York: Columbia University Press.
Potters, Jan. 2022. Conceptualising paradigms: On reading Kuhn's history of the quantum. *Annals of Science* 79:386–405.
Pulkkinen, Karoliina. 2023. On compatibility between presentism and anti-presentism in history of science. *Journal of the Philosophy of History* 17:310–327.
Quine, William V. O. 1969. Epistemology Naturalized. In *Ontological Relativity and Other Essays*, William V. O. Quine, 69–90. New York: Columbia University Press.

Shapere, Dudley. 1984. *Reason and the Search of Knowledge*. London: Springer.
Sharrock, Wes, and Rupert Read. 2002. *Kuhn: Philosopher of Scientific Revolutions*. Oxford: Polity.
Timmins, Adam. 2019. Between history and philosophy of science: The relationship between Kuhn's *Black-Body Theory* and *Structure*. *HOPOS* 9:371–387.
Weber, Max. 1949. *The Methodology of the Social Sciences*. Trans. and ed. Edward A. Shills and Henry A. Finch. New York: Free Press.

Chapter 4
Kuhn's Theory of Collective Rationality

Abstract One of the implications of the dynamic perspective is the shift from the traditional individualistic and 'algorithmic' theory of scientific rationality to a theory of collective rationality. Such a theory needs to specify who, exactly, is the rational agent of science. From *SSR* on, Kuhn stresses the importance of the scientific community and, by doing so, he can be credited with the development of an earlier form of social epistemology. At the same time, it is important to point out that he contemplated the existence of scientific communities of different levels of generality. This, in turn, may shed some new light on his history-based model of science. I conclude this chapter by discussing the relation between the history-oriented philosophy of science and the sociological approach to the study of scientific groups.

Keywords Collective rationality · Collective agency · Essential tension · Scientific communities · Social dimension of science · Historical studies of science · Social studies of science

4.1 Introduction

If one accepts that there exists an absolute and neutral foundation of scientific knowledge, being it experience or apodictic ideas, as well as a universal method for obtaining it, being it an inductive procedure to ascend from particular experiences to general claims or a deductive procedure to descend from universal laws to the explanation of particular cases, then one also accepts that every rational individual must 'tap' into the same foundational source and use the same method to obtain objective knowledge. In the static perspective, therefore, there can be no individual differences between rational individuals. If someone's makes decisions that are different from those of the other rational individuals, then it must be because of error or plain irrationality. The static perspective, in other words, holds an 'algorithmic' theory of rationality, which tells scientists to grab the right ingredients and follow

V. Politi, *Thomas Kuhn's Philosophical Perspective*, Boston Studies in the Philosophy and History of Science 348,
https://doi.org/10.1007/978-3-032-13540-7_4

the right instructions of the methodological receipt so that, in the end, everyone will get to the same conclusion. However, as Kuhn points out,

> The search for algorithmic decision procedures has continued for some time and produced both powerful and illuminating results. But those results all presuppose that individual criteria of choice can be unambiguously stated and also that, if more than one proves relevant, an appropriate weight function is at hand for their joint application. Unfortunately, where the choice at issue is between scientific theories, little progress has been made toward the first of these desiderata and none toward the second. Most philosophers of science would, therefore, I think, now regard the sort of algorithm which has traditionally been sought as a not quite attainable ideal. [...] Even an ideal, however, if it is to remain credible, requires some demonstrated relevance to the situations in which it is supposed to apply. (Kuhn 1977, in *ET*:326)

By rejecting the static perspective, Kuhn also rejected its unsuccessful theory of scientific rationality. This does *not* mean that he rejected the idea that science is a rational enterprise. Rather, it means that "[t]here is no neutral algorithm for theory-choice, no systematic decision procedure which, properly applied, must lead each individual in the group to the same decision" (*SSR*:200). In the dynamic perspective, scientific rationality is considered from the perspective of the scientific group, rather than from that of the (idealised) rational individual. From this perspective, disagreement among scientists may be rational, and even a desirable feature of well-functioning scientific groups. The rejection of the traditional theory of rationality, therefore, does *not* imply the rejection of the idea that science is a rational enterprise. In other words, Kuhn wanted to propose a new theory of scientific rationality, not to claim that science is irrational.

This is one of the most significant aspects of Kuhn's philosophy, but also one of the most misunderstood. On the one hand, some sociological and constructionists interpretations read *SSR* as defending an 'irrationalist' image of science. In several post-*SSR* writings, however, Kuhn stressed how he found such an idea "not only absurd but vaguely obscene" (Kuhn 1970, in *RSS*:159). On the other hand, Kuhn himself may have confused several readers when, in *SSR*, he described some episodes of scientific revolutions as if they had broken out through the intervention of some individual genius, such as Lavoisier, Newton, or Einstein. In reality, it would be enough to examine some of his pre-*SSR* writings, in particular 'The Essential Tension", as well as several of his post-*SSR* works, starting with the Postscript, to get a much more sophisticated picture of his view. This chapter, therefore, is about Kuhn's collective approach to scientific rationality.

In Sect. 4.2, I introduce Kuhn's theory of collective rationality, for which decisions are made by an internally diversified group. The theory of collective rationality poses questions about whether the agent of science is the individual or the group itself, which I address in Sect. 4.3. In Sect. 4.4, I explain how Kuhn considered the existence of scientific communities of different levels of generality, and how this contributes to a better understanding of his model of science. In Sect. 4.5, I compare Kuhn's theory of collective rationality to those held by some of his most notable critics. Finally, in Sect. 4.6, I discuss the analogies and differences between the history-oriented philosophy of science and the sociological approach to the study of the practice carried out by scientific groups.

4.2 Collective Rationality

A non-algorithmic theory of scientific rationality should possess at least two features. First, it should not dismiss disagreement as evidence that one of the disagreeing parties is simply being 'wrong' or 'irrational'. It ought to explain, in other words, how individual variations of judgment and 'rational disagreement' are possible. Second, the theory ought to explain how such individual variations and rational disagreements within a group do not conflict with group-rationality, but they rather constitute it.

Kuhn (1977) explains that scientific evaluation is not governed by universal criteria. In the absence of the Archimedean platform provided by an absolute and self-evident foundation, scientific evaluation is guided by a set of 'values', which are (though the list is not exhaustive): accuracy, consistency (both internal to a theoretical framework and in relation to the established knowledge), scope, simplicity, and fruitfulness. These values are 'constitutive' of science: theories that display them are scientific whereas, for example, inaccurate theories, or theories that go against every piece of established background knowledge, are not. Scientific theories may possess these values in varying degrees; for instance, one theory may be simpler than another one, which may however have a wider breadth of scope. The constitutive values, furthermore, are open to interpretation. Some scientists may define consistency mainly, or only, as 'internal consistency'; simplicity may mean different things to different individuals; and so on. Even though they agree on *what* the constitutive values of scientific theories are, scientists may have different views on *how* to define them. Moreover, they may also disagree about how to weigh and rank them. A scientist valuing simplicity over fruitfulness will choose a different theory than a scientist valuing fruitfulness over simplicity. None of them, however, could be deemed as 'irrational'. In short, the constitutive values of science, that in Kuhn's theory of rationality replace the universal rules of the static perspective, guide scientists in the evaluation process, but they do not determine its outcome.

Three things must be noticed. First, as recently noticed by De Benedetto and Luchetti (2023), while adopting a certain ranking of values drives the acceptance of certain theories over others, the acceptance of such theories, in turn, reinforces that ranking of values, which will therefore drive further research. Between values and theories, in short, there is a 'feedback-loop', "[a] dynamic by virtue of which the outcome of a certain system or process becomes the input for the same system or process at a later stage" (De Benedetto and Luchetti 2023:4; see also De Benedetto and Luchetti 2024).

Second, it is crucial to remember that the dynamic perspective looks at changes as they occur. The static perspective analyses fully fledged and complete theories. During theory change, however, scientists often find themselves choosing a theory that has not been fully developed yet, but that carries with itself the promise of being more fruitful, or more coherent, and so on. This process, which cannot be captured by the algorithmic approach, requires a study of the social dimension of science. Kuhn was accused of reducing epistemology to psychology. His plea to ground a collective theory of scientific rationality on sociological and psychological studies, however, could now be appreciated as one of the first attempts at a naturalized epistemology.

Third, Kuhn often speaks about 'theory choice', even though it is not entirely clear what he meant by 'scientific theory' to begin with. In any case, I take his considerations about the different ways in which values are defined and applied as valid not only for theory choice, but also for the evaluation of models, experimental methods, and other things scientists have to choose and decide about.

The theory of collective rationality, implied by the dynamic perspective, is one of the backbones of the history-based model of *SSR*. During scientific revolutions, differences in defining and ranking the constitutive values lead to the phenomenon of *methodological incommensurability* (see Sect. 1.4.3). This form of incommensurability does not create 'communication breakdowns' between opposing groups of scientists talking at cross-purposes. In fact, different groups of scientists may understand each other's theoretical languages, but they may still face irreconcilable disagreements over the application of values in the evaluative process. It is important to stress, however, that for Kuhn these differences are present at every stage of scientific development, even during periods of normal science. Judgment diversity, furthermore, is not a weakness or a sign of irrationality, but it is a mark of scientific rationality.

At a talk he delivered at the University of Utah in 1959, entitled 'The Essential Tension", Kuhn distinguishes two modes of reasoning in science: *convergent* and *divergent thinking*. The former is usually defined as a sort of dogmatic acceptance of the current tradition, while the latter is generally regarded as a mark of intellectual freedom, creativity, and an inclination to 'think out of the box'. Contrary to the common assumption that divergent thinking is the propulsive force behind the progress of science, Kuhn argues that convergent thinking is not only essential to scientific practice but, also, that it is the actual source of the discovery of major novelties.

Kuhn begins his description of convergent thinking by speaking about the process of education undergone by aspiring scientists. Such an education, based on special textbooks, aims at indoctrinating students into the current normal scientific tradition. Nevertheless, Kuhn insists that such a process of indoctrination is neither castrating nor does it constitute a weakness of the educational system. In fact, it is only after acquiring the dominant paradigm that scientists can discover its limits. As paradoxical as it may sound, tradition-bound work is the necessary condition for overthrowing the current tradition: only by applying and developing the existent paradigm it is possible to discover its weaknesses.

> [T]he practitioner of a mature science, from the beginning of his doctoral research, continues to work in the regions for which the paradigm derived from his education and from the research of his contemporaries seem adequate. He tries, that is, to elucidate topographical details on a map whose main outlines are available in advance, and he hopes—if he is wise

> enough to recognize the nature of his field—that he will someday undertake a problem in which the anticipated does *not* occur, a problem that goes wrong in ways suggestive of a fundamental weakness in the paradigm itself. In the mature sciences, the prelude to much discovery and to all novel theory is not ignorance, but the recognition that something has gone wrong with existing knowledge and beliefs. (Kuhn 1959, in *ET*:235, *original emphasis*)

The work of normal scientists has, therefore, two opposite and complementary functions: to confirm the current dominant paradigm, by trying to fit as many segments of nature as possible in its pre-constituted boxes, and to push the paradigm to a point in which it is disconfirmed. These two functions are not necessarily carried out by every scientist at the same time.[1] For Kuhn, normal science is pervaded by a *tension* between dogmatic and critical attitudes toward the dominant paradigm. Such a tension is indeed *essential* to the progress of science through revolutions. Some scientists may be so divergent in their thinking that they may end up endorsing unorthodox positions. Although dissidents of this sort rarely manage to overthrow the dominant paradigm, they carry out the fundamental task of maintaining, within normal science, a 'reserve tank' of alternative theories to be used in case of crisis, when the majority's consensus cracks and scientists are more open to consider new problem-solving approaches.

What emerges after taking stock of the idea of essential tension is a picture in which scientists' consensus and shared practice does not conflict with a dissent endemic to normal science. One of the few philosophers who has noticed this point is Joseph Rouse, who points out:

> There is room for considerable disagreement within [...] a research field. Even within normal science, scientists assess differently what is possible, plausible, or promising. [...] Yet such a divergence is always held in partial check, both by the use of common resources and by a more or less common sense of what is at issue in their field, why it matters, and what must be done to resolve these issues. The acceptance of a paradigm is thus not a matter of monolithic agreement within a community, but rather of sufficient common ground to make disagreement both intelligible and interesting. [...] Kuhn has often been misread as insisting that members of scientific communities are in substantial agreement about fundamental issues in their field. What he actually says is that normal science rarely engages in controversy about such fundamentals. A lack of controversy is quite consistent with extensive disagreement, however, if research can proceed coherently and intelligibly without having to resolve the disagreements. (Rouse 2003:109–110)

[1] It is also debatable whether scientists are *aware* that they are carrying out such functions. For instance, some of them may think that they are just attempting to solve yet another puzzle and to confirm the paradigm once again, yet they work ends up causing a crisis. However, it may also be the case that some scientists openly attempt to challenge the paradigm, because they would benefit from a revolutionary discovery or the opening up of a radically new research line. If no one aimed at overthrowing the paradigm, then the new generation of science should just be content with solving the tiniest puzzles and ticking the last boxes in an overcrowded field where the main discoveries have already been made. The point is that, from a collective perspective, the awareness or the intentions of individuals are not even too relevant for assessing the rationality of a group, as I will discuss towards the end of this section.

From a descriptive point of view, disagreement in science simply exists. It is in virtue of divergent thinking that new theories are created. That science ought always to encourage diversity and variability, even during periods of so-called normal science, to be ready to deal with an unexpected crisis—in other words, for its own 'survival'—is a prescriptive maxim.

> Most newly suggested theories do not survive. Usually, the difficulties that evoked them are accounted for by more traditional means. Even when this does not occur, much work, both theoretical and experimental, is ordinarily required before the new theory can display sufficient accuracy and scope to generate widespread belief. In short, before the group accepts it, a new theory has been tested over time by the research of a number of men, some working within it, others within its traditional rival. *Such a mode of development, however, requires a decision process which permits rational men to disagree, and such disagreement would be barred by the shared algorithm which philosophers have generally sought.* If it were at hand, all conforming scientists would make the same decision at the same time. With standards for acceptance set too low, they would move from one attractive global viewpoint to another, never giving traditional theory an opportunity to supply equivalent attractions. With standards set higher, no one satisfying the criterion of rationality would be inclined to try out the new theory, to articulate it in ways which showed its fruitfulness or displayed its accuracy and scope. I doubt that science would survive the change. What from one viewpoint may seem the looseness and imperfection of choice criteria conceived as rules may, when the same criteria are seen as values, appear an indispensable means of spreading the risk which the introduction or support of novelty always entails. (Kuhn 1977, in *ET*:332, *my emphases*)

If every one of its members places the same 'bet' (i.e., choses the same theory, adopt the same problem-solving strategy, and so on), the whole group would risk too much. In fact, a potentially successful and yet non-mainstream research strategy could remain underdeveloped or even completely abandoned, and it may turn out that all bets were put on a losing strategy. By 'spreading' the risk—i.e., by allowing members of the group to place different bets—the group as a whole can work on many tools and strategies at the same time (see also Rueger 1996; D'Agostino 2005, 2010). By promoting the survival of science, the essential tension, which allows the spreading of risk within scientific research groups, is the mark of collective rationality.

One problem that Kuhn overlooked is the potential conflict between individual and collective rationality. For a group, it is rational to invest most of its 'intellectual capital' on the current paradigm, which has proven its problem-solving effectiveness numerous times already, while keeping few scientists on the margins working on a 'reserve tank' of alternatives to be used in case of crises. For an individual scientist, however, it would not be rational to work on models, methods, or theories that are regarded as 'loosing bets' by most of their colleagues, thus remaining at the margins of their own community. How could individuals be persuaded to make such an irrational choice for the sake of maintaining the essential tension?

This is the problem that Philip Kitcher (1990) attempts to solve in his famous model of the *distribution of cognitive labour*. Kitcher develops an analytic model of the scientific community that represents scientists as self-regarding rational agents pursuing the maximisation of their own utility. The model assumes that the community aim at reaching an objective (i.e., the solution to a relevant scientific problem), that there are two available methods for reaching it, that one of the two has a higher probability of success and, finally, that all the agents know in advance the number of available methods, their probability of success, and each other' choices. In this scenario, most individuals will choose the method with a higher probability of success. However, joining an already overcrowded team would not make the team more likely to succeed, nor would make its members more likely to give some relevant contributions. This is why for some of the scientists it would be more rational to choose the method with a lower probability of success. Self-regarding scientists, in fact, do not aim at increasing the probability of success of the group but, rather, they choose the best strategy for maximizing their own success, which may come in terms of prestigious publications, career advancement, or public recognition. Choosing the research strategy with a lower probability of success, therefore joining a less crowded and less competitive group, may actually improve one's likelihood of making some relevant contribution and, therefore, maximising one's own personal success. This is why for some scientists it is rational to choose the less popular research strategy. In this way, the rational choices of individuals contribute sustain the essential tension and guarantee the rationality of the group.

Kitcher's model has inspired the development of more recent computer-based simulation that models the social organisation of research groups—see, for example, Zollman (2007, 2010) and Weisberg and Muldoon (2009).[2] All these models could be considered, in a sense, as the contemporary and formalized development of Kuhn's seminal idea about the essential tension, as I have argued in Politi (2021). It must be said that, in these models, all the agents agree on the relevance of the problem to be solved and on the existing problem-solving strategies. Therefore, they provide a formal representation of how normal science is carried by an internally diversified research group. It cannot be excluded that by adding or modifying some parameters they could also be used to represent scientific revolutions, which would surely involve a *redistribution* of cognitive labour within the group.

These models, however, may conflict with the Kuhnian perspective. Even though their rational agents do not apply some universal rule of theory choice, they follow the utility-maximising rules of classical Rational Choice Theory (in Kitcher's model), or the rules of behaviour established by the programmer (in the case of computer simulations). This is still too close to a rather 'algorithmic' and context-free view of scientific rationality. Furthermore, these models, like logic (see Sect. 2.3), are 'empty'. The modeller proceeds by establishing the set of rules, the number of agents, the range of possible choices, and the objective to be reached. After that, calculations are made, or simulations are run. The model itself, however, does not need to have any actual content, any real target. Only an interpretation of the model reveals that its agents represent scientists, its objective represent a scientific problem, and so on. And this, too, looks like a variation of the kind of formalization that Kuhn would resist.

---

[2] For a general introduction to contemporary computer-based models of the social structure of science, see Weisberg (2010) and Šešelja (2021, 2023).

Finally, it must be noticed that formal models of the scientific community rely on so-called *methodological individualism*, which holds that collective entities (such as states, institutions, or classes) do not have causal powers independent of the individuals who constitute them. Therefore, any adequate explanation of social structures or outcomes must ultimately be reducible to individual-level processes. Supporters of methodological individualism, in other words, holds that the behaviour of a group must be explained by looking only at the choices and behaviour of its individual members. In this way, they reduce the study of groups to the study of individuals and their choices. This approach has a solid tradition in philosophy as well as in the social sciences. It must be said, however, that foundational discussions about methodological individualism abound and that the doctrine itself has been reformulated in various ways, thus forming a spectrum that includes strong and moderate views. Here it would be impossible to provide an adequate review of the different versions of methodological individualism (for an introduction to the topic, see Udehn 2001).[3]

The point here is not whether methodological individualism is right or wrong in principle. Rather, what it must be stressed is that, for Kuhn, members of a scientific group belong to the same research culture, they have the same presuppositions, they share the same values, theories and metaphysical worldviews, as I will explain in the next chapter. Scientists are trained in light of the dominant paradigm and are influenced by it. Yet, at the same time, they do not make the same decisions, and they do not necessarily share the same attitude towards the paradigm. In short, from a Kuhnian perspective, thinking that the behaviour of the research group supervenes the behaviour of its individual members would be too simplistic. The relationship between the collective and the individual dimension within scientific groups is far more complex. Therefore, a Kuhnian approach to the study of the scientific communities should follow other routes.

4.3 Collective Agency

It has been suggested that, from how Kuhn describes both the periods of normal science and the episodes of revolutionary changes it seems that, in his philosophy, the scientific research group is not only the locus of science, that is the place where decisions are made and changes happen, but also "the agent of science" (Hoyningen-Huene 1993:8, 65, 82), that is the agent who makes the decisions and operates

[3] As discussed in the previous chapter, the dynamic perspective rejects what Kuhn calls 'methodological solipsism', which is different from methodological individualism. What Kuhn rejects is the idea that the same set of universal truths can be obtained by every rational individual by following the same rules of the scientific method. This idea implies that the only explanation of disagreement is error or irrationality. Methodological individualism, by contrast, allows for individual rational agents to make different choices and it explains how such different individual behaviours determine social macro-phenomena.

change. To say that the scientific community is the agent of science, rather than just the locus where science grows and undergoes changes, means to say that the scientific community produces and possesses scientific knowledge, and that it makes decisions when it comes to discard an old paradigm in favour of a new one. In short, from how he characterizes it, and for the role it plays in his model, it appears plausible to interpret Kuhn's scientific community as a *collective agent*.

It is worth stressing that a non-individualistic theory of collective agency does not require postulating the existence of metaphysically dubious entities, such as that of a 'collective mind', but only the specification of the group structures that facilitate collective reasoning and acting. The same Kuhn explicitly rejected the idea of a group mind and, with it, his own early use of the analogy between scientific revolutions, that involve scientific community, and Gestalt switches, that involve individuals' perceptions. The Gestalt analogy, in his view, would be an obstacle to understanding his collective approach. As Kuhn came to realize, in fact, attributing to a group properties and processes of individuals, like a mind, corresponds to a category mistake: "The mistake is to treat groups as individuals writ large or else individuals as groups writ small. It results, at its crudest, in talk of the group mind (or group interests)" (Kuhn 1993, in *RSS*:241–242). All a non-individualistic theory of collective agency needs is to specify the structure of a group that makes it possible for its behaviour not to be reducible to the behaviour of its individual members.

One of the theories of collective agency that does not presuppose the existence of a collective mind is developed by Margaret Gilbert (1990, 2013). In her view, what distinguishes random groups of people from proper collectives is the establishment of a *plural subject*, which is achieved when individuals share the same intentions and know that they share the same intentions. Members of a plural subject are bound together by their commitment to the group they are part of. For Gilbert, it is possible to attribute to plural subjects both collective intentionality and collective beliefs—that is, intentions and beliefs that are more than the sum of individuals' intentions and beliefs.

Gilbert applies her theory of plural subjects to scientific communities (Gilbert 2000). She also uses it to redescribe Kuhn's position (Weatherall and Gilbert 2015). In her view, the theory of the plural subject explains the scientists' resistance towards alternatives to the established paradigm, a phenomenon that characterizes normal science and that may lead to controversies between the supporters of the old tradition and those of the new paradigm during periods of extraordinary science. Since members of a scientific community form a plural subject and are therefore bound together by their group commitment to the old paradigm, they tend to treat everything that challenges their group identity as an alien menace.

Gilbert's account is not left unchallenged. For instance, it is possible to imagine a collection of individuals who share the same goals and objectives, who know about their shared intentions, who are also committed to reaching a common end, and yet act in isolation with one another. Some philosophers would not consider such a group as a proper collective. This is what Bratman (1992, 1993, 2013) argues: in his view, what characterizes a collective agent is not shared intentionality and beliefs, but *cooperative activities*. The problem, then, would be to investigate whether shared intentionality and beliefs are always necessary to cooperation. If, one the one hand, Pettit (2009) claims that joint commitment is necessary but not sufficient to the establishment of socially integrated groups, on the other hand, List and Pettit (2011) demonstrate how a group may make decisions that do not reflect the preference of any of its members in particulars, thus challenging the necessity of joint intentionality and joint beliefs for collectives.

When it comes to scientific research, although it is true that scientists may agree on many things, it is also true that disagreement permeates scientific communities. Some argue that scientific collaborations do not even require mutual agreement (Dang 2019), while others maintain that dissent, rather than a bug, is a desirable feature of scientific practice (Andler 2013). Contrary to some rather popular (mis) readings, and as explained in the previous section, Kuhn also maintained a view of the scientific community as internally diversified. *Contra* Gilbert's reading, therefore, the reason why, for Kuhn, scientists hold on to the established paradigm and show resistance towards new ideas is not because of their shared commitment, but to rationally 'spread the risk'.

It is true that, in several passages of *SSR*, Kuhn speaks about 'commitments'. In his post-SSR distinction between disciplinary matrix and exemplars, he even defines the previous in terms of 'group commitments' (see Chap. 5). However, he does not use 'commitments' in the same sense of Gilbert's. For Kuhn, scientists working on a paradigm are committed to a set of professional rules, or ethos of research, rather than to the protection of the paradigm at all costs. He says, for example, that "the relation between rules, paradigms, and normal science" can be discovered by isolating " the particular loci of commitment that have just been described as accepted rules" (*SSR*:43). In his view, scientists are committed to exploit and extend the paradigm, even at the cost of eventually reaching a crisis. In short, they are committed to practicing science in a certain way. And this sort of commitment is the same both during normal science, when scientists work on the agenda set by the paradigm, and during periods of extraordinary science, when they look for alternative paradigms (*SSR*:100–101).

This kind of 'commitment to a practice' does not even require that every scientist believe in the paradigm, have the same intentions, and know about each other intentions. As he explains, "Men [*sic*] whose research is based on shared paradigms are committed to the *same rules and standards for scientific practice*. That commitment and the *apparent consensus* it produces are prerequisites for normal science" (*SSR*:11, *my emphases*; see also Rouse 2003, 2013). Furthermore, if one moves beyond *SSR* and considers Kuhn's view on the essential tension and the risk-spreading argument, then it appears clear that Gilbert's theory of plural subject does not capture some central aspects of Kuhn's views.

Recently, the idea that the scientific community can be regarded as a collective agent has been gaining momentum. Several philosophers have begun to theorize about the fact that scientific knowledge is neither produced nor possessed by individual scientists, but rather by groups. For example, although with different arguments and from different perspectives, both Bird (2010, 2014, 2022) and Nersessian (2022) have recently characterised scientific groups as collective epistemic agents based on the 'distributed cognition' examined by Hutchins (1995). In Hutchins' framework, cognition is not something 'in the head' of individuals, but it is distributed across a group. For a group to display collective cognition, it is not necessary for its members to know what the others know, or to be all committed to the same things.

Claiming that Kuhn 'anticipated' the work of Hutchins and the views of Bird and Nersessian would be a stretch. However, his model of a scientific group pervaded by an essential tension that has the function of spreading the risk, without every one of its individual members necessarily sharing the same knowledge and intentions, seems at least compatible with Hutchinson's distributive approach. It must also be stressed that when Kuhn was taking a collective approach to science, discussions about the ontology of social groups were still in their infancy in analytic philosophy. Indeed, Kuhn's own view about the metaphysics of scientific groups is rather underdeveloped. Nevertheless, if one takes into consideration his view on collective rationality and the idea that individual variations are necessary for group rationality, then one could also interpret Kuhn as close, at least in spirit, to the recent theories on collective agency.

4.4 Scientific Communities

Kuhn's collective perspective cannot be underestimated. His various philosophical turns notwithstanding, Kuhn maintained two central elements of his philosophy throughout his career: the incommensurability thesis, which he attempted to refine and clarify in many of his post-*SSR* writings, and the idea that science is a collective enterprise carried on by a group, namely the *scientific community*. Many commentators of *SSR* have focused on the concept of paradigm. However, it is important to stress that paradigms and scientific communities are co-implicated (see Sect. 5.2). What makes normal science possible, in fact, is the adherence of the scientific community to a paradigm, while a scientific revolution is achieved when the scientific community endorses a new paradigm and abandons the old one. In other words, while the paradigm is the unit undergoing change during a revolution, "the locus of scientific change is the research community"; more specifically, "it is a research community that undergoes a revolutionary change, not an individual scientist" (Wray 2011:43). The community-based character of Kuhn's model is restated—and its importance, in a sense, reinforced—in some of his post-*SSR* writings. In 'Second Thoughts on Paradigms', Kuhn maintains that "the scientific enterprise is distributed and carried forward by communities" (Kuhn 1974:297). In 'Reflections on my Critics' (Kuhn 1970), he clearly states that he does not aim to explain how the individual scientist makes decisions; rather, what he is really interested in is "the normal (i.e., non-pathological) scientific group" (Kuhn 1970, in *RSS*:134). In the 'Postscript' he even says that, ideally, the model of *SSR* should be understood by looking at how scientific communities are structured first (*SSR*:176). Even in his very last writings, Kuhn never abandoned his focus on the scientific community. In relation to his more 'linguistic' concepts, he always makes it clear that kind terms, taxonomies, and lexicons are shared by the members of a scientific community, which resembles a 'linguistic community' (see Chaps. 6 and 7).

The importance of the concept of a scientific community for Kuhn's model is recognised in the Kuhnian scholarship. In general, however, the scientific communities Kuhn talks about are implicitly or sometimes even explicitly identified with the group of scientists working in the same discipline. Consequently, scientific revolutions, which are changes affecting a scientific community, are assumed to be a change within a whole scientific discipline—i.e., scientific revolutions in physics, in chemistry, and so on. Characterising Kuhn's scientific communities as 'discipline-communities', therefore, is linked to an interpretation of the Kuhnian model as being concerned solely with big paradigmatic changes affecting a whole discipline. Kuhn's thoughts about these issues were far more nuanced.

To begin with, he questioned the "implicit one-on-one identification of scientific communities with scientific subject matters" (*SSR*:179), which gives a misleading picture of the internal organization of science and does not stand historical scrutiny.

> What is today the subject matter for a single broad community has been variously distributed among different communities in the past. Other narrower subjects, for example heat and the theory of matter, have existed for long periods without becoming the special province of any single scientific community. Both normal science and revolutions are, however, community-based activities. To discover and analyse them, one must first unravel the changing community structure of the sciences over time. A paradigm governs, in the first instance, not a subject matter, but rather a group of practitioners. Any study of paradigm-directed or paradigm-shattering research must begin by locating the responsible group or groups. (*SSR*:179–180)

For example, until the beginning of the XX century, the so-called 'theories of matter' which were not the subject-matter of one specific scientific community or of one specific discipline (*SSR*:180). There was not a single discipline studying the theories of matter and the analysis of the content of the theories cannot help with the individuation of the actual scientific group(s) which endorsed them.

Kuhn also clarifies that there exist scientific communities of different sizes and at different levels of generality.

> The most global is the community of all natural scientists. At an only slightly lower level the main scientific professional groups are communities: physicists, chemists, astronomers, zoologists and the like. For these major groupings, community membership is already established except at the fringes. Subjects of highest degree, membership in professional societies, and journals read are ordinarily more than sufficient. Similar techniques will also isolate major subgroups: organic chemists, and perhaps protein chemists among them, solid-state and high-energy physicists, radio astronomers, and so on. It is only at the next lower level that empirical problems emerge. (*SSR*:177)

Kuhn's reference to the multi-level structure of scientific communities solves a problem raised by several critics. It has been said that Kuhn oscillates between *gradualism* (when he stresses small incremental changes in the scientists' activity) and *discontinuism* (when he speaks about revolutionary breaks). This double attitude towards scientific change does not help in understanding why some episodes of scientific change are revolutionary, whereas something like Maxwell's electromagnetic theory is regarded as a 'normal change' within the wider paradigm of classical mechanics (Nickles 2013:118).

Since scientific communities exist at different levels, scientific revolutions (which are caused by and occur within scientific communities) occur at different levels too. An example of a 'high-level revolution' is the Copernican revolution, which not only changed astronomy but also had shattering implications for the general metaphysical view of its time. If Kuhn was interested only in this type of revolutions, his model would only capture some extremely rare episodes in the history of science. This was not what he had in mind:

> [a] few readers of [*SSR*] have concluded that my concern is primarily or exclusively with *major revolutions* such as those associated with Copernicus, Newton, Darwin, or Einstein. *A clearer delineation of community structure should, however, help to enforce the rather different impression I have tried to create.* A revolution is for me a special sort of change involving a certain sort of reconstruction of group commitments. But it need not be a large change, nor need it seem revolutionary to those outside a single community. (*SSR*:180–181, *my emphases*)

By considering that revolutions can occur at different levels of the scientific community, one can see how, in a sense, Maxwell's electrodynamic theory both was and was not revolutionary. Before Maxwell's theory, there were indeed *two* distinct disciplines: electric physics and magnetic physics. After Maxwell, the electric and the magnetic forces, once believed to be different, became the 'electromagnetic force' and the two different sub-communities of scientists became a single sub-branch of classical physics. For electric and magnetic physicists, Maxwell's electrodynamic theory was indeed a revolution: they had to re-conceptualize old phenomena in new ways, the communities they once belonged to no longer existed, and the old division of knowledge they were accustomed to died off. At the high-level view of classical mechanics, however, there was not such a big change: electric and magnetic phenomena kept on being regarded as scientific problems pertaining to classical physics, both before and after Maxwell.

Kuhn's view on scientific communities and revolutions can be summarized as follows:

1. the normal scientific activity is carried out within a community and, similarly, a scientific revolution also happens within, and involves the members of, the community
2. there are high-level scientific communities (e.g., the communities of 'physics', 'chemistry', 'biology') and low-level scientific communities (e.g., the communities of 'quantum mechanics', 'organic chemistry', 'molecular biology')
3. from 1 and 2, there can be high-level revolutions (occurring in communities at the high-levels) and low-level revolutions (occurring in lower-level communities)
4. a revolutionary change occurring in a scientific community may not be noticed by the members of other scientific communities; or, if it happens at a low-level

community, or in a sub-community, it may not be perceived as revolutionary by all the members of the rest of the wider community

4.5 Theories of Rationality: Kuhn, Popper, Feyerabend, and Lakatos

How does Kuhn's theory of collective rationality score with respect to the views of some of his contemporaries? Notoriously, many philosophers saw in the concept of scientific revolutions a threat to scientific rationality. However, normal science was also criticized. For example, Popper (1970) feared that the normal scientist described by Kuhn was a rather dull individual, far from the ideal of a critical scientist capable of making bold conjectures and attempting to refute them fiercely.

> [Normal science] is the activity of the non-revolutionary, or more precisely, the not-too-critical professional: of the science student who accepts the ruling dogma of the day; who does not wish to challenge it; and who accepts a new revolutionary theory only if almost everybody else is ready to accept it—if it becomes fashionable by a kind of bandwagon effect. To resist a new fashion needs perhaps as much courage as was needed to bring it about. (Popper 1970:52)

Popper does not say that Kuhn is mistaken at the descriptive level: normal science does exist, and the normal scientist is simply someone who "has been taught in a dogmatic spirit: he is a victim of indoctrination. He has learned a technique which can be applied without asking for the reason why"; in short, the normal scientist "is a person one ought to be sorry for" (Popper 1970:53) For Popper, however, the uncritical acceptance of the existence of normal scientists means that Kuhn's philosophy simply fails at the normative level. In his view, and unlike Kuhn's, normative philosophy of science should not justify science as it is, but it ought to find the rules of rationality and promote the critical attitude.

Popper does not notice that, instead of threatening it, the idea of normal science compliments his own critical approach. In very general terms, Popper's *critical rationalism* rests upon two main ideas: (1) *fallibilism*—what we call knowledge is always conjectural and hypothetical, and it could be revised at any time; and (2) *criticism*—since a scientific theory can never be proven to be true inductively, but it can be proven to be false by new evidence, every scientific conjecture ought to be challenged and put through severe tests, in an attempt to falsify it. Together, (1) and (2) form the core of a critical rationalist attitude that, for Popper, is necessary for the development of bolder conjectures and new ways for attempting to refute them, thus leading to scientific progress (Popper 1959, 1963).

Kuhn surely does not deny (1). In a sense, he is as much a 'fallibilist' as Popper: in fact, even the longest period of normal science can be put to an end and even the most perduring paradigm can be overthrown. But Kuhn does not deny (2) either. He does not say that the paradigm cannot be criticized by normal scientists. What he says is that a paradigm can be rationally criticized *by working on it*. Without the work of normal scientists, it would be more difficult to make extraordinary discoveries and to encounter the anomalies which force scientists to revise their assumptions, theories and methods. In fact, it is not entirely clear why it would be 'rational' to attempt to overthrow a paradigm that has solved many problems so far in the absence of unexpected discoveries and anomalies, which in turn can emerge only during periods of normal science.[4]

Not only Kuhn's normal science does not conflict with Popper's critical attitude, but it also highlights some obvious practical limitations to Popper's project. If every hypothesis or conjecture is constantly challenged and criticized, and if the main aim of experimentation is the attempt at their falsification, problem resolutions and technological progress would be considerably slowed down. Scientists could just spend their lives criticizing and challenging each other's conjectures, without making any practical progress. What is not clear, in the Popperian model of 'bold conjectures' and 'severe attempts at refutations', is how so-called applied science would be possible without scientists accepting a conjecture and exploring its practical potentialities. Kuhn, by contrast, speaks of normal science as a period in which the dominant paradigm is developed and applied, often with the creation of new experimental devices. The fallible nature of scientific knowledge notwithstanding, normal science allows for the exploration and application of the parts of the paradigm which work.

Feyerabend (1975) defends a pluralistic methodology and the permanent proliferation of alternative scientific theories. For this reason, in his (1970) he criticizes the rigid 'dogmatism' contained in the very idea of normal science. However, normal science is not the monolithic activity Feyerabend is so afraid of. The fact that Kuhn puts a special emphasis on periods of consensus in science, in fact, does not mean that he rejects pluralism. What he rejects is the idea that science enjoys the same degree of pluralism all the time. Kuhn justly reminds us that, to be properly assessed, or even formulated, theories require time and effort. If several different theories receive the same share of such an effort all the time, the problem-solving process would be slowed down, since it will always take longer to decide which is the theory best suited for the resolution of the current problems. It is far more rational, instead, to allow periods in which most scientists work on one theory, to maximize its applicability, while consenting to the exploration and development of other theories 'on the side'.

Finally, Lakatos (1970) accuses Kuhn of being an elitist, since he depicts normal science as governed by the 'authoritarian' regime of a dominant paradigm, and scientists as governed by 'mob mentality'. However, the consensus bestowed upon a paradigm is not synonymous with unanimity and homologation of thought. The consensus is *achieved*, not imposed. Furthermore, a rational community must always cultivate internal diversity.

---

[4] Incidentally, and again *contra* Popper, it must also be stressed that the fervent activity taking place during periods of normal science—during which problems are selected, experiments performed, measurements taken, theories applied, hypotheses tested, devices designed, and so on—is far from a 'ticking-small-boxes' routine and requires much more ingenuity and creativity than its Popperian caricature.

Interestingly enough, and *contra* Lakatos, Kuhn's normal science could even be compared, with the due differences, to the ideal of democratic society. Marcum (2015) notices that the parallel between Kuhn's account of the collective nature of science and democracy was established in one of the first reviews of *SSR*.

> A review of *Structure* also appeared in *The Nation*, along with reviews of another dozen books on various topics related to science. The reviewer was Philip Siekevitz, a biochemist from the Rockefeller Institute. Siekevitz focused on the community structure of Kuhn's new image of science, especially a community structure that provides for open discussion and debate of issues germane to scientific advance. Interestingly, he linked Kuhn's analysis of the scientific community's paradigmatic structure with Gerard Piel's analysis of democracy and science in *Science in the Cause of Man*. According to Siekevitz, "Just as science is based on the 'paradigm' subject to change, so is our democracy based on the Common Law, subject to constant reinterpretation". (Marcum 2015:77)

As in every democracy, normal science is characterized by periods in which the most voted party is in power, but that does not mean that the other parties are silenced once and for all and cannot exercise political opposition. The dominant 'party' continues to get the highest percentage of 'votes' as long as it keeps its promise and solves problems, but political crises can always lead to new elections and the victory of a rival party. Not every dominant party has to be tyrannical, and some revolutions may even be bloodless, carried out with democratic instruments. Examining the political metaphors highlights another important aspect of science, namely the fact that its pluralism is always 'regulated'. In many democracies, the number of official parties is controlled; the expression of some political ideologies, such as fascism, is forbidden by law. Similarly, normal science can allow the rise of 'alternative minorities', but not every alternative idea will be considered 'scientific'. This means that at the basis of the plurality of views that can co-exist within normal science there is always a minimal agreement about what must count as 'scientific'—and that, therefore, it is not true that 'anything (whatsoever) goes', as some readers of Feyerabend may be tempted to say.

4.6 History, Sociology, and Philosophical Perspectives

Kuhn insisted that, within a dynamic perspective, disagreement is not evidence of the irrationality of one of the parties involved in a dispute. He stressed this to make it clear that his target was not the idea of scientific rationality per se, but the view of scientific rationality associated with the static image. He fleshed out his view by looking at some episodes of scientific change when scientists indeed disagreed. Perhaps he could have reached this or a similar conclusion had he kept pursuing a *sociological approach* to the study of the scientific community. In response to those who accused him of describing scientists as being under the spell of 'mob mentality', Kuhn responded that his descriptions of the scientific community were grounded on social psychology.

> My recourse has been exclusively to social psychology (I prefer 'sociology'), a field quite different from individual psychology reiterated n times. Correspondingly, my unit for purposes of explanation is the normal (i.e., the non-pathological) scientific group, account being taken of the fact that its members differ but not of what makes any given individual unique. (Kuhn 1970, in *RSS*:133–134)

In his latest writings, he returns to this point.

> It is now the group rather than the individual that guards the rationality of belief, and much of the belief it guards, because constitutive of the community's way of life, is simply not subject to justification. The perspectives of the psychologist, the sociologist, and, above all, the anthropologist become relevant to philosophy after all. (*SKHP*, in Kuhn 2022:11–12)

In Kuhn's dynamic perspective, the scientific community is a collective agent that holds beliefs, makes decisions, and constitutes its own (scientific) world. Understanding such a collective agent may require the use of research methods from sociology and anthropology. This is exactly what some philosophers of science have begun to do in the past couple of decades. Ethnographic and qualitative research methods, once used almost exclusively by sociologists of science and STS scholars pursuing a constructionist agenda, may be fruitfully used by philosophers of science interested in the study of non-idealized *in situ* distributed cognition (Nersessian and MacLeod 2022; Hangel and ChoGlueck 2023).

Kuhn even claims that he could have developed his model by looking at sociology rather than at the history of science.

> Actual experience in the practice of science would probably be a more effective bridge [across the gap between philosophers of science and science] than the study of its history. Sociology of science [...] might do as well. (Kuhn 1971, in *ET*:13)

> To most of us [philosophers rejecting the static perspective], history seemed primarily a convenient source of data about actual science, data which might in large part have been gathered without exhuming the past. (*SKHP*, in Kuhn 2022:1)

Furthermore, Kuhn (2022) defines the historians' work as akin to that of ethnographers: "historians are like ethnographers striving to understand and describe the apparently incongruous behaviour of the members of an alien culture (*SML*, in Kuhn 2022:29; see also 47, 71). These claims seem to reduce the differences between the 'historical' and the 'sociological approach' in philosophy of science and may even point out at the primacy of the latter over the previous (after all, it is the historian whose task is like that carried out by the ethnographer, not the other way around).

It is however important to stress that, in the same way in which Kuhn was not interested in the historical data per se, but on the nature of scientific progress and change through history, had he chosen to pursue a much stronger sociological approach, he would have likely been more interested in the *nature* of scientific groups, rather than on the fine details of an ever-growing body of sociological microstudies. This is consistent with the observations made in the previous chapter. Kuhn did not use the data coming from the study of actual scientific activity to develop his philosophical perspective. Rather, his philosophical perspective led him to look at the data coming from the study of actual scientific activity, no matter whether such data came from the history of science or from sociological and ethnographic laboratory studies.

Whether Kuhn could have developed a model of the progress of science through revolutions by looking at sociology and social psychology is not entirely obvious. Scientific revolutions, and scientific change in general, are processes extended through time and could hardly be captured by the sociological look at a scientific group in the 'here and now'. As explained in the previous chapter, however, one thing is to consider the relation between (historical or sociological) data and a particular model of the progress of science, another thing is to analyse a philosophical perspective (which may lead to consider different sources of data and may lead to the development of different models of the progress of science) and its implications.

4.7 Conclusions

A persistent 'textbook' image of Kuhn represents him as the philosopher who attacked the very idea of scientific rationality, claiming that normal science is a matter of dogmatic thinking and revolutions are akin to religious conversions. Without a doubt, Kuhn is guilty of using some rather bold and ultimately misleading analogies in *SSR*. At the same time, many philosophers are also guilty of focussing on the trees of some controversial passages in *SSR* while missing the whole forest of Kuhn's philosophy. Kuhn rejected the static image of science and, by doing so, he also rejected the algorithmic theory of scientific rationality. He then developed his own philosophical perspective and, by doing so, he developed a new theory of scientific rationality. In this theory, disagreements in science can indeed be rational. Furthermore, the scientific group, far from being characterized by conformism, must guarantee a tension between 'dogmatic' and 'iconoclast' scientists as a rational measure to spread the risk of error.

If there is no such a thing as a universal, context-insensitive and a-historical set of methodological rules, then, the philosopher holding the dynamic perspective must look at how context-sensitive decisions in science are taken at the group level. This, however, poses the challenge of accessing and describing the designated group. Whether and how Kuhn tackled this problem will be discussed in the next part of the book.

References

Andler, Daniel. 2013. Dissensus in science as a fact and as a norm. In *New Challenges to Philosophy of Science*, ed. Hanne Andersen, Dennis Dieks, Wenceslao Gonzalez, Thomas Uebel, and Gregory Wheeler, 493–506. Dordrecht: Springer.

Bird, Alexander. 2010. Social knowing: The social sense of 'scientific knowledge'. *Philosophical Perspectives* 24:23–56.

Bird, Alexander. 2014. When is there a group that knows? Distributed cognition, scientific knowledge, and the social epistemic subject. In *Essays in Collective Epistemology*, ed. Jennifer Lackey, 42–63. Oxford: Oxford University Press.
Bird, Alexander. 2022. *Knowing Science*. Oxford: Oxford University Press.
Bratman, Michael. 1992. Shared cooperative activity. *The Philosophical Review* 101:327–341.
Bratman, Michael. 1993. Shared intentions. *Ethics* 104:97–113.
Bratman, Michael. 2013. *Shared Agency: A Planning Theory of Acting Together*. Oxford: Oxford University Press.
Dang, Haixin. 2019. Do collaborators in science need to agree? *Philosophy of Science* 86:1029–1040.
D'Agostino, Fred. 2005. Kuhn's risk-spreading argument and the organization of scientific communities. *Episteme* 1:201–209.
D'Agostino, Fred. 2010. *Naturalizing Epistemology. Thomas Kuhn and the 'Essential Tension'*. New York: Palgrave Macmillan.
De Benedetto, Matteo, and Michele Luchetti. 2023. Theory choice as niche construction: The feedback loop between scientific theories and epistemic values. *Philosophy of Science* 2023:1–18. https://doi.org/10.1017/psa.2023.160.
De Benedetto, Matteo, and Michele Luchetti. 2024. Worlds, algorithms, and niches: The feedback-loop idea in Kuhn's philosophy. In *Rethinking Thomas Kuhn's Legacy*, ed. Yafeng Shan, 103–120. Cham: Springer.
Feyerabend, Paul. 1970. Consolation for the specialist. In *Criticism and the Growth of Knowledge*, ed. Imre Lakatos and Alan Musgrave, 197–230. Chicago: Chicago University Press.
Feyerabend, Paul. 1975. *Against Method: Outline of an Anarchistic Theory of Knowledge*. London: New Left Books.
Gilbert, Margaret. 1990. Walking together: A paradigmatic social phenomenon. *Midwest Studies in Philosophy* 15:1–14.
Gilbert, Margaret. 2000. Collective belief and scientific change. In *Sociality and Responsibility: New Essays in Plural Subject Theory*, 37–49. Lanham: Rowman and Littlefield.
Gilbert, Margaret. 2013. *Joint Commitment*. Oxford: Oxford University Press.
Hangel, Nora, and Christopher ChoGlueck. 2023. On the pursuitworthiness of qualitative methods in empirical philosophy of science. *Studies in History and Philosophy of Science* 98:29–39.
Hoyningen-Huene, Paul. 1993. *Reconstructing Scientific Revolutions*. Chicago: Chicago University Press.
Hutchins, Edwin. 1995. *Cognition in the Wild*. Cambridge: MIT Press.
Kitcher, Philip. 1990. The division of cognitive labour. *Journal of Philosophy* 87:5–22.
Kuhn, Thomas S. 1959. The essential tension. In *The Third University of Utah Research Conference on Identification of Scientific Talent*, ed. Charles Taylor, 162–174. Salt Lake City: University of Utah Press. Reprinted in *ET*:225–239.
Kuhn, Thomas S. 1970. Reflections on my critics. In *Criticism and the Growth of Knowledge*, ed. Imre Lakatos and Alan Musgrave, 231–279. Cambridge: Cambridge University Press. Reprinted in *RSS*:123–175.
Kuhn, Thomas S. 1971. The relations between history and history of science. *Daedalus* 100:271–304. Reprinted in *ET*:127–163.
Kuhn, Thomas S. 1974. Second thoughts on paradigms. In *The Structure of Scientific Theories*, ed. Frederick Suppe, 459–482, Urbana: University of Illinois Press. Reprinted in *ET*:293–319.
Kuhn, Thomas S. 1977. Objectivity, Value Judgment and Theory Choice. In *ET*:320–339.
Kuhn, Thomas S. 1993. Afterwords. In *World Changes*, ed. Paul Horwich, 311–341, Cambrige: MIT Press. Reprinted in *RSS*:224–252.
Kuhn, Thomas S. 2022. *The Last Writings of Thomas S. Kuhn*, ed. Bojana Mladenović. Chicago: University of Chicago Press.
Lakatos, Imre. 1970. Falsification and the methodology of scientific research programmes. In *Criticism and the Growth of Knowledge*, ed. Imre Lakatos and Alan Musgrave, 91–196. Cambridge: Cambridge University Press.

List, Christian, and Philip Pettit. 2011. *Group Agency: The Possibility, Design, and Status of Corporate Agents*. Oxford: Oxford University Press.

Marcum, James. 2015. *Thomas Kuhn's Revolutions. A Historical and an Evolutionary Philosophy of Science?* London: Continuum.

Nersessian, Nancy. 2022. *Interdisciplinarity in the Making*. Cambridge: MIT Press.

Nersessian, Nancy, and Miles MacLeod. 2022. Rethinking ethnography for philosophy of science. *Philosophy of Science* 89:721–741.

Nickles, Thomas. 2013. Some puzzles about Kuhn's exemplars. In *Kuhn's Structure of Scientific Revolutions Revisited*, ed. Vasso Kindi and Theodore Arabatzis, 112–133. New York: Routledge.

Pettit, Philip. 2009. The reality of group agents. In *Philosophy of the Social Sciences: philosophical theory and scientific practice*, ed. Chrysostomos Mantzavinos, 67–91. Cambridge: Cambridge University Press.

Politi, Vincenzo. 2021. Formal models of the scientific community and the value-ladenness of science, *European Journal for Philosophy of Science* 11: article 97. https://doi.org/10.1007/s13194-021-00418-w

Popper, Karl.1959. *The Logic of Scientific Discovery*, translation by the author of *Logik der Forschung* (1935). London: Hutchinson.

Popper, Karl. 1963. *Conjectures and Refutations: The Growth of Scientific Knowledge*. London: Routledge.

Popper, Karl. 1970. Normal science and its dangers. In *Criticism and the Growth of Knowledge*, ed. Imre Lakatos and Alan Musgrave, 51–58. Cambridge: Cambridge University Press.

Rouse, Joseph. 2003. Kuhn's philosophy of scientific practice. In *Thomas Kuhn*, ed. Thomas Nickles, 101–121. Cambridge: Cambridge University Press.

Rouse, Joseph. 2013. Recovering Thomas Kuhn. *Topoi* 32:59–64.

Rueger, Alexander. 1996. Risk and diversification in theory choice. *Synthese* 109:263–280.

Šešelja, Dunja. 2021. Exploring scientific inquiry via agent-based modelling. *Perspectives on Science* 29:537–557.

Šešelja, Dunja. 2023. Agent-based modelling in the philosophy of science. In *The Stanford Encyclopedia of Philosophy* (Winter 2023 Edition), ed. Edward N. Zalta and Uri Nodelman. https://plato.stanford.edu/archives/win2023/entries/agent-modeling-philscience/

Udehn, Lars. 2001. *Methodological Individualism: Background, History, and Meaning*. London: Routledge.

Weatherall, James, and Margaret Gilbert. 2015. Collective belief, Kuhn, and the string theory community. https://philsci-archive.pitt.edu/11413/

Weisberg, Michael. 2010. New approaches to the division of cognitive labour. In *New Waves in Philosophy of Science*, ed. P. D. Magnus and J. Busch, 250–269. London: Palgrave Macmillan.

Weisberg, Michael, and Ryan Muldoon. 2009. Epistemic landscapes and the division of cognitive labour. *Philosophy of Science* 76:225–252.

Wray, Karl B. 2011. *Kuhn's Evolutionary Social Epistemology*. Cambridge: Cambridge University Press.

Zollman, Kevin. 2007. The communication structure of epistemic communities. *Philosophy of Science* 74:574–587.

Zollman, Kevin. 2010. The epistemic benefit of transient diversity. *Erkenntnis* 72:17–35.

Part II
From Paradigms to Lexicons

Chapter 5
From Paradigms to Exemplars

Abstract A collective approach to science must define and describe the collective agent of science, namely the scientific community. In *SSR*, the concepts of scientific community and paradigm are co-defined: a paradigm defines the scientific community that holds it. This circularity may pose some problems. However, Kuhn maintained that it is possible to isolate a scientific community without prior recourse to its paradigm. If that is the case, then the philosopher describing a scientific community can 'discover' its paradigm. The concept of paradigm faced several criticisms, and, for this reason, in some post-*SSR* works Kuhn attempted to refine it. This chapter analyses how Kuhn went from a 'wide' to a 'narrow' conception of paradigm, by introducing the distinction between exemplars and disciplinary matrix. While exemplars have been recently rehabilitated by some philosophers, it is questionable whether they can be isolated from disciplinary matrices, the concept of which deserves deeper analysis.

Keywords Scientific community · Paradigm · Disciplinary Matrix · Exemplar · Analogical reasoning · Naturalised philosophy

5.1 Introduction

In the previous chapters, I have reconstructed Kuhn's dynamic perspective as well as one of its most salient elements, namely the non-algorithmic theory of collective rationality. Since there are not such things as absolute methodological rules, the philosopher adopting a dynamic perspective must describe actual episodes of scientific change and then make it explicit why the decisions taken at the group level were indeed rational in those contexts. This means that, once it is established that the scientific community is the *locus* and the *agent* of science, it remains to find a way to access and describe such a community and the collective activity its members carry.

V. Politi, *Thomas Kuhn's Philosophical Perspective*, Boston Studies in the Philosophy and History of Science 348,
https://doi.org/10.1007/978-3-032-13540-7_5

In *SSR*, Kuhn defines a scientific community based on its paradigm, and defines a paradigm based on the scientific community that holds it. To break the circularity, he claims that it is possible to isolate a scientific community without prior recourse to the paradigm that defines its activity. By looking at the activity carried out by the scientific community, the philosopher of science will eventually 'discover' its paradigm. Circularity, however, was not the only problem Kuhn had to deal with. Notoriously, several philosophers criticized the very concept of a paradigm. This may have led Kuhn to distinguish, in some of his post-*SSR* writings, between a 'wide' and a 'narrow' conception of paradigm, through the concepts of *disciplinary matrix* (DM) and *exemplar* respectively. The latter, in recent years, went through a philosophical revival, thanks to several scholars who read Kuhn's exemplar as anticipating case-based and model-based epistemologies.

The aim of this chapter is not only to tell the story of how Kuhn went from paradigm to DM and exemplar, and to define these concepts. Contrary to the recent wave of renewed interest in exemplars, in fact, I will also question the tenability of the very distinction between the 'wide' and the 'narrow' sense of paradigm.

In Sect. 5.2, I examine Kuhn's idea of isolating scientific communities without resorting to paradigms first, which he expressed in the Postscript. While it was heavily criticised when it was initially formulated, I find such an idea more forward-thinking than how it was portrayed by his critics. In Sect. 5.3, I introduce Kuhn's post-*SSR* distinction between DM and exemplars. In Sect. 5.4, I focus on exemplars and on how they have been recently reevaluated by some philosophers of science. In Sect. 5.5, I explain why considering exemplars in isolation with respect to a DM poses a problem to Kuhn's history-based model of science. Specifically, I discuss how exemplars on their own cannot help us discriminate between incommensurable scientific traditions. Therefore, they cannot play the same 'pivotal role' Kuhn had assigned to paradigms in *SSR*. In Sect. 5.6, I consider whether an 'expanded conception' of exemplars could solve this problem. In Sect. 5.7, I question the viability of discussing exemplars without considering the DM they are part of. I suggest that one of Kuhn's major contributions to philosophy of science is not the concept of exemplar but, rather, the unjustly neglected concept of a DM.

5.2 Communities and Paradigms

One of the difficulties with developing a philosophical theory of science that describes and explains the activity carried out within a scientific community is that it is extremely difficult to individuate a scientific community. The problem is that what needs to be described, that is the activity within the group, is what defines the group the begin with. In other words: would it be possible to individuate a scientific community without first individuating the paradigm which allows the members of such a scientific community to practise science? After all, in *SSR*, Kuhn uses the concept of a paradigm to define a scientific community and the concept of a scientific community to explain what a paradigm is. At the same time, in the Postscript to the second edition of *SSR*, he maintains that it is possible to break this circularity, even though the task presents several difficulties.

> A paradigm is what the members of a scientific community share, and, conversely, a scientific community consists of men who share a paradigm. Not all circularities are vicious [...], but this one is a source of real difficulties. Scientific communities can and should be isolated without prior recourse to paradigms [...]. If this book [*SSR*] were being rewritten, it would therefore open with a discussion of the community structure of science. (*SSR*:176)

Kuhn tries to solve the problem of isolating scientific communities without prior recourse to their paradigms by looking into some works in the sociology of science, confident that the discipline can develop promising tools for this task. In particular, he welcomes the method of examining scientists' communication networks and counting the citation linkages of their publications. The rough idea is that scientists citing each other's works form a scientific community. One does not need any special scientific knowledge to apply this technique: that is, one does not need to grasp what those scientists are talking about to understand that they are indeed talking about the same thing and about each other's publications.

In his review of the second edition of *SSR*, Musgrave (1971) examines the limits of this approach. He rightly points out that scientists from different communities may nevertheless cite each other's works for various reasons, and that therefore counting citations and compiling bibliometric indexes 'mechanically' is not sufficient to determine membership to one scientific group rather than to another. Therefore, the sociological methods used to isolate scientific groups suggested by Kuhn are incomplete at best.

> The sociologists to whom Kuhn refers us take pride in what they call the 'objectivity' of their methods. What they mean is that their methods can be applied without taking any notice of the scientific content of the activities investigated. [...] [However] the fact that one scientist cites another hardly guarantees that they share a subject-matter (he might, after all, cite him in distinguishing their subject-matters), or that their "professional judgement" coincides (he might, after all, cite him in disagreement). Citations, not to mention the papers containing them, will have to be read and understood to see if they provide evidence of common group membership in Kuhn's sense. But Kuhn's sociologists, as we have seen, appear to have a methodological aversion to reading what they so assiduously count, and therefore cannot solve Kuhn's problem. It seems that Kuhn's own sociological citations should be taken with a pinch of salt. (Musgrave 1971:288)

But Musgrave does not criticise only the 'incompleteness' of the method, but also the very idea of attempting to define scientific communities without analysing the special knowledge shared by its members (or, as Kuhn would say, without prior recourse to paradigms).

> [Kuhn] is worried about the circularity involved in defining a paradigm as that which the members of a scientific group share, and then defining the group by its shared paradigm. Now most philosophers of science, if they paid any attention at all to the sociological dimension of science, would presumably have a ready solution to this Kuhnian puzzle. They would insist that the content of science is primary, so that if scientists do organise themselves into different groups this must be a sociological reflection of their different problems, theories and techniques. Such philosophers of science would, therefore, define

> the group by the common scientific content of its activities, and not vice versa. (Musgrave 1971:287).

It must be said, however, that Kuhn does not take the sociological methods he talks about as complete and conclusive. He only suggests possible routes for future works. It must also be acknowledged how Kuhn's ideas on how to define scientific communities were way ahead of his times. It is enough to mention that the academic journal *Scientometrics* was founded in 1978, almost a decade after the publication, in 1970, of the Postscript. Kuhn looked at sociological ideas and methods that were still in their infancy but, even so, he could foresee how they could eventually be developed for philosophical purposes. That such (back then) 'immature methods' should be taken with a pinch of salt does not mean that they should be completely dismissed and rejected. Moreover, Kuhn's proposal appears to have been fully vindicated by the methodology of contemporary philosophy of science. Empirically informed philosophers of science do exactly what Kuhn suggests: they use a wealth of quantitative and qualitative research methods to analyse what scientists do, and, in this way, they end up discovering their assumptions, patterns of reasoning, research culture, and so on.[1]

Ultimately, Musgrave's criticism fails to appreciate the *philosophical* point of Kuhn's proposal. Although his suggestions for how to define scientific communities without looking at science's content first are underdeveloped, if not vague, Kuhn raises an important methodological reflection for the community-oriented philosophy of science. The point is not whether scientific communities can be completely and unambiguously defined without prior recourse to their paradigms *via* scientometric (or other) methods. Rather, the point is that to understand the practice of science one should begin by looking at scientists at work. In a sense, Kuhn turns the philosophy of science upside down: we do not understand what scientists do by knowing beforehand the paradigm they share; instead, we must first look at what scientists do and then we reconstruct the paradigm they share. In other words, paradigms must not be presupposed; instead, they should be "*discovered* by scrutinising the behaviour of a given community's members" (*SSR*:176, *my emphasis*).

Assuming that one can isolate a scientific community and discover its paradigm, the next step would be clarifying what such a paradigm consists of. To begin with, it is crucial to stress that a paradigm is *not* a theory. In the Kuhnian model of science, the concept of a scientific theory is both too restrictive and too liberal. It is too restrictive because, at least in his early writings, Kuhn is not interested only in theories, but also in the activity of science which involves, among other things, evaluative standards, instrumentation, methodological and ontological commitments (unevenly spread among the members of the scientific group), and so on. The concept of a theory is also too relaxed because it would allow far too many scientific revolutions. In *SSR*, a scientific revolution is defined as a change of paradigm. While scientific revolutions often involve changes of theory, not every change of theory is

[1] For an introduction to contemporary empirical philosophy of science, with a special focus on the use of qualitative research methods for philosophical purposes, see Wagenknecht et al. (2015).

also a scientific revolution. If every change of theory were also a revolutionary change of paradigm, then one would be led to say, for instance, that the Copernican Revolution was followed by a 'Keplerian Revolution'. Yet, although Kepler's theory was different from Copernicus', they still belonged to the same 'heliocentric paradigm'. Kuhn makes it clear that paradigms are not theories when, in the *Postscript*, he says that "[as] currently used in philosophy of science [...] 'theory' connotes a structure far more limited in nature and scope than the one required [by his model of science]" (*SSR*:182).

Granted that they are not theories, it is still not clear what paradigms are. As discussed by Hoyningen-Huene (1993:131–162), Kuhn speaks of paradigms in different ways in different writings, so that it is possible to distinguish the pre-*SSR* paradigm, intended as "concrete problem solutions that the profession has come to accept" (Kuhn 1959, in *ET*:229), the *SSR*-paradigm, and the post-*SSR* distinction between 'disciplinary matrix' and 'exemplars'. There is a clear *fil rouge* connecting the pre-*SSR* 'concrete problem solutions' to the post-*SSR* notion of 'exemplars', which is the idea that scientists solve new problems by looking at how older problems have been successfully solved. Alongside this general idea, however, *SSR*-paradigm has a plethora of many other different meanings.

That in *SSR* 'paradigm' is polysemous could be seen as a lack of analytical sharpness and precision. Some would even say that, because of this polysemy, the concept of paradigm is unclear (even though polysemy does not imply unclarity). The polysemy of 'paradigm' is highlighted by Masterman (1970), who famously notices how, in *SSR*, the term is used in at least twenty-one different ways. Masterman's aim, however, is *not* to criticise Kuhn on the ground of the polysemy of paradigm which, from her perspective, does not necessarily lead to unclarity. In fact, "not all the senses of 'paradigm' are inconsistent with one another: some may even be elucidations of others" (Masterman 1970:65). Rather, Masterman aims at criticising Kuhn's critics. Because of its various meanings, paradigm may represent a 'moving target' for the critics of Kuhn, to the point that it becomes difficult to understand who is criticising what exactly. In other words, "although Kuhn's whole general view of the nature of scientific revolutions depends on it, those who attack him have never taken the trouble to find out what [a paradigm] is" (Masterman 1970:61).

Masterman groups the different meanings of paradigm in three main families: the *metaphysical paradigm* (or 'metaparadigm'), that contains the set of speculations, beliefs, ontological commitments and ways of seeing the world; the *sociological paradigm*, that refers to those features of science that make it analogous to political institutions, such as inter-subjective deliberations, shared values and standards; and the *artifact paradigm* (or 'construct paradigm'), a more concrete sense of paradigm referring to tools and instrumentation for the scientific practice, but also to examples found in textbooks as well as to analogies.

Perhaps convinced by Masterman's analysis of the many meanings of paradigms, in some post-*SSR* writings Kuhn distinguishes between *disciplinary matrix* (DM) and *exemplar*, which somehow resemble Masterman's sociological and artifact paradigm respectively. Prima facie, this distinction seems to leave aside the metaphysical sense of paradigm. However, as I illustrate below, Kuhn considers metaphysical worldviews and ontological commitments as part of a DM (that is, of the 'sociological' paradigm). Yet, his decision to focus on the sociological and artifact sense of paradigm is even more remarkable when considering that, as Masterman points out, the 'metaphysical paradigm' was the favourite target of Kuhn's critics. Many philosophers, in fact, took the notion of scientific revolution as indicating a change of high-level worldviews, that is, as a change in the metaphysical picture of the world. Kuhn, of course, does not deny that some metaphysical elements play some function in science. For instance, metaphysical beliefs may play the role of 'heuristic models' and can supply analogies and metaphors which, in turn, play a central role in the articulation of a paradigm, by pointing out new avenues for further research. As with scientific theories, however, paradigms contain, but cannot be reduced to, a metaphysical worldview.

Granted that it cannot be reduced to either theories or metaphysical beliefs, it remains to discuss what exactly is in a DM, as well as its relations with exemplars.

5.3 Disciplinary Matrix

There are at least two reasons for examining the post-*SSR* distinction between DM and exemplars. First, Kuhn himself seemed to regard exemplar as the most original contribution of his own work and as a sort of ultimate definition of paradigm. Second, after making such a distinction he dropped the notion of paradigm altogether. Because this book looks at the different phases of Kuhn's developing philosophy, it is necessary to analyse the status of the concept of paradigm right before its vanishing from his writings.

The distinction between DM and exemplar kept Kuhn busy for several years. He began to delineate it less than a couple of years after attending Masterman's presentation at the 1965 London Colloquium (the same colloquium in which several philosophers, including Popper, Lakatos, and Feyerabend, criticised Kuhn's history-based model of science). In 1967, in the second of a lecture series at Swarthmore College, titled 'Paradigms and Theories in Scientific Research', Kuhn introduced the notion of 'professional matrix', consisting of: general statements (that is, theoretical and empirical laws), a collective metaphysics, methodology and instrumentation, solved problems. In 1969, he took part in a symposium about the structure of scientific theories, hosted by the University of Illinois in Urbana, where he read his "Second Thoughts on Paradigms". (Although this paper was read the same year in which Kuhn wrote the Postscript, and a year before its publication in the second edition of *SSR*, it was published only years later in Suppe 1974). In this paper, 'professional matrix' becomes 'disciplinary matrix': "'disciplinary' because it is the common possession of the practitioners of a professional discipline, 'matrix' because it is composed of ordered elements of various sorts, each requiring further specification" (Kuhn 1974, in *ET*:297). In the version of "Second Thoughts", a DM consists of: symbolic generalisations, models, and exemplars. Finally, there is the DM of the Postscript, defined as the "constellation of group commitments" (*SSR*:181) and containing: symbolic generalisations, metaphysical beliefs, values, and exemplars.[2]

The different reformulations of DM have in common two things. The first is the concept of exemplars. Even though in the Swarthmore lecture Kuhn spoke of 'solved problems', it is not too uncontroversial to suppose that what he had in mind was something like exemplars. Indeed, in the following papers, he defines exemplars as 'exemplary solved problems'. While the list of the other elements of DM underwent revisions, then, it seems that exemplars always played a stable and continuative role in Kuhn's post-*SSR* redefinitions of paradigm.

The second is the idea that a DM indeed consists of several different things. In all its three formulations, a DM is a composite entity that cannot be reduced to only one of its elements. This consideration leads to the conclusion that none of these elements is a sufficient condition for a DM to obtain. Further questions concern whether every single non-sufficient element of the DM is necessary, or whether some elements play a more prominent role, and are in a sense 'more' necessary, than others. For example, as said above, exemplars appear in every reformulation of DM, a lot of emphasis is put on them and, therefore, it seems that exemplars are necessary constituents of DM (i.e., there would be no DM without exemplars). However, are metaphysical beliefs necessary, or 'as necessary as' exemplars? One of the main issues tackled in the remainder of this chapter is whether it is possible to conceive exemplars in isolation from the other constitutive elements of DM.

5.4 Exemplars

In Kuhn's words, 'exemplars', or 'shared examples', represent the "most novel and least understood aspect of [*SSR*]" and consequently "require more attention than the other sorts of components of the disciplinary matrix" (*SSR*:187). In 'Second Thoughts on Paradigms', he specifies that 'exemplar' "provides a new name for [...] the most fundamental sense of paradigm" (Kuhn 1974, in *ET*:298).

What do exemplars exemplify? They suggest to scientists how normal research should be conducted: they exemplify the 'paradigmatic' way of solving problems. They are, therefore, good examples of successful scientific achievements. That past scientific achievements are constitutive of paradigms and play a fundamental role within normal science is already suggested in the first pages of *SSR*, where Kuhn states: "In this essay, 'normal science' means research firmly based upon one or more past scientific achievements, achievements that some particular scientific community acknowledges for a time as supplying the foundation for its further practice" (*SSR*:10).

---

[2] Marcum (2013) provides a detailed exegesis of Kuhn's evolving definitions of 'professional' and 'disciplinary matrix' during his post-*SSR* years.

Hoyningen-Huene (1993) points out that exemplars play a *normative role*. In fact, they are not accepted just as "particular solutions to particular problems" but, rather, "[future] research must refer itself to these problem solutions, proceeding by analogy from them" (Hoyningen-Huene 1993:159). He then distinguishes three normative functions of exemplars:

semantic function: learning an exemplar means learning how to 'attach' scientific concepts to nature and, ultimately, how to use them
identification of scientific puzzles: exemplars tell scientists which problems should be counted as scientific problems in need of a solution
puzzle-resolution standards: exemplars tell scientist what counts as an appropriate solution of a problem (Hoyningen-Huene 1993:159–162)

Bird (2000) individuates a fourth function which, in his opinion, should "fill a role that fits between function (ii) and (iii)", namely:

solution identification: exemplars "enable scientists to see the world in such a way that solutions to those puzzles become apparent" (Bird 2000:69)

Let's begin by examining the *semantic function*. As pointed out in the previous section, one of the elements of a DM is the set of 'symbolic generalisations'. While the early description of a DM, at the Swarthmore lecture, referred to 'general statements' of both a theoretical and empirical nature, later, in 'Second Thoughts' and the 'Postscript', Kuhn considers mathematical formulas as having a status on their own. In his view, such formulas taken by themselves are meaningless. To make this point, he discusses the meaning of Newton's Second Law of Motion that, in its simplified form, is written as:

$$F = ma \tag{5.1}$$

where F is 'force', m is 'mass' and a is 'acceleration'.

Kuhn's point is that scientists do not learn the meaning of Newton's Second Law by learning that 'force is equal to mass time acceleration'. To understand this definition, one should first understand what 'mass' and 'acceleration' mean. But one cannot learn any one of these concepts without all the others. The concepts expressed in the so-called laws of nature, in fact, are co-defined. They form a 'cluster' of interrelated terms that must be learned together. It is the exposure to exemplars, as well as the activity of solving problems that are similar to the ones already solved, that allow scientists to understand the meaning of symbolic generalisations.

In the case of the Second Law, to get the meaning of $F = ma$, the science student must face a series of 'problem situations'. For instance, she will have to solve problems concerning free fall, the pendulum, and interacting harmonic oscillators. To solve such problems, she will use the following equations respectively

$$mg = \frac{md^2s}{dt^2} \tag{5.2}$$

$$mg\sin\theta = -ml\frac{d^2\theta}{dt^2} \tag{5.3}$$

$$m_1\frac{d^2 s_1}{dt^2} + k_1 s_1 = k_2\left(d + s_2 - s_1\right) \tag{5.4}$$

By using (5.2)–(5.4), science students learn the meaning of (5.1). In short, exemplars give semantic content to symbolic generalisations: this is why they represent two distinct elements of a DM. This is also why, in a sense, exemplars play a more 'fundamental role' than symbolic generalisations: while the former without the latter are meaningless, one could learn how to solve specific problems without knowing the meaning of the general formula. In other words, one may learn how to use $mg = \frac{md^2 s}{dt^2}$ without necessarily knowing the meaning of $F = ma$.

The *identification of new scientific puzzles* occurs when, after practising with examples of past scientific achievements and learning the meaning of symbolic generalisations, the science student has acquired the ability to perceive new problems as being similar to the old ones.

> The [...] ability to see a variety of situations as like each other, as subjects for $F = ma$ or some other symbolic generalisations, is, I think, the main thing a student acquires by doing exemplary problems, whether with a pencil and paper or in a well-designed laboratory. After he has completed a certain number, which may vary widely from one individual to the next, he views the situations that confront him as a scientist in the same gestalt as other members of his specialists' group. For him, they are no longer the same situations he had encountered when his training began. He has meanwhile assimilated a time-tested and group-licensed way of seeing. (*SSR*:189)

To practise science, one needs to acquire a special way of seeing new problems as being similar to other problems. By practising with (5.2)–(5.4), one learns to see them as similar to one another and also to (5.1). In short, exemplars teach how to recognize and then solve new problems. As such, they teach us how to *identify solvable problems*, by teaching us to see the world in a certain way (Bird's 'intermediate' function iv); and, in virtue of this, they incorporate the *standards* of problem resolution (Hoyningen-Huene's function iii).

Being able to see new problems as similar to those already solved is a precondition of scientific practice. It is only by spotting similarities among various problem situations that scientists can solve new problems. But how do scientists acquire this skill? For Kuhn, similarity-relations are learned through ostension: science students learn to see similarity patterns with the help of their teachers, who expose them to exemplars and guide them through the acquisition of the relevant similarity relations. This process happens within the scientific community, which is where scientific skills are transmitted and acquired. This process does not rely on formal definitions and logical rules but, rather, on guided exposure to previous scientific achievement.

To explain the role of ostension for the acquisition of the similarity criteria, Kuhn (1974) discusses the example of a little boy, named Johnny, who learns to spot and distinguish different kinds of waterfowl during a walk with his dad at the zoological garden. Johnny has already learned what birds are, as well as the name of some kinds of birds, like robins. He then learns how to recognize swans because, whenever he sees one, his dad explains to him that that is indeed a swan. After being exposed to several exemplars of 'swan', Johnny feels confident about his knowledge of swans. However, when he sees a duck and a goose, he mistakenly takes them for swans, and his father corrects him. Through the guided ostension to exemplars and after a 'trial and error' process, Johnny will have learned how to recognize swans, ducks and geese.

Kuhn also stresses that Johnny does not learn to distinguish different kinds of waterfowl by memorising formal definitions. In fact, "the primary tool is ostension. Phrases like 'all swans are white' may play a role, but they need not" (Kuhn 1974, in *ET*:309). It is through ostension that Johnny learns that, despite their similarities, ducks don't have a long neck like geese and swans, which in turn are also white, unlike geese, and so on. At the same time, Johnny would have not really got what swans are had he not been exposed to instances of ducks and geese. The three subfamilies of waterfowls constitute a *contrast set* of similarities and dissimilarities. This is way Johnny must learn the meaning of swans, ducks, and geese together. The process of learning through ostension, furthermore, has a social nature: as Kuhn remarks, during his walk with his dad, Johnny has been "programmed to recognize what his prospective community already knows" (Kuhn 1974, in *ET*:312).

Kuhn establishes a parallelism between the Johnny example and the recognition of similarities among different problem situations, as in the case of Eqs. (5.1)–(5.4).

> Need I now say that the swans, geese, and ducks which Johnny encountered during his walk with father were what I have been calling exemplars? Presented to Johnny with their labels attached, they were solutions to a problem that the members of his prospective community had already resolved. Assimilating them is part of the socialisation procedure by which Johnny is made part of that community and, in the process, learns about the world which the community inhabits. [...] But the same technique, if in a less pure form, is essential to the more abstract sciences as well. I have already argued that assimilating solutions to such problems as the inclined plane and the conical pendulum is part of learning what Newtonian physics is. Only after a number of such problems have been assimilated, can a student or a professional proceed to identify other Newtonian problems for himself That assimilation of examples is, furthermore, part of what enables him to isolate the forces, masses and constraints within a new problem and to write down a formalism suitable for its solution. Despite its excessive simplicity, Johnny's case should suggest why I continue to insist that shared exemplars have essential cognitive functions prior to a specification of criteria with respect to which they are exemplary. (Kuhn 1974, in *ET*:313)

In recent years, Kuhn's theory of exemplars, often overshadowed by unending discussions about revolutions and incommensurability, has attracted the interests of several philosophers of science. To begin with, the idea that one learns to perceive salient similarities by practising with exemplars gives a new rationale to Kuhn's notoriously controversial incommensurability thesis. In *SSR*, incommensurability is 'the lack of a common measure between paradigms'. Of course, without a precise characterization of 'paradigm', such a definition risks being too vague. Bird (2007) suggests that incommensurability can be reformulated on the basis of the perceptual theory associated with exemplars, thus becoming a thesis about differences in the learned perceived similarities deriving from being exposed to and practising with different exemplars.

The pre- and post-revolutionary scientists see things in different ways; that is, they perceive different similarity-relations populating the world. Being exposed to different exemplars leads to perceiving the world in different ways, but since—as pointed out by Hoyningen-Huene and Bird—exemplars have also the functions of regulating values, standards of good practice, and so on, 'exemplar-incommensurability' is not less rich than the definition in *SSR*. In the exemplar-driven perspective, however, the 'richness' of the incommensurability thesis is given a foundation in terms of acquired cognitive skills.

In Bird's interpretation, the notion of exemplar also enables Kuhn to develop a case-based reasoning view of science, in which new problems are solved by spotting salient similarities with some previously solved problems. In this view, scientific reasoning is largely based not on rules, but on *analogical reasoning* (see also Nickles 2003). The main points of this interpretation can be summarised as follow:

1. selecting puzzles, solving them, and assessing the quality of a proposed puzzle solution are driven by perceived similarity to exemplary puzzles and their solutions (paradigms [as exemplars])
2. the ability to perceive such similarities is acquired by training with [exemplars]
3. that ability is primarily an ability to recognize patterns and relevant similarities
4. it is therefore not an ability that is mediated by following rules (Bird 2005:102)[3]

Bird (2000, 2005) reads the theory about the acquisition of perceived similarities through exemplars as an expression of Kuhn's early naturalistic approach. In his view, one of Kuhn's major contributions is his use of ideas and empirical findings from the special sciences. For instance, in *SSR* Kuhn does not speak about scientific observation by recurring to some a priori theory but, rather, his ideas about scientists' perception are informed by, and somehow continuous with, the actual experimental results on vision obtained by the psychologists Bruner and Postman (1949). For Bird, Kuhn's early naturalism culminates with the discussion of the functions of exemplars: Kuhn not only explains the functioning of science in terms of perceptual and cognitive mechanisms which can be (and are) the object of disciplines such as experimental psychology, but his theory of exemplars can be empirically tested. As

[3] Point (4) interprets Kuhn as maintaining that scientists do not solve problems by applying a predetermined set of formal rules, but simply by practising and by applying old solutions to new problems. This is generally seen as one of Kuhn's most important (and audacious) contributions. "Kuhn compared normal science to solving puzzles such as crossword puzzles. One feature of puzzle-solving to which Kuhn sought to draw our attention is the fact that one can learn to do crossword puzzles quicker and more easily simply by practising them. Although one can write very effective computer programmes that play chess by implementing an algorithm, good human players recognize patterns in positions in a non-algorithmic way that is acquired by past exposure to many similar positions. [...] [What] makes a difficult problem more difficult will be not just increased complexity but also the fact that it is less clearly similar to a problem the student has seen before" (Bird 2005:102–103).

Bird reminds us, Kuhn even mentions the project of building a computer capable of developing by itself some ‘transformation functions’ to categorise stimuli on the bias of their similarity and, therefore, to ‘learn’ (Kuhn 1974, in *ET*:310).

For Bird, the new frontiers of cognitive science vindicate Kuhn’s rule-less case-based analogical model of scientific reasoning, which can be re-defined in terms of *neural networks*. Roughly described, a neural network is a net in which the ‘neural units’ are joined together through interconnections of different levels of strength. There are three types of neural units: the input, the output and the ‘hidden’ units. The connectionist model of intelligence started to attract the general attention of cognitive scientists with the implementation of the ‘learning algorithm’ of *back-propagation*, which allows the modification of the weight of the neural connections to readjust the output of the system. In this way, the system ‘learns’ which output must be produced (Rumelhart and McClelland 1987). Without delving into technicalities, it is enough to point out how connectionism does not model knowledge through the ruled computation of digital symbols. It is the repeated exposure to the same stimuli that reinforces some hidden neural paths and, consequently, neural networks ‘learn’ through a process which is distributed throughout the system.

5.5 Exemplars and Scientific Change

Nickles (2003) moves some criticisms against Kuhn’s notion of exemplars. In particular, he argues that: the exemplars used in learning concepts and those used to solve problems are not always of the same type; it is not obvious that junior scientists learn how to solve scientific problems by using the same set of exemplars used by senior scientists, as Kuhn claims sometimes; Kuhn mentions only ‘positive exemplars’ (examples of what works), whereas a more realistic model of how science is learned should also include a discussion of ‘negative exemplars’ (examples of what was thought would have worked but actually did not). In what follows I focus on one of Nickles’ criticisms that, in my opinion, highlights some weaknesses with the notion of exemplar within Kuhn’s model of science.

Nickles argues that exemplars cannot function as the ‘pivotal concept’ of Kuhn’s model, since they are not sufficient for discriminating between normal and revolutionary science. Nickles’ criticism has to do with the ambiguities of the notion of exemplars. Kuhn (perhaps) dropped the notion of paradigm to provide a more refined definition of what guides scientists during periods of normal science. Like paradigm, however, exemplar means different things. There are at least four different ways in which Kuhn speaks of exemplars:

1. as concrete problem solutions
2. as textbook reformulations of concrete problem solutions
3. as contemporary textbooks
4. as classic books

Although (5.3) and (5.4) may come as a surprise, intuitively enough textbooks and classic books are exemplary in the sense that they 'contain' (5.2). In other words, they "expound the body of accepted theory, illustrate many or all of its successful applications, and compare these applications with exemplary observations and experiments" (*SSR*:10). Furthermore, since textbooks are a relatively recent invention, for the historical study of science the items of (5.4)—like Aristotle's *Physics*, Ptolemy's *Almagest*, Newton's *Principia* (*SSR*:10)—have the same function as (5.3). In short, (5.3) and (5.4) can be defined as repositories of (5.2).

The problem is that the 'actual' and 'concrete' scientific achievements of (5.1) are not the same things as (5.2)–(5.4), the latter being "sanitised, idealised versions of those achievements, written up from a later point of view" (Nickles 2003:168). Kuhn too recognized that "past scientific achievements [...] are recounted [by science textbooks], though seldom in their original form" (*SSR*:10). Yet, he did not see any problem with such 'sanitised versions' of the actual problem solutions presented in textbooks. As Nickles reminds us, Kuhn's own historical approach led him to criticise the very fact that 'textbook science' is quite 'whiggish'. It may come as a surprise, therefore, that he then based his epistemology of science textbook-exemplars, which often offer a non-faithful characterisation of those very concrete achievements they are supposed to represent.[4]

In my view, that Kuhn first rejects Whig historiography of science and then puts Whig textbooks at the centre of his epistemology of science is not a contradiction. While, at a descriptive level, scientists are trained through Whig textbooks, it may also be the case that, to learn how to master the central exemplars of their field, scientists ought to be trained in such a Whig way. In other words, Kuhn criticises a Whig *image* of the development of science from the perspective of historians and philosophers; such an image, however, does not need to be discarded by the practising scientists. For Kuhn, it can be the case that scientists' training *has to be* Whig to be effective.[5] Leaving aside scientists' Whiggish education, the historically oriented philosopher can still analyse the development of science in non-Whiggish terms.

The real problem here, for me, is that textbook-exemplars, as Nickles rightly points out, are 'static' and rigid representations of actual, concrete exemplars.

> Fitting your current problem to available exemplars is a two-way street, since matching is a symmetric relation. It is a question of mutual fit, after all. You not only deform your problem so that it better fits an exemplar; you may also find a way to deform the exemplar in order to achieve a better fit with your problem. Successful work that stretches the exemplar in this way may result in a gradual reinterpretation of the exemplar itself. Eventually, the reinterpretation may be stretched practically out of recognition [...]. On the whole, however,

[4] Are the exemplars of the old scientific books as idealised and sanitised as the exemplars of contemporary textbooks? It seems so. Besides, it is not obvious that the old scientific books present the actual scientific achievements as they had been actually achieved.

[5] As said in Sect. 3.5, the fact that scientists ought to be trained in such a way is one of the motivations that lead Kuhn to re-evaluate Whig history in his last writings. In that section, however, I questioned whether scientists need any history at all. A sanitised exemplar is just an idealised version of what originally occurred in the past, *not* a historical narrative. Science students do not even need to know *when* the original problem was solved, or *by whom*.

> Kuhn leaves us with the idea that exemplars are fixed historical achievements, permanent models, whereas they, too, historically evolve as normal science progresses. (Nickles 2003:169)

The plasticity of actual exemplars (as opposed to the idealised fixity of textbook exemplars) may cause some problems for Kuhn's model of scientific development. If 'paradigm' is restricted to exemplars, a period of normal science would be defined by the scientists' adhesion to the same set of exemplars, while scientific revolutions would be individuated based on when the set of exemplars is changed. But if, as Nickles says, exemplars are not static and historically fixed entities, but can be modified even during periods of so-called normal science, then we are back to one of the main difficulties faced by *SSR*, namely: how is it possible to distinguish between changes occurring during periods of normal science and revolutionary changes? This looks like a revamped version of Toulmin's (1970) criticism, which threatens to blur the lines between normal and revolutionary science.

Nickles is right to argue that exemplars belonging to the same normal tradition get modified through time and, therefore, may cease to 'look similar'. To Kuhn's further detriment, I would add that similar-looking exemplars may nevertheless belong to incommensurable research traditions. In the previous section, I discussed Kuhn's own example of exemplars: Eqs. (5.2)–(5.4) look similar to one another as well as to the symbolic generalisation expressed by (5.1). It is by learning how to use (5.2)–(5.4) that the science student also learns the meaning of (5.1). At this point, one can assume that the more equations similar to (5.2)–(5.4) science students are exposed to, the more they refine their understanding of Newton's Second Law, that is one the fundamental laws in classical mechanics.

Let's now look at Eq. (5.5), known as Ehrenfest's theorem:

$$F(r) = \frac{md^2r}{dt^2} \tag{5.5}$$

where r is a quantum mechanical operator and $< r >$ its expectation value. Ehrenfest's theorem looks similar to (5.1)–(5.4). The problem here is that, while (5.2)–(5.4) are all exemplars belonging to the tradition of classical mechanics, Ehrenfest's theorem was developed within the paradigm of quantum mechanics. On the basis of notational similarities alone, one could be tempted to say that quantum mechanical expectation values obey Newton's Second Law and that, therefore, classical and quantum mechanics are continuous—a conclusion endorsed by the same Eherenfest (1927) as well as by a number of scientists and philosophers (see, for example, Landsman 2007:475–481), but that Kuhn, the philosopher of scientific revolutions, would not be very happy with.

As Max Jammer (1966) rightly points out, things look different when one goes beyond the level of similarities among equations.

> A [...] general and direct line of connection between quantum mechanics and Newtonian mechanics was established in 1927 by Ehrenfest, who showed 'by a short elementary calculation without approximation' that the expectation value of the time derivative of the

> momentum is equal to the expectation value of the negative gradient of the potential energy function. Ehrenfest's affirmation of Newton's second law in the sense of averages taken over the wave packet had a great appeal to many physicists and did much to further the acceptance of the theory. For it made it possible to describe the particle by a localised wave packet which, though eventually spreading out in space, follows the trajectory of the mechanical motion. [...] Ehrenfest's theorem [...] does not conceptually reduce quantum dynamics to Newtonian physics. [It] merely establish[es] an analogy - though a remarkable one in view of the fact that, owing to the absence of a superposition principle in classical mechanics, quantum mechanics and classical dynamics are built on fundamentally different foundations. (Jammer 1966:184)

For Jammer, although it establishes an analogy between classical and quantum mechanics, the Ehrenfest's theorem is not sufficient ground to justify any claims of intertheoretical reduction between the two. Notational similarity of some of their exemplars notwithstanding, classical and quantum mechanics are *conceptually* different.[6]

Classic functions are continuous and differentiable at every point of their domain. This means that their real variables can take on any *definite value* between their minimum and maximum value, and their exact value is determined by other variables, such as time. Unlike real variables, the Hermitian (or 'self-adjoint') operators used in quantum mechanics possess a *range of possible values* at a given time. These mathematical differences are grounded on different metaphysical world-views. The scientists who solve classic functions represent the entities of the world as being knowable in a definite way, whereas those using quantum operators describe the world as probabilistic. Although (5.2)–(5.5) 'look similar', they are 'conceptually' different. One could even venture as far as saying that, while (5.2)–(5.4) allow scientists to grasp the meaning of $F = ma$, one cannot fully get the Ehrenfest theorem unless one has *already* learned parts of the conceptual and mathematical structure of quantum mechanics. In short: is one accept, with Kuhn, that

[6]The similarity between Ehrenfest's theorem and some classical equations is mentioned by Ladyman (1998), who re-develops the main idea, originated by Worrall (1989), of structural realism. In this view, although the episodes of revolutionary theory change displayed by the history of science threaten the realists' idea that scientific knowledge gets closer to the truth, it is still possible to find a 'structural continuity' between old and new theories. Such a continuity is established by some equations, which are retained by the new theories. (Whether structural realism should be interpreted as an epistemic or as an ontic thesis—which is Ladyman's concern—is not an issue to be examined here.) Another interesting topic which I have no time to develop any further is whether Kuhn, based on his ideas about the similarities between exemplary equations, was a structural realist after all. The point I want to make here is that structural realism is consistent with Kuhn's model of science—and, indeed, with any model taking seriously the discontinuities of the history of science. Structural realism says that there can be structural continuity and equation retention, not that there are no revolutions and discontinuities in science. In this view, there is structural continuity *despite* revolutions. This means that structural continuity cannot be used to determine when a revolution occurred and what it consisted of. In short, the retention of some equations, or the structural similarity between the equations of the old theory and the equations of the newly accepted theory, is consistent with the idea of revolutionary conceptual change in science—an idea which is central in Kuhn's philosophy and yet does not seem to be captured by the idea that normal science is established on structurally similar equations.

classical and quantum mechanics are incommensurable (or that the previous can be reduced to the latter), that one ought to be able to explain why Eq. (5.5) does not belong to the same 'cluster' of Eqs. (5.2)–(5.4) and generalisation (5.1). Clearly, such an explanation cannot to be found at the level of similarities/dissimilarities.

Without denying the necessity of exemplars, it seems that, alongside the ability to perceive similarities between new and old problems, scientists also must learn a 'way of reasoning', in virtue of which they can understand why similar-looking exemplars may belong to different problem-solving traditions. And different ways of reasoning may, in turn, be grounded in different metaphysical worldviews. While there is an undeniable notational similarity among the equations discussed in this section, similarity by itself is not sufficient to discriminate exemplars belonging to different, or even incommensurable, scientific traditions.

5.6 Expanded Exemplars

Exemplars belonging to the same research tradition may be modified to the point of becoming less similar to each other, while similar-looking exemplars may belong to incommensurable research traditions. This problem perhaps could be solved if one conceives exemplars as being something more than collections of similar equations. Shan (2020), for example, proposes an 'expanded' account of Kuhn's exemplars. For Shan, Kuhn's novel contributions to the philosophy of science are fundamentally two: the concept of a paradigm, especially in its narrow sense of exemplars, and the incommensurability thesis. He agrees with Bird (2002) that, by abandoning the concept of exemplars and by restricting incommensurability to its semantic aspects, Kuhn undermined his own major philosophical insights. To straighten such a 'wrong turn', Shan, like Bird, rescues Kuhn's concept of exemplars.

The difference between Bird's and Shan's interpretation of Kuhn's exemplars is the following. Bird re-reads the concept from a cognitive/perceptual perspective: in his view, learning an exemplar, or a family of exemplars, means learning to 'see the world' in a certain way. Incommensurability thus amounts to deep differences in how scientists see the world. Shan targets some questions surrounding exemplars that, in his view, Kuhn left unanswered. For instance, how are exemplars established and how are they then applied in practice? Granted that exemplars are successfully solved problems, where do problems come from, and why are some of them deemed relevant? When is a solution to a problem also a good solution—indeed, an exemplary solution? A good theory of exemplars ought to answer these questions.

Shan's approach has the merit of stressing the relation between exemplars and the problem-solving activity of science. This requires, among other things, an understanding of what 'scientific problems' are—an issue that has not received enough attention in philosophy of science, with some notable exceptions (see Nickels 1978, 1980, 1981). Solving problems requires not only the ability to see them as similar to the ones that have been already solved, but also a whole set of other skills, such as the ability to manipulate either mathematical symbols or experimental tools. These skills, for Shan, are exemplar dependent. He therefore defines exemplars in the following way:

> An exemplar has five main components: a vocabulary, which is a set of the concepts employed in the problems and solutions; a set of well-defined research problems; a set of practical guides, which specify all the procedures and methodology as means to solve the problems; a set of hypotheses or models, which are proposed to solve the problems; and a set of patterns of reasoning, which indicate how to use other components to solve the problems. (Shan 2020:392)

These various components are intertwined and lay the foundations for an 'exemplary scientific practice'. As an example, Shan describes the work of Gregor Johann Mendel as establishing an exemplary practice that was followed by other scientists and that led to the establishment of genetics. In a sense that will be discussed in Chap. 10, Shan's conception of exemplars also plays a role in his rehabilitation of the so-called functionalist view on scientific progress (Shan 2019).

To begin with, it must be noticed that some of the components of an exemplar Shan discusses had already been examined before in terms of 'functions' of an exemplar. The "set of well-defined research problems" and the "practical guides" for solving them amount to the functions ii, iii and iv that Hoyningen-Huene (1993) and Bird (2000) attribute to exemplars (see Sect. 5.4). It is also worth stressing that, for Shan, the conceptual vocabulary is part of an exemplar. In his view, moreover, "the vocabulary of an exemplar does not suggest that all the concepts in the vocabulary are first introduced by the exemplars, though it is not unusual that the vocabulary of an exemplar has some pre-defined concepts" (Shan 2020:392). As already discussed, Kuhn regards symbolic generalisations and exemplars as two distinct parts of a DM, with exemplars providing the 'semantic content' to the symbolic generalisations. Scientists, therefore, learn the meaning of the theoretical language of their DM through exemplars. This is different from saying that an exemplar carries with itself its own conceptual language. What I am stressing here is just the difference between Kuhn and Shan on how they conceive the relation between conceptual language and exemplars, without necessarily defending the position of neither.

One could not help but notice, however, that Shan 'expands' exemplars to the point of making them co-extensive, if not indistinguishable, from DMs. In his account, apart from a conceptual vocabulary, a set of problems, and the guides for solving them, exemplars also include "patterns of reasoning" and "hypotheses or models". How, then, are Shan's 'expanded exemplars' any different from DMs? As already explained, after *SSR* Kuhn distinguished between a 'wide' and a 'narrow' meaning of paradigm. The originality of the narrow paradigm, that is exemplars, consisted in showing how scientists do not need to know a conceptual language and a set of rules beforehand but that, rather, they would learn these things by practising with exemplary problems. Shan takes Kuhn's narrow paradigms, but he widens them so much that, in the end, they are not narrow anymore.

Finally, it must be noticed that Shan does not address directly the issue of incommensurability. Of course, a philosopher who focuses on some of Kuhn's concepts does not need to examine all of Kuhn's ideas. The absences of an explicit discussion of incommensurability, therefore, would not be necessarily a problem for Shan, had he not said that the most fundamental elements of Kuhn's philosophy are the concept of paradigm and the incommensurability thesis, and that all his other contributions are "dependent" on these two (Shan 2020:386). By not addressing incommensurability, Shan does not show how his 'expanded' exemplars could solve the problem discussed in the previous section. His case study shows how the establishment of genetics was the result of a 'chain' of exemplary practices that began with Mendel and then continued with other scientists, whose way of posing and solving problems was analogous to Mendel's. But can Shan explain what exactly must change in a chain of exemplary practice for a revolution to occur? In short, how can we distinguish between two incommensurable exemplary practices?

Even though Shan does not answer these questions, he could. From his perspective, he could say that two exemplary practices are incommensurable if they employ conflicting conceptual vocabularies, or conflicting methods, or conflicting ontological commitments. This would be, in other words, the incommensurability between the wide and multi-composite paradigms of *SSR*.

5.7 Reclaiming the Centrality of the DM

Going back to the problem pointed out by Nickels, either Kuhn recognizes that exemplars can be modified and stretched 'out of recognition' even during periods of normal science, which would make normal and revolutionary science undistinguishable; or he should allow for the existence of more revolutionary episodes in the history of science, but how many? The first horn of this dilemma poses a problem only if exemplars are regarded in isolation with respect to the other elements of a DM. I have explained that notational similarity among exemplars cannot help with the discrimination among different normal traditions, since similar-looking exemplars can belong to incommensurable scientific frameworks. If exemplars are considered as part of, and from within, the DM they belong to, then the lines between two different scientific traditions become neater. A period of normal science is not established only upon exemplars, but upon exemplars embedded in a matrix of values, concepts, and metaphysical assumptions. Consequently, two different scientific traditions will not be distinguished *only* based on their exemplars. However, if exemplars are not capable of demarcating between different normal traditions without the other components of DM, then they cannot function as the 'pivotal' element of the Kuhnian model of science.[7] I therefore agree with Kindi (2013) when she argues that Kuhn should have not introduced a too strict distinction between the wide and narrow sense of a paradigm, since the two have complex relations. As

[7] This is only part of the argument. The problem discussed by Nickles can be solved by challenging *both* the idea that exemplars can be considered in isolation with respect to the other elements of their DM, *and* the focus on similarity relations, rather than on dissimilarities. This point will become clearer in Sect. 6.4, where I will introduce Kuhn's late view on 'differentiae'.

shown in previous sections, this becomes more evident by looking at the relation between exemplars and symbolic generalizations. Equations used in solving problems provide the meaning of the laws of nature expressed by symbolic generalizations, which in turn constraint the kind of equations that instantiate them and help discriminate them from the equations belonging to other research traditions and instantiating different laws of nature. Exemplary achievements may help establish a DM, but a DM in turn dictate to which new problems the exemplars will be applied, or which new problem solution can be counted as new exemplars. At the same time, since exemplars are part of DMs, contrary to what Kuhn himself maintained, I regard the somehow neglected concept of a DM to be one of Kuhn's central contributions to the philosophy of science.

The neglect of Kuhn's concept of DM is surprising, if not baffling. Sociologists and STS scholars consider Kuhn as the founder of the social studies of science. In their studies, however, they tend to develop externalist explanations of science and to focus on how social forces (such as politics and power balances) drive and, in their view, determine scientific research. This is different from the approach Kuhn had in mind. A DM—or, to use Masterman's terminology, a 'sociological paradigm'—is the 'common possession' of a scientific community. I suggest, moreover, that 'disciplinary' could be interpreted not only as 'pertaining to a scientific discipline' (assuming that Kuhn really wanted to talk about scientific disciplines, and not just about scientific communities), but also as establishing the 'code of conduct' (or, if we wish, the 'rules of the game') of an organised group, that must be followed to acquire and maintain membership to that group, and that has been developed to help the group to achieve its aims. The concept of a DM, therefore, helps understand precisely why Kuhn's look at the collective nature of science does not lead to the conclusion that science is subject to all sorts of external 'social forces'. Scientific research is a social enterprise, in the sense that it is not carried out by isolated individuals. At the same time, such individuals cannot share a constellation of commitments, values, metaphysical views. As discussed in the previous chapter, they do not need to have the same beliefs and to behave in the same way (actually, internal diversity of a scientific community is one of the requirements of collective rationality), but they nevertheless must follow the same (implicit) rules of the others. Failing to follow such rules will lead to the exclusion from the group. Those provided by a DM, furthermore, are the rules governing the *scientific* activity itself. This is why the results of the social enterprise of science are not 'socially constructed', but they are reached at and justified by scientific norms applied and interpreted within the social group of the scientific community.

A DM provides the background for understanding why some exemplars instead of others give meaning to a symbolic generalisation. In a sense, it also provides to scientists with an 'identity', in the sense that allows them to recognize themselves as part of a scientific community. The concept of a DM should have been in the forefront of the analysis of Kuhn's philosophy and, in particular, of the discussions about its relationship with social constructionist. But it has not. For a long time, most philosophers of science simply assumed that whatever is 'social' is not rational, cognitive, and scientific, and that scientific rationality must be found in the logical structure of scientific theories or at the level of the idealised rational individual. Contemporary philosophers have become more open towards the study of the social dimension of science. In the past few decades, philosophy of science has undergone a 'practice turn' and, also, a 'social turn'. Social epistemology of science is a respected sub-field of philosophy of science and some even regard Kuhn as one of its earlier proponents (Wray 2011). And yet, now that the times would be ready to re-examine Kuhn's DM, many philosophers seem to dismiss it.

For instance, Chang (2011, 2012a, 2014) puts a major emphasis on the importance of studying scientific practice: "Serious study of science must be concerned with what it is that we actually *do* in scientific work [...][which consists [of] actions, carried out by agents. An agent carries out intentions. A scientist is not a passive receiver of facts or an algorithmic processor of propositions. [...][Therefore,] in serious study of science, we need to consider human capabilities (capacities and skills) in performing epistemic activities" (Chang 2014:70). He then introduces the notion of *systems of practice*, that are coherent sets of epistemic activities held by a group of scientists and performed to achieve epistemic objectives. Despite the similarities between the two concepts, Chang claims that his own notion of systems of practice is better than Kuhn's DM, because Kuhn did not clarify how the elements of a DM hang together in a coherent way. This is a far too harsh criticism. As I have discussed in this chapter, Kuhn as well as some of his commentators have explained very well the relations between at least two elements of a DM, namely exemplars and symbolic generalisations. Furthermore, one could level the same objection to Chang, asking him to show why and how the elements of his systems of practice cohere *more* and *better* than those of a DM. Indeed, few philosophers regard Kuhn's and Chang's concepts as complementary, instead of different. For example, Boon and Van Baalen (2019) express this position: "we suggest to consider Kuhn's notion of 'disciplinary matrix' and Chang's notion of 'system of practice' as complementary to a better understanding of scientific practices. Additionally, contrary to Chang, we claim that there exists coherence between the elements of the matrix in the sense that these elements mutually support and reinforce each other, which is why it functions as a paradigm" (Boon and Van Baalen 2019:15, ft.14). In a sense, Chang's work could even be regarded as a further development of Kuhn's DM (no matter how Chang decides to call it), aimed at further explicating its constitutive elements and their relations.[8] Ultimately, however, one may even question the need for con-

[8] There are some clear Kuhnian elements in Chang's work. For instance, Chang accepts Kuhn's incommensurability thesis, especially in its methodological sense (see Chang 2012b). This kind of incommensurability can be regarded as indicating irreconcilable differences in the way in which problems are conceptualised and solved (i.e., through a different ranking and application of the same epistemic values). It could be therefore regarded as an incommensurability between DMs, rather than theoretical languages (which, anyway, are part of a DM). Yet, Chang does not use the term 'disciplinary matrix' and prefers 'system of practice' instead. One may wonder whether using different words for referring to the same Kuhnian concept, or to concepts very similar to Kuhn's, is motivated by the worry that one's work could be negatively received because of many philosophers' prejudices about Kuhn's philosophy. If that is the case, then it could be that there are more Kuhnian philosophers than one could guess, although many of them may be 'covert Kuhnians'.

cepts such as Chang's system of practice, or for Shan's extended exemplars, when one could simply look, analyse and further develop Kuhn's DM.

Ankeny and Leonelli (2016), as mentioned in Sect. 1.1, propose a 'post-Kuhnian' approach to the study of science, on the ground that Kuhn's philosophy cannot capture many important features of contemporary scientific research, such as the nature of 'big science' projects and interdisciplinary collaborations.[9] Instead of considering wide-ranging revolutionary changes affecting a whole discipline, as they maintain Kuhn does, they look into the finer structure of contemporary scientific practice, which is characterised, among other things, by the use of *repertoires*. A repertoire is *both* a list of items that can be performed *and* the set of abilities for performing them.[10] In the biological sciences, repertoires are 'animal models' used by different research groups and in different projects. The more a repertoire is used, the more it generates results and its own literature, the more it establishes itself as a reliable tool for research and as a recognisable mark of excellence. The 'recognisability' of repertoires influences the direction science takes. For example, articles referring to well-known and established repertoires are more likely to be published and cited, research projects that use them are more likely to get funded, and so on. Ankeny and Leonelli's insightful analysis of repertoires brings a fresh perspective on some elements of contemporary research. My problem is not with their much-needed study on repertoires, but with their characterisation of the philosophical perspective repertoires provide as 'post-Kuhnian'.

A possible way to respond to Ankeny and Leonelli is to reflect, as Rouse (2024:23–24) does, about the changes that science, its practice, and its funding mechanisms have undergone in the past few decades. Kuhn developed his history-based model of science around the 1960s and with the aim of explaining how so-called 'mature' sciences (i.e., physics, chemistry, and biology) make progress. One should not expect his model to be able to account contemporary science, which is more interdisciplinary, innovative and technology driven. However, this does not mean that Kuhn's model is wrong or that there is no use in it anymore. Furthermore, in the same way in which the Kuhnian model is limited to the progress of the mature sciences, so Ankeny and Leonelli's does not aim at universal validity. In fact, they explicitly address the complex relationship between repertoires and scientific research groups, including the fact that not every research group possesses a repertoire or needs it to obtain innovative results. While Rouse is surely right in pointing

---

[9] That Kuhn's philosophy cannot make sense of interdisciplinary collaborations is debatable. For example, Boon and Van Baalen (2019) use a Kuhnian framework to develop an epistemology of interdisciplinarity. See also Sect. 8.4.

[10] To make a musical example, 'Edith Piaf's repertoire' refers both to the song Edit Piaf used to perform and to the ability of performing them. Edith Piaf's repertoire, therefore, may be performed by other singers, as long as they possess the adequate skills and scenic presence. To say, for example, that Italian singer Milva knew 'Edith Piaf's repertoire', means to say that Milva knew Edith Piaf's songs but also that she had the right technical skills and scenic presence to convey the raw emotions the French diva was known for.

out the limited scope of both Kuhn's and Ankeny and Leonelli's proposals, I have further comments.

To begin with, Ankeny and Leonelli seem to criticise a 'textbook' and misleading version of Kuhn. As explained in Sect. 4.4, Kuhn was not interested only in high-level communities and in high-level revolutions. Nor Kuhn's paradigm is the rigid and inflexible monolith discussed by Ankeny and Leonelli. Apart from this, I do not see why repertoires cannot be regarded as *part* of DMs. To some extent, repertoire seems to function like exemplars. Ankeny and Leonelli rightly insist that the same repertoires are used across different specialties and research groups. This observation, however, could help enrich Kuhn's philosophy, rather than overthrowing it. I have already discussed how similar exemplars could be used in different research traditions, and that this will pose a problem to Kuhn's model of scientific change only if one considers exemplars in isolation with the other (conceptual and methodological) elements of the DM. Similarly, from a Kuhnian perspective, one would not only notice that the same repertoires appear in different research groups but would also investigate whether they are also 'performed' differently, or how they are 'adapted' when they get embedded in different DMs.

Moreover, and following Ankeny and Leonelli's own account, some repertoires become successful because in virtue of their power to attract funding. This motivates scientists to write grant proposals for research that will make use of such successful and fund-attracting repertoires. In other words, thanks to their fund-attracting power, some repertoires may acquire an ever-growing prominence. Does this mean that repertoires risk becoming as rigid and fixed as what Ankeny and Leonelli think a paradigm is? Or, rather, does this mean that repertoires, like Kuhn's *actual* paradigms, simply perdure through time, even though they can undergo several changes?

Finally, a point that deserves to be made concerns normativity. Kuhn's model is *both* descriptive *and* normative. The scientific community *is* pervaded by an essential tension between scientists with a dogmatic attitude towards the paradigm and more iconoclastic and 'out of the box' scientists, who develop a reserve tank of alternatives to be used in case of crisis. But the scientific community also *ought* to be pervaded by such an essential tension, which is the mark of the collective rationality of science. During normal science, scientists *do* extend and exploit the paradigm, and they also *ought* to do so, because it is by using the paradigm that scientists discover its limits. And so on. Surely, Ankeny and Leonelli highlight some crucial elements of contemporary research, such as the use of repertoires in data-driven biology. They provide a rich description of how such repertoires possess epistemic, social, and institutional dimensions. However, it is not clear whether they also defend their normativity. Scientists use repertoires, but are repertoires good for reaching the epistemic aims of science?

However, Kuhn also left a wealth of concepts, such as the concept of a DM, as well as a philosophical perspective which are yet to be fully appreciated, even by those that, in my view, are developing such concepts and perspective, rather than overcoming them.

5.8 Conclusions

By looking at the activity carried out by a scientific community, the philosopher of science will discover its paradigm, that includes a theoretical language, laws of nature, models, methodological views, metaphysical presuppositions, and so on. Much has been said about the polysemy of 'paradigm' in *SSR*, with most philosophers simply assuming that if a word is polysemous than its underlying concept is imprecise. However, since a paradigm complex and multifaceted, the word that refers to it is inevitably polysemous.

Nevertheless, to reply to his critics, Kuhn attempted to refine the concept by distinguishing between a DM, that captures the broad sense of a paradigm, and exemplars, that reflect its narrow sense. Exemplars are fundamental to explain how scientists learn to solve new scientific problems by analogy with previously solved problems. However, it is important to stress that exemplars are *part* of a DM. By contrast, some contemporary philosophers have focused on exemplars without examining their relationship with the other elements of a DM. Such a relationship is dynamic and multi-lateral. For instance, while exemplars give meaning to the laws of nature expressed by symbolic generalisations, the interpreted symbolic generalisations in turn establish the similarity/dissimilarity criterion for interpreting some exemplars as part of a DM rather than another. Ultimately, Kuhn's paradigm and DM were among the first concepts in philosophy of science that attempted to capture what makes *scientific practice* possible. Some of the concepts proposed by contemporary philosophers of scientific practice may even be regarded as further developments of Kuhn's DM.

In the end, perhaps Kuhn was mistaken in his attempt to distinguish between a wide and a narrow sense of paradigm. He should have sticked to his ground and defend the usefulness of polysemous words expressing complex concepts. Had he done so, however, he would have still needed to solve the methodological problem of how to describe the complex activity carried out by the scientific community. In his mature writings, perhaps he solved this problem by regarding scientific communities as akin to 'linguistic communities'.

References

Ankeny, Rachel, and Sabina Leonelli. 2016. Repertoires: A post-Kuhnian perspective on scientific change and collaborative research. *Studies in History and Philosophy of Science* 60:18–28.

Bird, Alexander. 2000. *Thomas Kuhn*. Princeton: Princeton University Press.

Bird, Alexander. 2002. Kuhn's wrong turning. *Studies in History and Philosophy of Science* 33:443–463.

Bird, Alexander. 2005. Naturalizing Kuhn. *Proceedings of the Aristotelian Society* 105:109–127.

Bird, Alexander. 2007. Incommensurability naturalised. In *Rethinking Scientific Change and Theory Comparison*, ed. Lena Soler, Howard Sankey, and Paul Hoyningen-Huene, 21–39. Dordrecht: Springer.

Boon, Mieke, and Susan Van Baalen. 2019. Epistemology for interdisciplinary research – Shifting philosophical paradigms of science. *European Journal for Philosophy of Science* 9:16. https://doi.org/10.1007/s13194-018-0242-4.
Bruner, Jerome, and Leo Postman. 1949. On the perception of incongruity: A paradigm. *Journal of Personality* 18:206–223.
Chang, Hasok. 2011. The philosophical grammar of scientific practice. *International Studies in the Philosophy of Science* 25:205–221.
Chang, Hasok. 2012a. *Is water H2O? Evidence, Realism and Pluralism*. Dordrecht, The Netherlands: Springer.
Chang, Hasok. 2012b. Incommensurability: Revisiting the chemical revolution. In *Kuhn's the Structure of Scientific Revolutions Revisited*, ed. Vasso Kindi, Vasso and Theodore Arabatzis, 153–177. New York: Routledge.
Chang, Hasok. 2014. Epistemic activities and systems of practice: Units of analysis in philosophy of science after the practice turn. In *Science After the Practice Turn in the Philosophy, History, and Social Studies of Science*, ed. Lena Soler, Michael Lynch, Sjoerd Zwart, and Vincent Israel-Jost, 75–87. London: Routledge.
Eherenfest, Paul. 1927. Bemerkung über die angenäherte Gültigkeit der klassischen Mechanik innerhalb der Quantenmechanik [Remark on the approximate validity of Classical Mechanics within Quantum Mechanics]. *Zeitung der Physik* 45:455–457.
Hoyningen-Huene, Paul. 1993. *Reconstructing Scientific Revolutions. Thomas Kuhn's Philosophy of Science*. Chicago: University of Chicago Press.
Jammer, Max. 1966. *The Conceptual Development of Quantum Mechanics*. New York: McGraw-Hill.
Kindi, Vasso. 2013. Kuhn's Paradigm. In *Kuhn's Structure of Scientific Revolutions Revisited*, ed. Vasso Kindi and Theodore Arabatzis, 91–111. New York: Routledge.
Kuhn, Thomas S. 1959. The Essential Tension: Tradition and innovation in scientific research. In *The Third University of Utah Research Conference on the Identification of Scientific Talent*, ed. C. W. Taylor, 162–174. Salt Lake City: University of Utah Press. Reprinted in *ET*:225–239.
Kuhn, Thomas S. 1974. Second thoughts on paradigms. In *The Structure of Scientific Theories*, ed. Frederik Suppe, 459–482. Urbana: University of Illinois Press. Reprinted in *ET*:293–319.
Ladyman, James. 1998. What Is Structural Realism? *Studies in History and Philosophy of Science* 29:409–424.
Landsman, Nicolaas. 2007. Between classical and quantum. In *Handbook of the Philosophy of Science: Philosophy of Physics*, ed. Jeremy Butterfield and John Earman, 417–554. Amsterdam: Elsevier.
Marcum, James. 2013. From Paradigm to Disciplinary Matrix and Exemplar. In *Kuhn's Structure of Scientific Revolutions Revisited*, ed. Vasso Kindi and Theodore Arabatzis, 41–63. New York: Routledge.
Masterman, Margaret. 1970. The nature of a paradigm. In *Criticism and the Growth of Knowledge*, ed. Imre Lakatos and Alan Musgrave, 59–89. Cambridge: Cambridge University Press.
Musgrave, Alan. 1971. Kuhn's second thoughts - Review of the *Structure of Scientific Revolutions* by Thomas S. Kuhn. *British Journal for the Philosophy of Science* 22:287–297.
Nickels, Thomas. 1978. Scientific problems and constraints. *PSA: Proceedings of the Biennial Meeting of the Philosophy of Science Association 1978*:134–148.
Nickels, Thomas. 1980. Scientific problems: Three empiricists model. *PSA: Proceedings of the Biennial Meeting of the Philosophy of Science Association 1980*:3–19.
Nickels, Thomas. 1981. What is a problem that we may solve it? *Synthese* 47:85–118.
Nickles, Thomas. 2003. Normal Science: From logic to case-based and model-based reasoning. In *Thomas Kuhn*, ed. Thomas Nickles, 142–177. Cambridge: Cambridge University Press.
Rouse, Joseph. 2024. Has there been a second kuhnian transformation in the philosophical image of science?. In *Rethinking Thomas Kuhn's Legacy*, ed. Yafeng Shan, 11–30, London: Springer.
Rumelhart, David, and James McClelland. 1987. *Parallel Distributed Processing: Explorations in the Microstructure of Cognition*. Cambridge: MIT Press.

Shan, Yafeng. 2019. A new functionalist approach to scientific progress. *Philosophy of Science* 86:739–758.
Shan, Yafeng. 2020. Kuhn's "wrong turning" and legacy today. *Synthese* 197:381–406.
Suppe, Frederick. 1974. *The Structure of Scientific Theories*. Urbana: University of Illinois Press.
Toulmin, Stephen. 1970. Does the distinction between normal and revolutionary science hold water? In *Criticism and the Growth of Knowledge*, ed. Imre Lakatos and Alan Musgrave, 39–47. Cambridge: Cambridge University Press.
Wagenknecht, Susan, Nancy Nersessian, and Hanne Andersen. 2015. *Empirical Philosophy of Science: Introducing Qualitative Methods into Philosophy of Science*. London: Springer.
Worrall, John. 1989. Structural Realism: The best of both worlds? *Dialectica* 43:99–124.
Wray, K. Brad. 2011. *Kuhn's Evolutionary Social Epistemology*. Cambridge: Cambridge University Press.

Chapter 6
Kind Terms and Taxonomies

Abstract In his more mature years, Kuhn became increasingly interested in the conceptual structure of conceptual taxonomies and in the meaning of kind terms. Contrary to those who claim that such an interest is evidence of a 'linguistic turn' he underwent to be accepted by the philosophy community of his time, in this chapter I examine the philosophical reasons that drove Kuhn to the problem of meaning, and how he attempted to solve such problems by focussing on taxonomies. Then, I analyse Kuhn's crucial shift from the 'principle of similarities' to the 'principle of *differentiae*', and how the latter sheds a new light on other parts of his philosophy. Finally, I discuss the problems that Kuhn's concept of taxonomy faces.

Keywords Meaning · Reference · Kind terms · Conceptual taxonomies · No-overlap principle · Prototype theory · Differentiae · Developmental psychology

6.1 Introduction

Around the 1980s, Kuhn gave up the concepts of paradigm, exemplars, and DM and begun to write about issues concerning the meaning of scientific terms. For Bird (2002), Kuhn decided to focus on the linguistic elements of scientific theories to be accepted by the community of philosophers, who regarded him as a historian at best, or as a naïve and unsophisticated thinker at worst. By doing so, however, he turned his back to his earlier and more original concept of exemplar that, if properly developed, could have been the basis for a fully naturalised philosophy of science. Bird's explanation of Kuhn's linguistic and anti-naturalistic turn is not entirely plausible, however. It is true that, after not getting tenured at Harvard, he went to work at the University of Berkley, where he became Professor of History of Science in 1961. However, in 1964 he became Professor of History and Philosophy of Science at Princeton and Laurance S. Rockefeller Professor of Philosophy at MIT in 1979. Meanwhile, he kept being invited to present in major philosophy conferences, such as the Urbana conference on the structure of scientific theories, organised by Patrick

V. Politi, *Thomas Kuhn's Philosophical Perspective*, Boston Studies in the Philosophy and History of Science 348,
https://doi.org/10.1007/978-3-032-13540-7_6

Suppe, where he read his "Second thoughts on paradigms". It appears that, by the time he underwent his 'turn', Kuhn was *already* part of the philosophy community of his time, at least from a strictly institutional and academic point of view. Of course, this does not mean that he also *felt* being accepted as a good philosopher by his colleagues and peers. The point, however, is that it is impossible to know what Kuhn's 'real feelings' were, and I also have doubts about explaining the development of his views based on how he personally felt about his place within the philosophy community of his time. What I suggest, then, is to analyse the *philosophical reasons*, that is the reasons arising from within his philosophy, that led him to tackle the problem of meaning.

In the previous chapter I explained some of the difficulties faced by exemplars. Still, such difficulties are not sufficient to motivate a 'turn' from the study of scientific practice carried out by a community to the analysis of meanings. For instance, Kuhn could have gone 'full sociological' by focussing on DMs, which contain exemplars alongside other elements. Even so, it is not clear from which element of a DM philosophers should (or could) begin their analysis. The study of the scientific communities of the past adds a further layer of complexity since, in this case, historically oriented philosophers of science cannot observe them directly. The description of past scientific activity is, indeed, a reconstruction. Kuhn bypassed these issues by considering scientific communities as akin to *linguistic communities*. Philosophers observing the behaviour of practicing scientists will easily come across their shared language, the analysis of which represents the 'entry point' for understanding what they do. When it comes to the history of science, the language of past scientific communities must be interpreted. It is by studying past scientific communities and their languages that Kuhn develop his view on meaning change which, in turn, requires a theory of meaning to begin with.

In Sect. 6.2, I explain Kuhn's dissatisfaction with the theories of meaning of his time. In Sect. 6.3, I explain how taxonomies provided the theory of meaning Kuhn needed to make sense of his view on meaning change. In Sect. 6.4, I explore one of Kuhn's crucial but unanalysed insights: the idea that kind membership is not established through similarities with the other members of that kind but, rather, through *differentiae*. In Sect. 6.5, I discuss one of the main problems with Kuhn's account of taxonomies, that is the existence of crosscutting kinds within the same scientific theories. Finally, in Sect. 6.6, I discuss an even bigger problem, namely the inadequacy of the concept of taxonomy to account for many important conceptual changes in science.

6.2 The Problem of Meaning

Even accepting the idea that exemplars ought to be understood from within a DM, which defines the community that holds it, it remains to see *how* to describe and analyse a DM. This is a serious methodological problem that remains surprisingly under-discussed by many philosophers of scientific practice. In Kuhn's view, philosophers who study a scientific community are akin to ethnographers. For instance, to those who accused him of 'overpopulating' philosophy of science with too many concepts, such as exemplars, symbolic generalizations, and so on, he replied in the following way:

> Imagine yourself as Quine's imaginary linguist learning a language, but now you are examining a scientific, rather than a language, community. It is vastly easier to discover that scientists say to each other 'force equals mass time acceleration' than to discover what the devil they mean or are talking about when they say it. Though the expressions are interpreted by the [scientists] who use them, discovering the signs and discovering the interpretation are partially separate operations, particularly since the same deduction follows with or without interpretation. (Discussion on 'Second Thoughts on Paradigms', in Suppe 1974: 516–517)

For practicing scientists, there is no neat distinction between symbolic generalizations and their meanings; for the philosopher studying their community, however, understanding the elements of a DM begins with a process akin to language learning. Whether one is studying the constellation of commitments held by a scientific community or the constellation of meanings making up an alien culture, the analysis of language seems unavoidable, at least as a first step. Such an analysis leads the historian and the historically oriented philosophers to experience incommensurability when facing seemingly senseless passages in old scientific texts. Semantic incommensurability faced many criticisms, while its 'local' version allowed Kuhn to clarify that it does not imply neither incommunicability nor incomparability (see Sect. 1.4.3). To talk about incommensurability of meaning, however, it is necessary to have a theory of meaning to begin with. As Dudley Shapere notices, however, even though both Kuhn and Feyerabend speak of meaning change through history, "neither author gives us a criterion for determining what counts as a part of the meaning of a term, or what counts as a change of meaning, even though these notions are central to their portrayals of science" (Shapere 1966, in Shapere 1984: 82).

For Kuhn, the 'received view of meaning' cannot explain the kind of meaning changes he discovered in the study of old scientific textbooks (see *PW*, in Kuhn 2022: 169 and ff). In the received view, the meaning of a term consists of its *extension*, that is all the things that that name names, and of its *intension*, that is the concept associated with that term, its 'sense'. Some words or expressions may have the same extension (that is, they refer to the same things), but different senses, as in the case of 'planet Venus', 'morning star', and 'evening star'. Intension determines extension, not in the sense that intension determines the 'size' of extension (i.e., how many entities are the referents of a word), but in the sense that it establishes the conditions for being the referent of a word. Such conditions amount to the set of features shared by the entities falling into the extension of a word. For example, to know the meaning of 'swan' would mean to know the traits that all the swans share (i.e., wings of a certain amplitude, a beak of a certain shape, the plumage of a certain colour, and so on). These features can be divided into two categories: 'essential' features, that all the referents of a term *must* possess, and 'accidental' features. The intension of a word is made up only of essential features. Therefore, in the traditional view, to know the meaning of a word is to know the essential features of the members of its extension, that is the list of the necessary and sufficient conditions something must satisfy to be a referent of that word.

Kuhn mentions some of the main limits of the traditional view of meaning. To begin with, actual linguistic competence does not require full knowledge of necessary and sufficient conditions: speakers know how to use words even though they don't know their formal definitions. When it comes to scientific language, Kuhn believes that theoretical terms are not acquired through formal definitions, but by practicing with exemplars (see Sect. 5.4). In both ordinary and scientific language, the usefulness of formal definitions is questioned, and it is not even clear whether such definitions are used at all to begin with. Furthermore, attempts to establish the full list of the necessary and sufficient conditions shared by all the referents of a term has proven to be a much challenging task.

Kuhn relates the crisis of the traditional conception of meaning to Quine's criticism of the analytic/synthetic distinction. The division between 'essential' and contingent features, in fact, is reflected on the variety of descriptive statements that can be formulated about the referents of a term. Statements that make explicit some of the essential features of a concept (i.e., "all bachelors are unmarried men") are true 'by definition'. They express, that is, analytic or tautological truths. The truth value of empirical or 'synthetic' statements, by contrast, depends on experience. As already discussed in Chap. 2, Quine argued that the analytic/synthetic distinction is untenable: even so-called analytic statements may be revised based on new evidence. This means that, apart from the statements in logic and mathematics (though there is room for debate in these cases too), there exists no statement 'true by definition'. Quine concludes that "Taken collectively, science has its double dependence on language and experience; but this duality is not significantly traceable into the statements of science taken one by one" (Quine 1953: 42).

Kuhn questions Quine's 'extreme holism', for which "The totality of our so-called knowledge or belief […] is a man-made fabric which impinges on experience only along the edges" (Quine 1953: 42). Indeed, the 'local' version of the incommensurability thesis is a departure from Quinean holism. At the same time, however, Kuhn reflects on one of Quine's insights.

> But it is important to recognize that, as Quine's phrase "double dependence" suggests, the removal of the [analytic/synthetic] distinction has two aspects which need, so far as possible, to be separated. The first is a threat to a neutral foundation for science, a foundation held to be independent of language and culture. The second is a threat to a long-standing understanding of meaning, in particular to the distinction between the intension and extension of words and phrases. These two aspects are closely interrelated […]. But seeing them as two makes it easier to recognize the alternative paths which the demise of the analytic/synthetic distinction opens. (*PW*, in Kuhn 2022: 173–174)

Those who take the first path opened by Quine preserve the idea of a neutral foundation for knowledge but abandon the notion of meaning as intension. For Kuhn, those who follow this path believe that "Words, phrases, and statements are to be understood extensionally, in terms of their referents alone. What a world means is simply the set of things to which it refers" (*PW*, in Kuhn 2022: 174).

For Kuhn, Quine (1960) himself took the first of the two paths he opened through his rejection of the analytic/synthetic distinction. In his famous example, English-speaking ethnographers studying the (imaginary) tribe of Arunta will face non-trivial difficulties when trying to learn its language. One of such difficulties concerns the establishment of the referents of the Arunta words. For instance, whenever they point at a running rabbit, Arunta people exclaim "Lo, gavagai!". The ethnographers would be tempted to translate that sentence with "Look, a rabbit!". However, one cannot be entirely sure that gavagai simply means rabbit. If one of the sources of nourishment of the tribe is indeed rabbit meat, in pointing at a running rabbit the Arunta people may refer to 'undetached rabbit parts' or simply to 'food' that they must hunt. Of course, to clarify the meaning of gavagai, ethnographers could ask questions to the Arunta native-speakers. This would be possible, however, only if they had already learned Arunta, in which case they would know what gavagai means. Reference, in other words, is *imperscrutable*.

For Quine, 'radical translators', who learn and translate a completely alien language, must proceed by developing analytical hypotheses about the grammar and lexicon of the object-language. These hypotheses are tested through continued and sustained interactions with the native speakers, for instance by observing the kind of behaviour associated with the utterance of some sentences and by relying on other contextual clues. The imperscrutability of reference notwithstanding, Quine believes that radical translators may learn the correct meaning of observational sentences. These are sentences that contain words whose referents can be observed by the speakers. For translating this kind of sentences, the ethnographers/translators ask themselves which sentence they would have uttered if they had been in the same place of the alien-language speaker and received the same kind of stimuli. This view, however, opens further philosophical problems. It is not entirely clear whether Quine believes that everybody perceives the world in the same way because we all have access to the same mind-independent world of stimuli, or because we all have the same system of receptors. And even though, early in his career, he was open to the idea that observationality admits different degrees, in his more mature years he maintained that empirical experience, expressed by observational sentences, is the fundamental and stable 'checkpoint' of science (see Quine 1995).

Leaving aside the issues concerning stimuli and receptors, it must be noticed that abstract terms pose their own set of problems. Testing analytical hypotheses about sentences containing abstract terms is an extremely complex and virtually unending affair, abstract terms lacking any tangible observational referent, no matter how initially imperscrutable that could have been. For Quine, then, while it is possible to produce correct translations of observational sentences, when it comes to abstract or theoretical sentences translation is *indeterminate*. The absence of an 'objective' referent to determine the meaning of abstract terms, in fact, can result in a range of distinct, yet equally valid, translations.

Hilary Putnam (1975) is also one of the champions of Quine's first path. One of the main representatives of this strategy. He challenges the traditional assumption for which "knowing the meaning of a term is just a matter of being in a certain psychological state" (Putnam 1975: 219) and claims that, instead, "meanings ain't just in the head!" (Putnam 1975: 227). At the same time, he substitutes intension, traditionally conceived as the set of necessary and sufficient features shared by the referents of a term, with a description based on a 'stereotype'.

A stereotype is "a standardized description of features of the kind that are typical, or 'normal', or at any rate stereotypical. The central features of the stereotype generally are criteria—features which in normal situations constitute ways of recognizing if a thing belongs to the kind or, at least, necessary conditions (or probabilistic necessary conditions) for membership in the kind" (Putnam 1975: 230). Speakers do not even need to know all the features of a stereotype to confidently individuate its referents (and, therefore, to know the meaning of a word): for Putnam, such a 'specialized' knowledge is possessed only by few members of a speakers' community, that is therefore characterized by a "division of linguistic labour" (Putnam 1975: 227).

With his theory, Putnam aims at establishing a new conception of linguistic competence, which is based not on the knowledge of formal definitions but, rather, on the actual interactions with the world and the entities it contains.[1] One of such interactions is the 'naming ceremony', through which an object receives its name. During the ceremony, the referent of a term is fixed once and for all. Even though the stereotype associated with a word may be significantly altered through time, in the ways permitted and sanctioned by the linguistic experts, its extension will not be affected. For example, even though different 'properties' have been historically attributed to 'tigers', 'gold', 'water' or 'humans', the classes of entities these words refer to have remained the same. This is so because the referents of these terms all share the same hidden 'essence' with the sample of the kind used in the original naming ceremony. Putnam maintains that his 'causal' theory of reference applies to both ordinary and scientific languages. Reference stability of theoretical terms can therefore be a remedy to the issues raised by the incommensurability thesis. Reference, not intension, is what constitutes meaning, while its stability through theory change may play the role of a neutral and unchangeable foundation of scientific knowledge.

For both Putnam and Quine, meaning is fixed by reference. For Quine, the reference of observational terms can be determined by developing and testing analytical

[1] "Linguistic competence and understanding are not just knowledge. To have linguistic competence in connection with a term it is not sufficient, in general, to have the full battery of usual linguistic knowledge and skills; one must, in addition, be in the right sort of relationship to certain distinguished situations (normally, though not necessarily, situations in which the referent of the term is present). It is for this reason that this sort of theory is called a 'causal theory' of meaning" (Putnam 1975: 199).

hypotheses, and by assuming a homogeneity of perceptual experience. While theoretical terms may present additional challenges regarding their meaning, observational sentences help maintain consistency within our knowledge system. In Putnam's view, it is possible to fix reference once and for all and this guarantees stability across language changes. In short, both philosophers assume that there exists a neutral and culture-independent reality that contains the referents of the words; or, at least, of some words.

Such an assumption conflicts with the dynamic perspective (see Sect. 3.2). Kuhn, therefore, takes the 'second path' opened by Quine which, in opposition to the first one, rejects the idea of a neutral foundation for knowledge and preserves the notion of meaning as intension.

Kuhn notices a crucial yet missing step in Quine's theory of radical translation. After learning the alien object-language, the ethnographers/translators must communicate it to an audience of speakers of their own mother-language. It is at this point that the translator, *contra* Quine, face problems even with observational sentences. Relying on the example by the linguist John Lyons, Kuhn explains how the unequivocally observational sentence 'the cat is on the mat' could be translated into different French sentences. The English word 'mat', in fact, may refer to 'paillasson', 'descente de lit', or 'tapis'. As Lyons explains,

> There is a set of lexemes in English, 'mat', 'rug', 'carpet', etc., and a set of lexemes in French, 'tapis', 'paillasson', 'carpette', etc.; and none of the French words has the same denotation [reference] as any one of the English lexemes. Each set of lexemes divides, or categorizes, a certain part of the universe of domestic furnishings in a different way; and the two systems of categorization are incommensurate. [...] The denotation of a lexeme is limited by the relations of sense which hold between it and other lexemes in the same language. The denotation of 'mat' is limited by its contrast in sense with 'rug' and 'carpet'; the denotation of 'paillasson' in French is limited by its contrast in sense with 'tapis' and other lexemes. We could not reasonably say that 'mat' has two meanings because it is translatable into French by means of two non-synonymous lexemes, 'tapis' and 'paillasson'; or that 'tapis' has three meanings because it can be translated into English with three non-synonymous lexemes, 'rug', 'carpet', and 'mat'. The meanings of words (their sense and denotation) are internal to the language to which they belong. (Lyons 1984: 238; cited in *PW*, in Kuhn 2022: 186; for similar examples, see also Kuhn 1983)

A skilled English-to-French translator, aided by contextual clues, would surely know when it is more appropriate to use 'paillasson' or 'tapis' for each occurrence of 'the cat is on the mat'. But this is not the problem. For Kuhn, rather, the problem is that there cannot be "any universal French generalization that applies to all and only situation in which a cat sits on a mat. For practical purposes (buying a new floor covering, for example), that untranslatability may be unimportant. *But universal generalizations are constitutive of much of science, and their untranslatability can there be deeply consequential*" (*PW*, in Kuhn 2022: 187, *my emphasis*).

Different languages sort things and the words that expressed them into different classes and different lexeme-relations of sense. For Kuhn, in the history of science there have been changes in scientific language that has led to a redefinition of groups of inter-related terms and, consequently, of how things were sorted into categories. While Putnam believe that changes in the stereotype/description of a term do not affect its referent, Kuhn argues that the history of science shows instances in which *both* the intension *and* the extension of groups of words changed in non-trivial ways.[2] And while Quine maintains that empirical experience may led to the modification of even so-called analytic statements, for Kuhn change may also work the other way round, with changes in the 'core' of our scientific system of beliefs leading to a change of how we interpret what we observe and experience.

In short, in virtue of the rejection of a neutral foundation of knowledge dictated by the dynamic perspective, and after the examination of historical data about episodes of meaning change in science, Kuhn needs to establish a theory of meaning that (a) accounts for *intension/extension changes*; (b) and accounts for how such changes affect *groups of words* through the alteration of lexeme-relations. He believed that such a theory could be found by characterizing the conceptual structure of scientific theories as if they were organized in taxonomies.[3]

6.3 Taxonomies

A taxonomy is the systematic classification of things (in its original sense: of biological organisms) in groups or classes, put in a hierarchical order that goes from the high level of generality of the genus to the lower levels of species and subspecies. Individuals are classified as belonging to one group or another, based on the characteristics shared with the other members of the group. Kuhn uses the notion of taxonomy to describe the conceptual structure of scientific theories: the kind terms of

---

[2] To be more precise: Kuhn does not reject the idea of a 'rigid designation' for names whose referents are individuals, as in the theory of Kripke (1980). He is concerned with the application of such causal theories for names referring to groups of individuals, such as the names referring to natural kinds.

[3] Kuhn's theory of meaning is criticized by Bird (2000, 2002, 2004a, b), who finds it philosophically weak and ultimately unsuccessful in its attempt to reject Putnam's causal theory. By contrast, Sharrock and Read (2002) and Kuukkanen (2010) praise Kuhn's theory of meaning and lament the little attention it has received in contemporary analytic philosophy. I tend to side with Sharrock and Read and with Kuukkanen and surely they have delved into the details of Kuhn's view on meaning with far more depth than what I have done in this section. My main aim here, however, was not to provide a full assessment of Putnam's, Quine's and Kuhn's theories. Rather, I explained how the problem of meaning was relevant for Kuhn's philosophy, not something he decided to tackle just to appeal to the mainstream analytic philosophers of his time.

a scientific theory are taxonomic in the sense that they can be systematized in a taxonomic fashion.

As Hacking (1993) comments, taxonomic kind terms have three fundamental properties:

1. they respect the *no-overlap principle*, which states that the members of a kind C_1 can also be members of a kind C_2 if and only if C_1 and C_2 are in a genus-to-species relation, or K-relation; that is, if and only if either C_1entirely subsumes C_2 or it is entirely subsumed by it. This means that if an individual is a member of C_1 and, also, a member of C_2 but C_1 and C_2 are not k-related, then the no-overlap principle is violated. The genus/species relation, or k-relation, is intransitive: while every individual belonging to the same species also belongs to the same genus, the opposite does not necessarily obtain. For example, while every dog is a mammal, not every mammal is also a dog. Individuals classified by two kind terms which are not k-related represent a transgression of the principle, as in the case of an individual which is classified both as a dog and as a cat
2. they have *infimae species*, in the sense that the taxonomic hierarchies terminate with a 'lowest level' of kinds which do not split in further sub-species
3. they are *projectible*, in the sense that they allow us to make generalizations or to have expectations about the properties and behaviours of classified individuals—i.e., if we classify an individual as a dog, we can expect it to bark and be playful, whereas if we classify it as a cat we will have different expectations

A few more words need also to be spent on an important characteristic of taxonomic categorizations, that is the *principle of similarity*. In a taxonomy, things are grouped into the same genus, species, and subspecies if they share some characteristics, that is if they are similar to one another with respect to some specific traits. Grouping things based on how similar they are is not sufficient, since there may be uncountable traits shared by diverse individuals. The risk is that everything may appear to be similar to anything else (in one way or another). This is the so-called 'problem of the open texture' which can be resolved by complementing the principle of similarity with the principle of *dissimilarity*. In other words, taxonomic kinds group similar things together and segregate dissimilar things, based on some salient traits (Andersen 2000).[4]

The introduction of the notion of taxonomy has significant consequences for the notion of a scientific revolution. What was once indicated as a change of paradigm in Kuhn's late philosophy becomes a change in the taxonomic relations of scientific kind terms. This is not only a mere variation of terminology. In his mature work, Kuhn redescribes revolutions as being triggered by some problems with the dominant taxonomy: some of the individuals classified as belonging to a kind display some unexplainable behaviour or, even worse, newly discovered individuals possess

[4] Andersen (2000) rightly insists on the need of introducing the principle of dissimilarity, alongside that of similarity, to complement or even 'correct' Kuhn's initial account. As I will argue in the next section, Kuhn progressively abandons the principle of similarity to focus mainly on dissimilarities, or *differentiae*.

the characteristics of two unrelated kinds, thus threatening the no-overlap principle. Revolutions occur when, to solve these classificatory anomalies, the taxonomy undergoes a significant change.

It must be stressed that while every revolution is a change of taxonomy, not every change of taxonomy is a revolution. Taxonomies can be in fact modified in non-revolutionary ways. For instance, a new kind can be discovered and added the new kind to the pre-existing taxonomy, without such an addition threatening the rest of the classification. This would surely count as a 'taxonomic change', but not revolutionary. It can also occur that, through time, some classifications must be 'corrected'. For example, after new and more careful observation, it may result that individuals mistakenly classified as belonging to a kind (for example, 'planets') and found out to belong to another kind (for example, 'dwarf planets'). These cases are not counted as revolutionary, since they involve a redistribution of the individuals classified as belonging to kinds, not the criteria of kind membership.

In Kuhn's mature philosophy, what changes during a scientific revolution is not just what the taxonomy looks like, through additions or adjustments, but the very *criteria* of classification. A scientific revolution, in other words, is an alteration of the similarity/dissimilarity criteria at the basis of classification.

> In normal change, one simply revises or adds a single generalization, all others remaining the same. In revolutionary change one must either live with incoherence or else revise a number of interrelated generalizations together. [...] Normal science, too, alters the way in which terms attach to nature. What characterizes revolutions is not, therefore, simply change in the way referents are determined, but change of a still more restricted sort. [...] What characterizes revolutions is [...] change in several of the taxonomic categories prerequisite to scientific descriptions and generalizations [...]—a change in one's sense of what is similar to what, and of what is different. (Kuhn 1987, in *RSS*: 29–30)

The notion of a taxonomy of scientific concepts is linked to a theory of learning that is not dissimilar from that linked to exemplars. As seen in the previous chapter, Kuhn argues that scientific concepts are not learned by learning rules and explicit definitions, nor in isolation. In his view, learning the meaning of a concept requires learning a 'cluster' of inter-defined. In Kuhn's mature account, the 'cluster' of concepts is nothing more than some of the 'branches' of a taxonomy. This does not seem to be dramatically different from what Kuhn said in his earlier works. After all, the example discussed in "Second Thoughts of Paradigms" (Kuhn 1974) can be easily re-described as if Johnny, who learns to recognize swans by also learning to distinguish them from geese and ducks, is learning the taxonomy of waterfowl.

Andersen, Barker, and Chen (2006; from now on: ABC) argue that some findings in cognitive science vindicate Kuhn's late insights about concept acquisition. In recent decades, especially after Eleanor Rosch's impactful work in the study of cognition and categorization, the 'classic view' on concepts has been deemed untenable. The requirements of the classic theory of concepts are simply too demanding: in many cases, making a list of conditions shared by all the instances of a concept would be impossible. What Rosch and her collaborators have shown, instead, is that people classify instances of a concept based on their similarity to a *prototype* (Rosch 1973, 1978; Rosch and Mervis 1975). In this view, concepts have a 'graded structure': the instances of a concept may possess various grades of similarity to a representative type for that concept. This means that concepts are not learned by learning lists of necessary and sufficient conditions, but by grouping things based on the similarity they have with respect to each other and with respect to a *prototype*.

Both Kuhn's and Rosch's theories revolve around the idea of grouping instances of concepts based on their similarity relations. These two theories, in turn, resemble the theory of 'family resemblance' developed by Wittgenstein (1953) who, despite being often ignored or dismissed by most analytic philosophers, but not by psychologists, can be considered as an *ante litteram* critic of the classic view on concepts. Rosch (1987) explicitly acknowledges the connection between her theory of prototypes and Wittgenstein's family resemblance. Some philosophers interpret Kuhn's philosophy from a Wittgensteinian perspective (Kindi 1995a, b). Whether Kuhn was influenced by Wittgenstein, however, is open to debate.[5] What it is established, however, is that Kuhn was familiar with the work of Rosch, with whom he had a correspondence. Andersen (2013: 184), for example, cites a letter of Kuhn to Rosch, dated 7th December 1976, alluding to the 'Johnny example'.

ABC, then, read Kuhn's mature theory of kind concepts *via* Rosch's theory of prototypes and re-explain Kuhn's mature definition of scientific revolutions in terms of taxonomy changes by using the *dynamic frame model*, developed by Barsalau (1992) and Barsalau and Hale (1993), who expand on Rosch's work. In a dynamic frame, 'superordinate concepts' (occupying a higher place in the hierarchy) and 'subordinate concepts' (standing at lower levels) are connected by a series of 'attributes'. Not every subordinate concept possesses the same attributes in the same way; in other words, the attributes possessed by the subordinate concepts may have different 'values'. Interpreting Kuhn's taxonomies as frames, in which genus and species are superordinate and subordinate concepts, clarifies how revolutionary taxonomic changes are possible: the changes of taxonomic criteria are in fact described as changes of attributes connecting superordinate and subordinate concepts.

For example, the Copernican Revolution can be described as the transition between two taxonomy-frames. The conceptual structure of heliocentric astronomy would be then represented as a frame in which the subordinate instances of the superordinate concept 'celestial body' possess the following attributes (each with the possible values here indicated in brackets): orbit centre (star, planet, none), orbit shape (elliptic, parabolic, other), distance (far, medium, near), luminance (self-luminous, reflected) and size (large, medium, small). In this frame, the subordinate concept 'planet' is defined as an instance of the superordinate concept 'celestial body' which is characterized by reflected luminance, by having a star as the centre of its orbit, and by (generally) having an elliptic orbit. In older astronomical conceptual frames, by contrast, the subordinates of 'celestial body' have only one attribute: that is, celestial path. This is just one of the several differences between older and modern astronomical conceptual systems. The reason why modern frames classify

[5] See, Sect. 1.2 and the end of Sect. 3.2.

Earth as 'planet'—as a celestial body of reflected luminance having the Sun as the centre of its elliptic orbit—while the older does not is because of the different attributes and values used for defining superordinate and subordinate concepts.

This example also fits nicely with Kuhn's late notion of 'local incommensurability'. The Ptolemaic and the Copernican taxonomies are indeed incommensurable because there is no *lingua franca* in which, for example, the Sun is both a planet and a star, or the Earth both is and is not a celestial body. As already explained, Kuhn maintained that incommensurability involves only a relatively small sub-set of interrelated kind terms while, at the same time, vast parts of the scientific (as well as ordinary) language remain unchanged after a revolution, thus allowing for the possibility of inter-theoretic comparison. In Kuhn's more mature view, the interrelated kind-terms that change during a revolution could be those that are grouped together at some level of a taxonomy. In other words, the local incommensurability describes the differences among some of the 'branches' of two alternative taxonomies.

To sum up, Kuhn's notion of taxonomy clarifies what is the conceptual unit undergoing change during revolutionary periods, helps to individuate the sub-set of interrelated terms affected by local incommensurability and focuses on the language of a community as a way of understanding how the group classifies things. Kuhn's late insights seem to be vindicated by cognitive theories of concepts and can be reinterpreted through models of concept representation used by cognitive scientists.

6.4 From Similarities to *Differentiae*

In the transition between incommensurable conceptual taxonomies, *both* the meaning of some kind terms *and* their referents are altered. This view does not require knowledge of necessary and sufficient conditions for kind membership, as in the classical conception of meaning, and it is in stark contrast with Putnam's reference stability thesis. At the same time, it relies on a 'local', non-Quinean form of holism: the phenomenon of meaning change affects only a restricted group of inter-defined kind terms. ABC's interpretation, furthermore, vindicates Kuhn's ideas about the structure of conceptual taxonomies by reading it through the lenses of Rosch's theory. But would Kuhn himself agree with such an interpretation?

In some of his last writings, he clearly distances himself from the prototype theory of concepts. As explained in the previous section, Rosch's theory defines kind membership through the similarity to a prototype. There are some (often unnoticed) similarities between Putnam's stereotypes and Rosch's prototypes (Geeraerts 2008). For instance, they are both useful abstractions for determining meaning without recourse to the knowledge of necessary and sufficient conditions. However, there are also important differences. For example, while Putnam insists on reference stability across theory change, Rosch's approach does not seem to exclude in principle that changes in the prototype may also change the class of its referents. In fact, Kuhn does not criticize the prototype theory because of its similarities with

Putnam's. Rather, his complaint is that such a theory does not make justice to the local holism characteristic of scientific languages.

> Kind terms do not attach to nature individually. [...] Ironically, prototype theory, the theory of meaning and concept formation most like my own, misses the need for this bit of local, non-Quinean holism. Like those who believe in characteristic features, prototype theorists conceive classification and meaning as attaching kinds to nature one at a time. The distance they are concerned with is that separating an individual from a prototypical example of the kind in which it should be placed. No role is played by the distance to neighbouring kinds. (*PW*, in Kuhn 2022: 241–242)

It is not clear whether Kuhn was acquainted with Barsalou's dynamic frame theory, which represents a more sophisticated version of the prototype theory and may overcome this sort of criticism.[6] Barsalou introduces a more nuanced way to make sense of the relationship of members of a kind with a prototype *as well as* of the relationships between neighbouring kinds and sub-kinds. For example, his dynamic frames may represent how the concepts of stars and planets, together with the classes of their referents, were modified in response to the introduction of the concepts of satellites and comets. Therefore, such an approach is not limited to the study of kinds as attaching individually to nature, and it is consistent with local holism.

There is, however, another criticism that Kuhn moves against the prototype theory. He does so by discussing the example of face recognition: "The 'normal face' [...] is a product of many [...] comparisons, not something that the learner requires in advance. Reliance on a normal face, *like reliance on prototypes more generally*, still too much resembles the traditional shared features approaches to problems of identification and meaning" (*PW*, in Kuhn 2022: 211, *ft.*21, *my emphasis*).

Kuhn's complaint can be summarized as follows. The traditional conception of meaning requires speakers to know in advance the list of necessary and sufficient conditions for kind membership. The prototype theory requires speakers to know in advance at least some of those features that make the members of a kind similar to a prototype. In both cases, speakers are required to know in advance the meaning of words, being it in the form of formal definitions or prototypes, before they are able to use them.[7] Scientific change, however, resembles language acquisition: both in

---

[6] The previous quotation was from the fifth chapter of *PW* (Kuhn 2022: 224–259) that, as for the date appearing at its end, was completed on September 24, 1995. For prototype theory in general, Kuhn refers to a text from the 1980s (namely, Smith and Douglas 1981). Much of Barsalou's work on dynamic frames began to be published in the 1990s on. Even though this work was already circulating when Kuhn was attempting to finish the manuscript of his last book, it may be the case that he had not read it.

[7] Textual evidence, examined by Melogno and Giri (2023), demonstrates how Kuhn thought about this issue as early as 1980. In 'The Nature of Conceptual Change', a series of lectures delivered at the University of Notre Dame in November 1980, he rejects the traditional view of meaning, but also the 'cluster theory', namely Wittgenstein's theory which defines kind membership as a matter of 'family resemblance', and which, as discussed in the previous section, influenced the development of Rosch's prototype theory. Both Wittgenstein's cluster theory and Rosch's prototype theory rely on the idea of similarity. In many of his writings, starting at least from "Second thoughts on

the case of scientists developing a new conceptual vocabulary and in that of historians reconstructing a past scientific language, learners can rely on *neither* ready-made definitions *nor* ready-made prototypes. Even describing scientific revolutions by comparing different dynamic frames does not say much about how scientists went from one frame to the other, especially if the latter was not fully developed yet. The comparison of two fully formed and complete conceptual frames would not be that different from the comparison of the logical structure of two fully developed theories that philosophers from a static perspective engage in, but that would not be admitted from a dynamic perspective. It is worth reminding that Kuhn does not want to analyse fully formed and 'fixed' systems of beliefs but, rather, the *changes* necessary to go from one of such systems to the development of a new one; and such a process, in his view, is akin to language acquisition. The cognitive parallel of Kuhn's philosophy is not a theory of concept representation through prototypes but, rather, *developmental psychology*.

Both in *SSR* and in other writings, Kuhn often acknowledges his debt towards Jean Piaget, the French psychologist who explained psychological development as consisting of various 'stages' (for a discussion of Kuhn's own reading of Piaget, see Levine 2000). By the time he began to refine his position with respect to the theory of meaning, he had delved into the more updated literature on the empirical studies about the development of infants' cognition. Returning to the case of face recognition, Kuhn summarizes some experiments, like those of Bushnell, Sai, and Mullin (1989) and Field et al. (1984), in which newborns are shown the faces of two adults simultaneously, one of whom is their mother. As early as 2 days after being born, infants observe their mother's face for a longer time. Only after they have established their preference by focussing on their mother, they turn their sight to the stranger's face. These results are interpreted as evidence that infants learn to recognize faces very early. However, an interesting refinement of the theories about face recognition comes, rather unexpectedly, from other experiments: those about gesture imitation.

Like face recognition, infants acquire the ability to imitate facial gestures during their very first days of life.[8] In many of these experiments, infants are made to interact with some adults. When the adults open their mouths or stick their tongues out, the infants repeat the gesture. In the experiment run by an experiment ran by Meltzoff and Moore (1992), infants interact with and imitate the gestures of two adults, one of whom is their mother, but not at the same time. Specifically, they first observe their own mother making a facial gesture and then immediately leaving

paradigms", Kuhn himself argued that categorization requires the spotting of similarities. However, as discussed in this section, he gradually distanced himself from such a view to the point of abandoning it, therefore turning his back to Wittgenstein and Rosch.

[8] Whether infants really 'imitate' adults' facial gestures or whether their behaviour is a sort of mechanical reaction to the arousal caused by visual stimuli is still debated in contemporary developmental psychology (see Vincini et al. 2017). No matter how or whether this debate will be settled, it is important to stress that Kuhn is not interested in the actual nature of infants' response to the adults' facial gestures, but in their ability of discriminating between different 'kinds of adults'.

their sights. The place of the mother is than taken by another adult. What happens is that infants repeat their mother's gesture back to the stranger, without any significant pausing time to readjust their behaviour to the new person. They behave, that is, as if their mother has left for a moment and then she reappears shortly after. This result shows that while infants can perceive the differences in facial features when they are exposed to two faces at the same time, they cannot recognise that different faces belong to different people when shown in succession.

What does Kuhn make out of these experiments in developmental psychology? In his view, the ability to recognise *individual objects* and *kinds* is pre-linguistic and common to humans and animals. Like Aristotle's substances, an object is "a bounded region that moves through space over time, [...] a trajectory tracer" (*PW*, in Kuhn 2022: 203). In the experiments of postnatal imitation of facial gestures, infants look at the adult who has appeared *right after* their mother left as if the same object returned. However, they will treat their mother and the other adult as two different individuals if they first interacted with the two of them at the same time, and then only one of them exits their field of vision. They would do so because they have the pre-linguistic intuition that individual objects cannot occupy two different places at the same time, nor can they follow 'bifurcating trajectories'. What infants still do not have at this stage, however, is the ability of considering those two different objects as belonging to two different kinds. In other words, for them, "It is as though mother were herself a kind, capable of numerous different exemplifications. Or [...] it is as though mother-present were a frequently recurring situation" (*PW*, in Kuhn 2022: 206).

Only with time, infants can 'individuate' (that is, differentiate) their mother's face from anyone else's. For Kuhn, experiments in developmental psychology demonstrate that infants do not learn to recognise the face of a particular individual, A, by memorizing some of A's 'salient features'. Rather, they start by learning how A is *different* from X. However, "[they] may be able to distinguish A from X but not from Y, or from X and Y but not from Z, or from X, Y, and Z but not from W, and so on. That series can be continued, and at each step along the way, [the infant] will have learned one or more additional discriminations. [...] Learning to recognize an individual face is thus a process that can, in principle, continue indefinitely. Why should it have an end?" (*PW*, in Kuhn 2022: 210).

Once the ability of discriminating between different individual objects is acquired, a similar mechanism allows for the discrimination of different kinds. The ability of sorting groups of individuals into different kinds, in other words, does not require the recognition of similarities among the members of the same family, let alone of resemblance to a prototype. All it takes is the recognition of differences *between* members of different groups.

> Viewed in this way, face recognition suggests with special clarity that neither identification of a particular individual nor of its kind requires knowledge of some special constellation of qualities, properties, or features shared by all presentations of that object or by all members of that kind. One need only know features that can differentiate the substance or kind in question from others with which it might, *in the world as it is*, be confounded. Recognition is not [...] the only process by which objects or kinds can be identified. But it is the basic

> one, and understanding that it proceeds from differentiae rather than shared features has a number of significant advantages. (*PW*, in Kuhn 2022: 210)[9]

Cutting the world at its joints may not always be an easy task, especially if the world's joints are not so neat or easily discernible. There are individuals that are clearly different from others. In such cases, it is not difficult to see which one belongs to one group rather than to another. In between kinds, however, there is a 'space of *differentiae*' where things can get fuzzier. Such a space could be populated by individuals who do not clearly fall into one of the two neighbouring kinds. Shifts in how the space of differentiae is perceived may lead to re-categorisation and, therefore, revolutions. For example, it may turn out that the space of differentiae is populated by individuals who belong to neither of the neighbouring kinds, but to a kind of their own; and, with this discovery, it may happen that some of the individuals who previously belonged to the pre-existing kinds will be classified as belonging to the new kind.

For Kuhn, there are several advantages in focussing on differentiae rather than on similarities. Differentiating two groups of individuals does not require assessing those individuals' similarity to an abstract prototype. Furthermore, differentiation allows for individual variations within the same community. Going back to ABC's cognitive reinterpretation of taxonomies, one may worry that such a view implies that every heliocentric scientist ought to know that every planet must possess certain features with certain values. This approach risks being a too little departure from a conception that defines language competence in terms of knowledge of necessary features. By contrast, for Kuhn, one does not even need to know which features a speaker is choosing to differentiate individuals and kinds.

It may even be that different speakers use different features, even though they end up differentiating kinds in the same way. To explain this point, Kuhn proposes a variation to the Johnny experiment, based on something that occurred to one of his friends.[10] In the new example, Alexander learns how to recognise between different vehicles through ostension. Every day on the way to school his father points at different concrete 'exemplars' of truck, bus, and car and, after a while, through ostension Alexander has learned the meaning of these concepts and how to discriminate their instances. So far, the new example is akin to the Johnny example. One day, while playing in the backyard, Alexander hears the noise of a vehicle and exclaims "Truck!". Even though the father though that the learning of the 'taxonomy of vehicles' was based on visual features, Alexander's own conceptualisation of trucks and cars relied on further features, such as sounds. And his father could have kept his idea Alexander acquired kinds of vehicle through visual features only virtually forever.

[9] It is impossible to know what Kuhn meant by "in the world as it is" (emphasized in the original). I will not even attempt an interpretation of such a rather cryptic, and surely unexpected, expression.

[10] Kuhn speaks about receiving "a letter from a friend and sometimes collaborator resident in Zurich" (*PW*, in Kuhn 2022: 242). The letter he refers to is one by Paul Hoyningen-Huene, dated April 1994.

Knowing what features speakers use to segregate kinds is not required to successful communication, as long as everybody sort things in the same way. What both Johnny and Alexander learn through the interaction with their father, then, is the *community*'s way of sorting things into kinds. Such knowledge is passed through socialisation. At the same time, individuals may categorise based on different sets of features. All it takes for a linguistic community to persist through time and not to 'break' is consistency in differentiating kinds, no matter how such a differentiation is made by its members. As seen in Chap. 4, for Kuhn individual variation within a group is essential for the group's 'survival'. Individual differences in how kinds are differentiated are vital when the discovery of new individuals threaten the way in which the inter-kind space of differentiae is conceived. In such cases, what a speaker holds as the most important features for differentiating kinds may be easily given up by another speaker, who will then be in a better position to build a new taxonomy.

At this point, it is important to summarize how this detour on developmental psychology and differentiae relates to Kuhn's philosophy. Kuhn must deal with the problem of meaning to explain how meaning change is possible. Taxonomic kinds suit the theory of meaning Kuhn's is after: within a conceptual taxonomy, some of its branches and sub-branches form a cluster of inter-related kind terms, that must be learned together; when the criteria of classification of a taxonomy are altered, *both* the intension (the meaning of a kind term) *and* the extension (the class of referents of that term) change. The historian who must reconstruct a past conceptual taxonomy needs to learn how to reconceive the inter-kind space of differentiae, following a process akin to language learning. The historian, that is, must learn how the conceptual taxonomy of the past sorted things into taxonomy whose kinds possess meanings that are intensionally and extensionally different from our own. This theory of meaning is vindicated by developmental psychology, which shows how we learn to differentiate individuals and classes of individuals, not by learning definitions nor by relying on pre-made prototypes, but through pre-linguistic cognitive abilities that are available since we are born and that allow the development of further abilities.

Finally, one may wonder whether Kuhn truly managed to establish a new theory of *meaning*. While his ideas about re-categorisation and shifts in the space of differentiae clarify how changes in the extension of some terms happen in a (locally) holistic way, it seems that any mention of intensions expressed by definitions has vanished completely. Kuhn is not looking for definitions. In his view, the process of differentiation and categorization are more fundamental than definitions. It may even be the case that scientists may create a new kind without even agreeing on how to define it but only agreeing on what differentiate it from pre-existing kinds. This is the case, for instance, of the discovery of *virus*, which was established through differentiation of a group of entities from the pre-existing kind of *bacteria* but *without* a unanimous agreed upon definition of what viruses actually are (see Sect. 8.5). In a sense, for Kuhn the intension of a term is the whole conceptual web that contains that term and that is governed by the principle of differentiae.

6.5 Cross-Cutting Kinds

The no-overlap principle plays a central role in Kuhn's re-definition of scientific revolutions, since revolutionary changes to a taxonomy are required when the principle is violated. Yet this view faces the problem of the existence of so-called *cross-cutting categories*. An example of cross-cutting kinds is *poison*: some instances of both *vegetable* and *mineral* can be classified as also being instances of *poison*; yet *poison* is not in a genus-to-species relation with either *vegetable* or *mineral*, nor are *vegetable* and *mineral* in a genus-to-species relation with one another. Saying that there are objects which can be classified as both poisonous and vegetable (or mineral) means that there are objects which belong to two kinds which are not in a K-relation—which thus violate the no overlap principle.

The same Hacking (1993), who delves deeply into Kuhn's use of the notion of taxonomy and the no-overlap principle, is aware of the problems that cross-cutting kinds pose to taxonomic hierarchies. This issue cannot be solved by simply saying that kinds like *poison* are not natural kinds. He therefore argues as follows: there are natural kinds, some are taxonomic and some are not taxonomic; scientific kinds are 'taxonomic natural kinds'; therefore, we need a principled way to distinguish scientific taxonomic kinds from the rest of the natural kinds.

To operate such a distinction, Hacking resurrects J.S. Mill's definition of kinds. Elaborating on William Whewell's idea that science must classify the entities of the world in epistemologically relevant rather that random collections, Mill proposed a criterion for distinguishing between arbitrary classes and real kinds. The members of an arbitrary class share only one characteristic, or just a few of them. For instance, both a piece of paper and an aspirin can be grouped in the class of 'white things'. Apart from their being white, however, they would not share any other property. This means that we would not expect a piece of paper to alleviate headache, nor that aspirins would be good for writing letters. By contrast, a natural kind "is distinguished from all other classes by an indeterminate multitude of properties not derivable from one another" (Mill 1843: 99). For example, members of the kind 'planet' do not share only the characteristic of being 'objects in space', but also the characteristic of rotating around a star, of following a definite orbit, of having a specific size, and so on. Following Mill, then, Hacking argues that the entities grouped under the concept of poison, apart from being poisonous, like poison are natural but do not share any other characteristics; therefore, poison is not a real kind. As such, it poses no problems to scientific taxonomies.

On the hand, however, and as pointed out by the same Hacking, there are kinds which are scientific but that do not satisfy Mill's condition for real kinds.

> Mill cites *horse* and *phosphorus* as real kinds. What about Clydesdales and 32P? There are certain marks, criteria, that determine the kind of horse and the isotope of phosphorus. Are there indefinitely many things to find about Clydesdales? Perhaps we happily conclude that Clydesdales don't form a scientific kind. What about the radioactive isotope? Are there indefinitely many things to find out about it? I'm not sure it is a real kind. Kuhn might be troubled here too. (Hacking 1993: 301)

On the other hand, there exist ordinary (i.e., non-scientific) kinds which that nevertheless have an inexhaustible number of characteristics. As Dupré (1981) explains, ordinary language distinguishes between prickly pears and chollas, which in scientific language belong to the same genus of *Opuntia*; the ordinary category of hawks crosscuts at least three different families of *Falconiformae*; lilies, onions and garlic all belong to the scientific kind of *Liliaceae*. Therefore, if we accept that only scientific kinds are real kinds, then we would also need to accept that "onions had turned out to be lilies" (Dupré 1981: 75). Now, real kinds are defined as possessing an indefinite number of characteristics; but *for whom* are such characteristics are indefinite? For people interested in cooking, onions and garlic have many more characteristics than lilies. Dupré concludes that the difference between the ordinary and scientific classifications consists in the *function* they serve, not in their being more or less 'real'.

The appeal to real kinds with an indefinite number of properties exacerbates some difficulties with the notion of taxonomy instead of solving them. As discussed by Massimi (2005), these difficulties were already known to Medieval philosophers. Following Porphyry's neo-Platonic interpretation of Aristotle's doctrine of categories, it is possible to systematize entities into a tree-like hierarchy by grouping them based on the essential properties, or predicables, they share. What some philosophers realized, but Porphyry apparently did not, is that the high number of predicables shared by entities leads to the so-called 'problem of symmetric hierarchy'.[11]

Massimi discusses Hacking's example of *poison* in the light of the Medieval problem of symmetric hierarchy. As already said, the genus *poison* can be divided into *vegetable* and *mineral*, with hemlock belonging to the former and arsenic belonging to the latter. However, the genus *pharmaceutical* can also be divided into *vegetable* and *mineral*. In turn, we could consider *vegetable* to be a genus which divides into *pharmaceutical* and *poisonous* (hemlock being a vegetable that is also poisonous), or we could consider *mineral* to be a genus which divides into *pharmaceutical* and *poisonous* (arsenic being a mineral that is also poisonous). Massimi shows how it is possible to construct four different taxonomic trees by considering the two pairs of properties 'poisonous/pharmaceutical' and 'vegetable/mineral', with arsenic belonging to two of those taxonomies. What happens if we stick to Hacking's solution of the problem of crosscutting kinds? "Should the properties of arsenic be virtually inexhaustible", Massimi argues, "there would be a virtually

[11] "In *De divisione* Boethius remarked how differences could be reorganized in alternative taxonomic trees: we can divide the genus of white things into liquid (milk) and hard (pearl), but we can also divide the genus of liquid things into white (milk) and black (ink). Abelard in the *Editio super Porphyrium* made a similar point: the genus 'animal' can be divided into rational/irrational (so as to distinguish 'man' from 'horse'), and then into mortal/immortal; but it can also be divided into mortal/immortal first (so as to distinguish 'man' from 'god') and then into rational/irrational. These taxonomic rearrangements highlight how the specific differences that lexical taxonomies consist of do not form an asymmetric hierarchy of interlocking properties, each embedding the subsequent one. […] Given any pair of differences, there is no warrant that the first contains the second any more than the second contains the first" (Massimi 2005: 91–92).

inexhaustible number of trees it could potentially be classified under" (Massimi 2005: 90–91).

A similar case can be made for *parasite*, discussed by Khalidi (1998: 41). This category crosscuts the biological classification—i.e., living organisms belonging to different species are classified as parasites. Khalidi argues that *parasite* is indeed a scientific kind, so much so that the whole field of parasitology has been built around the study of such a kind. However, it is not entirely clear whether the scientific kind *parasite* is also a real kind in the sense that the entities defined as such share an indefinite number of characteristics. Whether parasite is a real kind or not, the problem is that if *living organism* is indeed real kind, its members sharing an indefinite number of properties, that crosscut taxonomic trees—for example, parasitology and the Linnaean classification may group the same living organisms in different ways.

Surely the problem of symmetric hierarchy, as explained by Massimi and further illustrated by Khalidi's example, poses non-trivial difficulties to the idea that there could be only *one* hierarchical classification. This, however, was not what Kuhn had in mind. Kuhn, in fact, allows for the existence of different ways of classifying things: he introduces the notions of kind terms and taxonomies to redefine incommensurability (between different ways of classifying things). In other words, what Massimi and Khalidi speak about is a problem for the essentialist ideal of one universal and fundamental classification of natural kinds. However, this is not a problem for Kuhn, for whom different Porphyrian trees are indeed incommensurable—i.e., the tree of parasitology crosscuts the Linnaean classification because the two are incommensurable (and one must choose either one of the two but cannot use them both at the same time).[12] The real problem for Kuhn arises with cross-cutting categories *within the same* scientific field.

Following ABC, some scientific classifications best represent the inviolability of the no-overlap principle. For instance, the periodic table of the elements is such an exclusive division: no atom can be both iron and carbon. Likewise, for astronomers in the sixteenth century, there are no fixed stars which are also planets, nor celestial objects which are also sublunar objects (ABC: 67). Curiously enough, some philosophers regard those fields which ABC take as exhibiting 'exclusive classifications'—chemistry and astronomy—as chief examples of classificatory systems containing cross-cutting kinds. For instance, McDonough (2003) notices that the table of chemical elements is anything but a 'taxonomic tree': while a taxonomy is composed by a 'head' from which various branches and sub-branches descend and form a hierarchy, the table of elements is composed by a sequence of parallel vertical columns, each forming a different family of chemical kinds—the first vertical column on the left hand-side of the table groups together the members of the family

[12] The problem of symmetric hierarchy can be tackled by abandoning Porphyry Aristotelianism and by reverting to another great Medieval tradition, namely nominalism. In a nominalist perspective, only individuals exist. Our ways of grouping them do not reflect real divisions in the world. This is the 'nominalist solution' of Hacking (1993) for the problem of the post-revolutionary 'new world': what changes after a revolution is not the actual world (i.e., the individuals in the world) but the way we conceive it (i.e., our classifications of individuals).

of *Alkali metals*, the second groups together the members of *Alkaline earth metals*, and so on. Moreover, some members of some families are grouped with other members of other families, therefore forming a series of cross-cutting kinds. The kind of *non-metals*, for example, includes all the members of the family of *Noble gasses*, some members of the families of *Chalcogens* and *Halogens*, and *Hydrogen* (*H*), which also belongs to the first column, together with the *Alkali metals*. Similarly, the kinds of *metals* and *semimetals* crosscut throughout the whole table, together with other kinds (*ion*, *acid*, *base*, and so on). McDonough concludes that "the structure of chemical kinds more closely resembles a quilt of overlapping strips and patches than a hierarchical tree" (McDonough 2003: 342).

The classification of stellar objects poses further problems. Following Ruphy (2010), the astronomical field admits alternative ways of classifying its objects.

> Take one of the brightest stars of the Northern Hemisphere, Vega in the Lyra constellation, and ask astrophysicists what kind of star Vega is. Commonly known as an "A0V" star, that is, a relatively hot, slightly bluish "main sequence dwarf", Vega is also classified by astrophysicists observing in the far infrared part of the electromagnetic spectrum as a "1n-18" star, that is, a star with no remarkable spectral feature in this domain of wavelength, and for those studying how the light emitted by a star varies, Vega is known as a "Delta scuti" type of star, that is, a kind of pulsating variable star. (Ruphy 2010: 1110).

Hacking (2007) has also recognized that the periodic table is "the permanent refutation of the idea that natural kinds have to be organized into a tree-like hierarchy" and has given yet another example of cross-cutting kinds:

> It has been repeatedly argued that natural kinds must, as a matter of logic, be arranged in a tree-like hierarchy. Not so. Bosons, isotopes and elements are generally regarded as natural kinds. But since rubium-47 is a species both of boson and of rubidium, but rubidium is not a species of boson or vice versa, you cannot put these on a branching tree. (Hacking 2007: 214)

Trying to avoid the violation of the no-overlap principle caused by the existence of cross-cutting categories leads to a dilemma. On the one hand, one can claim that crosscutting categories are not real kinds and, therefore, not scientific; however, the claim that some chemical and stellar kinds are 'more scientific' than others surely appears problematic: it seems uncontroversial to maintain that the kind *metals* is as scientific as the kind *Halogens*. On the other hand, one can say that cross-cutting kinds exhibit a violation of the no-overlap principle and, therefore, belong to incommensurable ways of classifying things; but this is simply not true, since there are examples cross-cutting categories within the same scientific field. A classificatory system that contains incommensurable taxonomies is simply incoherent.

Perhaps there is another way to look at this issue. Kuhn says that a taxonomy must respect the no-overlap principle. He does *not* say that a scientific theory, or even a whole research field, must rely on only one taxonomy. Kuhn speaks about the possibility of constructing different taxonomies, that categorize the same entities in different ways, that are driven by different values and interests, and that serve different purposes.

> Individuals tend to use more than one [hierarchical tree], depending on context. And whichever one they use, it leads them to the same identification in cases that make a difference. It would be hard to maintain that [...] the differences between [hierarchical trees][are disagreements] about matters of fact: that one is right, the others wrong. [Rather,] one [hierarchical tree] is perhaps more efficient when used for one purpose, another for another [purpose], but all lead to the same conclusions in everyday practice (*PW*, in Kuhn 2022: 248)

In the quote above, Kuhn is talking about taxonomies of the everyday language spoken by a linguistic community, making the examples of the categorization of fish and birds. However, we should remember that Kuhn uses everyday languages and linguistic communities as analogies for scientific conceptual frameworks and scientific communities. Therefore, I explain how I read this quote by returning to the example of the periodic table.

As already discussed, the periodic table arranges chemical elements in 'groups' (vertical column) and in 'trends' (that cut the table diagonally). Groups and trends do not form a neat 'grid' but, rather, intersect in complex ways. So, for example, although it is not a metal, *H* is put in the group of *Alkali metals* (first column). At the same time, it also belongs to the group of *nonmetals*. Kuhn, however, would not consider *H* as belonging to two kinds which are not K-related; rather, he would regard groups and trends as forming different systems for categorizing elements. In the periodic table, columns group together chemical elements with the same number of valence electrons (that is, electrons in the outer shell of an atom), whereas trends group together elements based on their behaviour. One considers the nonmetal *H* as belonging to the same group as the *Alkali metals* or, alternately, to the *nonmetal* trend depending on what one is interested in or aiming to explain. The classification in groups and trends are not incommensurable, because they are *not* competing to explain the same things. Rather, they are used to investigate different aspects of the same elements. Therefore, *H* does not really crosscut across two incommensurable conceptual systems, but it belongs to two different systems.

If this is what Kuhn had in mind, then he should have clarified that the conceptual system of a scientific field may include several different conceptual taxonomies, a 'forest' of tree-like hierarchies. This would have forced him to explain how such a forest of taxonomies is learned. Since different taxonomies classify things differently depending on what they want to explain, it seems that to learn the many taxonomies of the conceptual structure of a research field one should also learn which values and interests drive research in that field.

As for exemplars, it appears that conceptual taxonomies cannot be easily understood in full isolation with respect to the other elements of a DM.

6.6 Non-taxonomic Revolutions

A revolutionary taxonomic change involves a change of the very criteria of taxonomic classification. But does every scientific revolution involve such a kind of change? To answer this question (in the negative), I will use the example of the

Chemical Revolution.[13] Without discussing the rich historical details of this famous case of scientific change, I focus on the type of conceptual changes brought about by such a revolution as presented some years ago by Thagard (1992: 39–50), who analyses the conceptual structure of Stahl's phlogiston theory and of Lavoisier's oxygen theory. Thagard is not averse to taxonomic hierarchies himself: indeed, he represents both Stahl's and Lavoisier's conceptual systems as tree-like taxonomic structures (Thagard 1992: 41, 46; *Fig. 3.2, 3.6*). In such taxonomies, however, concepts are not connected only through kind-relations. There are also other concept-relations, represented by arrow-lines connecting sub-kinds belonging to different kinds. Thagard defines this sort of relations in terms of *rules*.

For instance, in Stahl's taxonomy there is a rule connecting *phlogiston* with *compound*, stating that when phlogiston is with a compound (i.e., sulphur) there will be some possibility of burning. In Lavoisier's taxonomy there is a rule connecting *oxygen* with *oxidable metallic*, stating that the two together will produce *oxides*; and another rule connecting *oxygen* and *non-metallic*, stating that the two combined together will burn to release *light* and *caloric*. According to Thagard, these 'rules' are necessary to understand the phlogiston and the oxygen theories. If the structure of a scientific conceptual system consists only of genus-to-species relations, then understanding rules such as 'it will produce' or 'it will burn with' becomes more difficult.

Learning a taxonomy means acquiring a 'semi-ontological knowledge': in virtue of their projectability, knowledge of kinds comes with beliefs and expectations about the regularities governing the behaviours of their individuals (ABC: 29). An ontology, however, is made of kinds of entities but also of processes. A semi-ontological knowledge should therefore be about *both* kinds of entities *and* processes. While a taxonomy can easily represent the former, it seems to leave aside the latter. Unless, of course, one can construct both a 'taxonomy of entities' and a 'taxonomy of processes'. This suggestion faces at least two problems.

---

[13] The Chemical Revolution is one of the central examples of scientific revolutions discussed in *SSR*. Hoyningen-Huene (2008) has even argued that, since it fits Kuhn's model so well, it is not just an 'empirical example' confirming the model but, rather, one of its 'constitutive parts'. However, some have counter-argued that, unless one reinterprets the historical facts in a distorted way, the Chemical Revolution can hardly be accommodated within Kuhn's rigid model (Blumenthal 2013). Both Pyle (2000) and Chang (2012) argue that the Chemical Revolution exhibits not too much 'semantic incommensurability' but some incommensurability of problems and methods. Finally, some have questioned whether the phlogiston and the oxygen theories were, respectively, the starting and the ending point of the Chemical Revolution. To begin with, there existed different reformulations of phlogiston theory, and the same Lavoisier, the French chemist credited for the discovery of oxygen, had put forward alternative versions of the phlogiston theory before formulating the first oxygen theory. Perhaps, such a revolution consisted in the transition from the old chemical ontology based on the idea of governing 'principles' to a new 'building-block' ontology of chemical composition. From this perspective, Lavoisier's theory is one phase of a much longer revolutionary process, the end of which should be postponed to Dalton's atomism—see Siegfried and Dobbs (1968), Chang (2010: 70–71).

First, it is not entirely clear whether it conflicts with Kuhn's own idea of what taxonomies should be about. As he clarified in his mature writings, what he means by 'kind terms' share an important characteristic:

> this characteristic can be traced to, and on from, the evolution of neural mechanisms for re-identifying what Aristotle called "substances": things that, between their origin and demise, trace a lifeline through space over time. [Substances are] not only kinds of physical objects (e.g., elements, fields, forces), but also kinds of furniture, of government, of personality, and so on. [...] Among English words, they can be identified by grammatical criteria: for example, most of them are nouns that take an indefinite article either by themselves or, in the case of mass nouns, when conjoined with a count noun, as in '*gold* ring'. (Kuhn 1993, in *RSS*: 229–230)

The variety of processes that play a crucial role in a scientific ontology can be hardly identified as Aristotelian substances. Besides, some terms indicating processes do not take an indefinite article. Kuhn's 'grammatical criterion' for the identification of kinds suggests the possibility of constructing kind-taxonomies only for entities (*an* electron, *a* chemical element, *a* gene) and materials (*a* gold ring, *a* silver plate, *a* wooden table). The situation with processes is less clear: some of them can (perhaps) take both the definite and indefinite article (i.e., *combustion* or *fermentation*), others only the definite (i.e., *meiosis* or *entropy*). But apart from the grammatical criterion, which Kuhn could have been mistaken about, it is not clear how 'kind-relations' among different processes would be established (i.e., are 'combustion' and 'evaporation' the sub-kinds of a more general kind of process?).

Incidentally, Chen (of the ABC team) has reconsidered the initial conception of taxonomies and frames. Chen (2007) argues that the focus on scientific concepts expressing entities or objects produces a 'biased' philosophy of science, incapable of doing justice to types of scientific changes at the level of processes and events, rather than involving the reclassification of entities. In his view, for example, one of the major elements of the shift from geocentric to heliocentric cosmology was the transition from the concept of *stellar path*, which ancient cosmology conceived as an actual entity, to the concept of *orbit*, which is not a 'thing' but, rather, the ideal line traced by the process of planets going around a star.

By relying on a series of theoretical and empirical studies, Chen maintains that the object-to-process changes involve different cognitive mechanisms than the simpler object-to-object changes. As such, this type of conceptual change cannot be represented by Barsalau's dynamic frames but, rather, by another model introduced by Barsalau in cognitive science, namely the *script*, capable of capturing the procedural dimension of processes and events (Chen 2003a, 2005). Incommensurability, Chen contends, may often be caused by the fact that the agents involved in episodes of revolutionary theory change do not understand that what in an old theory is an object corresponds to an event in the new theory, or vice versa (Chen 2003b).

Chen's considerations on the differences between the cognitive mechanisms behind the categorization of objects and events surely take ABC's original account to a new level of historical precision and sophistication. His efforts to do so are mentioned by Barker (2011), whereas Andersen (2013) seems to fully endorse the exclusively entity-centred Aristotelian conception of kinds Kuhn referred to in some of his mature works.

Finally, it must be stressed that Kuhn himself provided some good examples of scientific revolutions that are remarkably non-taxonomic, such as the transition from Newtonian Mechanics to Einstein's Relativity.

> What had previously been meant by space was necessarily flat, homogeneous, isotropic, and unaffected by the presence of matter. If it had not been, Newtonian physics would not have worked. To make the transition to Einstein's universe, the whole conceptual web whose strands are space, time, matter, force, and so on, had to be shifted and laid down again on nature whole. (*SSR*: 148)

Clearly, concepts like *space*, *time* and *matter* are not 'taxonomic': they are not subordinate instances of some superordinate concept, they do not divide into sub-kinds, they do not occupy a place in a tree-like hierarchical structure. Yet Einstein's revolution was brought about when these non-taxonomic concepts were altered.

Another example comes, again, from the scientific concepts of mechanics. Kuhn (1989, in *RSS*: 72) argues that Newtonian mechanics cannot withstand revision of the Second Law, of the Third Law, or of Hooke's law without such a revision being a revolutionary change. Let's consider Newton's Second Law, $F = ma$. Apart from *force*, which perhaps can be thought as splitting in sub-kinds, it seems safe enough to claim that *mass* and *acceleration* are not taxonomic kinds. How would it be possible, then, to represent $F = ma$—and its modifications—through a taxonomy? Describing revolutions as being caused by an anomalous violation of the no-overlap principle seems naive at best, if not falsified by the numerous historical examples of revolutionary scientific changes which did *not* involve taxonomy changes (see Bird 2012).

6.7 Conclusions

Kuhn aim at describing scientific communities, which he considered analogous to linguistic communities. When confronting scientific communities of the past, philosophers must reconstruct their scientific languages, that are incommensurable with respect to the contemporary one. Such a reconstruction is akin to the process of language acquisition, which in turn requires the two abilities of discriminating between individuals and of grouping them in kinds. Experiments in developmental psychology provide a description of how infants carry these tasks. Kuhn also believes that describing the conceptual language of science in terms of taxonomic categories, and to redefine scientific revolutions as changes in the criteria of categorisation, show how scientific terms must not be learned one at a time, but in clusters, and how meaning change affects both the intension and extension of groups of inter-related terms. In this way, he can ground his incommensurability of meanings on a theory of meaning that accounts for changes in both the intention and the extension of terms.

Kuhn's last writings contain a great wealth of ideas, and this chapter has only begun to scratch its surface. For example, while Kuhn's always criticised both the classical theory of meaning and Putnam's causal theory, his shift from the criterion of similarity to the criterion of differentiae for establishing kind membership allows him to also distance himself from the prototype theory. Had he had the time, perhaps he would have also developed further the ideas about the pre-linguistic requirements for categorization. But at what point does *scientific* categorizing becomes different from our natural ability to sort things into groups? Is there a continuity between the two processes, or are they marked by some substantial differences? Unfortunately, Kuhn could not answer these questions.

At the same time, the concept of taxonomy turns out not to be as helpful as Kuhn thought. The conceptual structure of a scientific language may not be a taxonomic hierarchy after all, and the redefinition of scientific revolutions as changes of taxonomies does not account for the numerous instances of non-taxonomic revolutions. The question, then, is the following: did Kuhn really need the concept of taxonomy for developing his theory of meaning and meaning change?

References

Andersen, Hanne. 2000. Kuhn's account of family resemblance: A solution to the problem of wide-open texture. *Erkenntnis* 52:313–337.

Andersen, Hanne. 2013. Scientific concepts and conceptual change. In *Kuhn's Structure of Scientific Revolutions Revisited*, ed. Vasso Kindi and Theodore Arabatzis, 180–204. New York: Routledge.

Andersen, Hanne, Peter Barker, and Xian Chen. 2006. *The Cognitive Structure of Scientific Revolutions*. Cambridge: Cambridge University Press.

Barsalou, Lawrence. 1992. Frames, concepts and conceptual fields. In *Frames, Fields and Contrasts: New Essays in Semantic and Lexical Organization*, ed. Adrienne Lehrer, Eva Kittay, and Richard Lehrer, 21–74. Hillsdale: Erlbaum.

Barsalou, Lawrence, and Christopher Hale. 1993. Components of conceptual representation: From feature-list to recursive frames. In *Categories and Concepts: Theoretical Views and Inductive Data Analysis*, ed. Ivan Van Mechelen, James Hampton, Ryszard Michalski, and Peter Theuns, 97–144. New York: Academic Press.

Bird, Alexander. 2000. *Thomas Kuhn*. Chesham, Princeton: Acumen, Princeton University Press.

Bird, Alexander. 2002. Kuhn's wrong turning. *Studies in History and Philosophy of Science* 33:443–463.

Bird, Alexander. 2004a. Kuhn, naturalism, and the positivist legacy. *Studies in History and Philosophy of Science* 2:337–356.

Bird, Alexander. 2004b. Kuhn on reference and essence. *Philosophia Scientiae* 8:39–71.

Bird, Alexander. 2012. What can cognitive science tell us about scientific revolutions? *Theoria* 75:293–321.

Blumenthal, George. 2013. Kuhn and the chemical revolution: A re-assessment. *Foundations of Chemistry* 15:93–101.

Bushnell, Ian, Fatima Z. Sai, and James Mullin. 1989. Neonatal recognition of the mother's face. *British Journal of Developmental Psychology* 7:3–15.

Chang, Hasok. 2010. The hidden history of phlogiston: How philosophical failure can generate historiographical refinement. *HYLE - International Journal for Philosophy of Chemistry* 16:47–79.
Chang, Hasok. 2012. Incommensurability: Revisiting the chemical revolution. In *Kuhn's The Structure of Scientific Revolutions Revisited*, ed. Vasso Kindi and Theodore Arabatzis, 153–177. New York: Routledge.
Chen, Xian. 2003a. Object and event concepts: A cognitive mechanism of incommensurability. *Philosophy of Science* 70:962–974.
Chen, Xian. 2003b. Why did Herschel fail to understand polarization? The differences between object and event concepts. *Studies in History and Philosophy of Science* 34:491–513.
Chen, Xian. 2005. Transforming temporal knowledge: Conceptual change between event concepts. *Perspectives on Science* 13:49–73.
Chen, Xian. 2007. The object bias and the study of scientific revolutions: Lessons from developmental psychology. *Philosophical Psychology* 20:479–503.
Dupré, John. 1981. Natural kinds and biological taxa. *The Philosophical Review* 90:66–90.
Field, Tiffany, Debra Cohen, Robert Garcia, and Reena Greenberg. 1984. Mother-stranger face discrimination by the newborn. *Infant Behaviour and Development* 7:19–25.
Geeraerts, Dirk. 2008. Prototypes, stereotypes, and semantic norms. In *Cognitive Sociolinguistics: Language Variation, Cultural Models, Social Systems*, ed. Gitte Kristiansen and René Dirven, 21–44. Berlin: DE Gruyter.
Hacking, Ian. 1993. Working in a new world: The taxonomic solution. In *World Changes: Thomas Kuhn and the Nature of Science*, ed. Paul Horwich, 275–310. Cambridge: MIT Press.
Hacking, Ian. 2007. Natural kinds: Rosy dawn, scholastic twilight. *Royal Institute of Philosophy Supplement* 82:203–239.
Hoyningen-Huene, Paul. 2008. Thomas Kuhn and the chemical revolution. *Foundations of Chemistry* 10:101–115.
Khalidi, Muhammad. 1998. Natural kinds and crosscutting categories. *The Journal of Philosophy* 95:33–50.
Kindi, Vasso. 1995a. *Kuhn and Wittgenstein: Philosophical Investigations of the Structure of Scientific Revolutions (in Greek)*. Athens: Smili Edition.
Kindi, Vasso. 1995b. Kuhn's *The Structure of Scientific Revolutions* revisited. *Journal for General Philosophy of Science* 26:75–92.
Kripke, Saul. 1980. *Naming and Necessity*. Cambridge: Harvard University Press.
Kuhn, Thomas S. 1974. Second Thoughts on Paradigms. In *The Structure of Scientific Theories*, ed. Frederik Suppe, 459–482. Urbana: University of Illinois Press. Reprinted in *ET*:293–319.
Kuhn, Thomas S. 1983. Commensurability, comparability, communicability. In *PSA: Proceedings of the Biennial Meeting of the Philosophy of Science Association 1982*, ed. Philip Asquith and Thomas Nickels, 669–688. Reprinted in *RSS*:33–57.
Kuhn, Thomas S. 1987. What are scientific revolutions? In *The Probabilistic Revolution*, ed. Lorenz Krüger, Lorraine Daston, and Michael Heidelberger, 13–32. Cambridge: MIT Press. Reprinted in *RSS*:13–31.
Kuhn, Thomas S. 1989. Possible Worlds in the History of Science. In *Possible Worlds in Humanities, Arts and Sciences*, ed. Sture Allén, 9–32. Walter de Gruyter and Co. Reprinted in *RSS*:58–89.
Kuhn, Thomas S. 1993. Afterwords. In *World Changes: Thomas Kuhn and the Nature of Science*, ed. Paul Horwich, 311–341. Cambridge: MIT Press. Reprinted in *RSS*:224–252.
Kuhn, Thomas S. 2022. *The Last Writings of Thomas Kuhn*, ed. Bojana Mladenović. Chicago: Chicago University Press.
Kuukkanen, Jouni-Matti. 2010. Kuhn on essentialism and the causal theory of reference. *Philosophy of Science* 77:544–564.
Levine, Alex. 2000. Which way is up? Thomas S. Kuhn's analogy to conceptual development in childhood. *Science & Education* 9:107–122.
Lyons, John. 1984. *Semantics*. Vol. I. Cambridge: Cambridge University Press.

Massimi, Michela. 2005. *Pauli's Exclusion Principle. The Origin and Validation of a Scientific Principle.* Cambridge: Cambridge University Press.
McDonough, Jeffrey. 2003. A *Rosa Multiflora* by any other name: Taxonomic incommensurability and scientific kinds. *Synthese* 136:337–358.
Melogno, Pablo, and Leandro Giri. 2023. Towards a genealogy of Thomas Kuhn's semantics. *Perspectives on Science* 31:385–404.
Meltzoff, Andrew, and Keith Moore. 1992. Early imitation within a functional framework: The importance of personal identity, movement and development. *Infant Behaviour and Development* 15:479–505.
Mill, John S. 1843. *A System of Logic*. New York: Harper & Brothers.
Putnam, Hilary. 1975. *Mind, Language and Reality: Philosophical Papers*. Vol. 2. Cambridge: Cambridge University Press.
Pyle, Andrew. 2000. The rationality of the chemical revolution. In *After Popper, Kuhn, Feyerabend: Recent Issues in Theories of Scientific Method*, ed. Robert Nola and Howard Sankey, 99–124. Dordrecht: Kluwer Academic Publisher.
Quine, Willard V. O. 1953. *From a Logical Point of View*. Cambridge: Harvard University Press.
Quine, Willard V. O. 1960. *Word and Object*. Cambridge: MIT Press.
Quine, Willard V. O. 1995. *From Stimulus to Science*. Cambridge: Harvard University Press.
Read, Rupert, and Wes Sharrock. 2002. Thomas Kuhn's misunderstood relation to Kripke-Putnam essentialism. *Journal for General Philosophy of Science* 33:151–158.
Rosch, Eleanor. 1973. On the internal structure of perceptual semantic categories. *Cognitive Psychology* 4:328–350.
Rosch, Eleanor. 1978. Principles of categorization. In *Cognition and Categorization*, ed. Eleanor Rosch and Barbara Lloyd, 27–48. Hillsdale: Erlbaum.
Rosch, Eleanor. 1987. Wittgenstein and categorization research in cognitive psychology. In *Meaning and the Growth of Understanding: Wittgenstein's Significance for Developmental Psychology*, ed. Michael Chapman and Roger Dixon, 151–166. Berlin: Springer.
Rosch, Eleanor, and Carolyne Mervis. 1975. Family resemblance: Studies in the internal structures of categories. *Cognitive Psychology* 7:573–605.
Ruphy, Stephanie. 2010. Are stellar kinds natural kinds? A challenge in the monism/pluralism and realism/anti-realism debate. *Philosophy of Science* 77:1109–1120.
Shapere, Dudley. 1966. Meaning and scientific change. In *Mind and Cosmos: Essays in Contemporary Science and Philosophy*, ed. Robert Colodny, 41–55. Pittsburgh: University of Pittsburgh Press.
Shapere, Dudley. 1984. *Reason and the Search of Knowledge*. London: Springer.
Siegfried, Robert, and Betty J. Dobbs. 1968. Composition: A neglected aspect of the chemical revolution. *Annals of Science* 24:275–293.
Smith, Edward, and Medin Douglas. 1981. *Categories and Concepts*. Cambridge: Harvard University Press.
Suppe, Frederick. 1974. *The Structure of Scientific Theories*. Urbana: University of Illinois Press.
Thagard, Paul. 1992. *Conceptual Revolutions*. Princeton, New Jersey: Princeton University Press.
Vincini, Stefano, Yuna Jhang, Eugene Buder, and Shaun Gallagher. 2017. Neonatal imitation: Theory, experimental design, and significance for the field of social cognition. *Frontiers in Psychology* 8:1323. https://doi.org/10.3389/fpsyg.2017.01323.
Wittgenstein, Ludwig. 1953. *Philosophical Investigations*. London: Blackwell.

Chapter 7
Lexicons

Abstract In his last writings, Kuhn discussed not only strict taxonomic kinds but also 'kinds of materials' and 'singletons'. This suggests that, for Kuhn, scientific lexicons are richer and more complex than simple conceptual taxonomies. At the same time, the distinctions among various components of lexicons may obscure the primary objective of their study: to elucidate the belief system of a scientific community through linguistic analysis. This original purpose can be restored by setting aside concerns related to meaning acquisition and by viewing a scientific lexicon as a representation of the inferential patterns employed by members of a scientific/linguistic community. As I will show, Kuhn appears to take this approach in some of his mature writings. Finally, given how his philosophy connects scientific language to the activities of scientific communities, I will dispute claims that Kuhn underwent a 'linguistic turn'.

Keywords Lexicon · Singletons · Systems of beliefs · Styles of reasoning · Thomas Kuhn's linguistic turn

7.1 Introduction

Many analyses of Kuhn's post-*SSR* writings focus on his theory of taxonomic kinds. So, for example, ABC have provided a cognitive reading of Kuhn's views on the taxonomic structure of scientific concepts, but without questioning the tenability of such views. Wray (2007, 2011) also accepts the idea that scientific revolutions are changes in the criteria of taxonomic classification, without critically assessing it. Neither ABC nor Wray acknowledges that numerous scientific concepts do not possess a taxonomic nature, or that several of Kuhn's examples of scientific revolutions do not entail violations of the no-overlap principle within segments of a prevailing conceptual taxonomy.

Even though he kept talking about taxonomies and the no-overlap principle, however, in his last writings Kuhn shows awareness of the richness and complexity

V. Politi, *Thomas Kuhn's Philosophical Perspective*, Boston Studies in the Philosophy and History of Science 348,
https://doi.org/10.1007/978-3-032-13540-7_7

of scientific language. For instance, alongside the terms referring to kinds of entities, he begins to investigate those referring to *kinds of materials* as well as a class of terms that he calls *singletons*. On the one hand, his insights about the generalisations these terms allow are surely interesting. On the other hand, however, one may wonder whether the increasing number of distinctions between different kinds of scientific terms, the ways they are learned, and the generalizations they allow risk making the analysis of scientific language too convoluted. This is why, in this chapter, I propose to go to the very root of Kuhn's need to analyse scientific language, which consists in studying the belief system of a scientific community from the analysis of the language shared by its members.

As already explained in previous chapters, in virtue of his dynamic perspective, Kuhn takes a collective approach to scientific rationality. In his view, science is an activity carried out by communities of scientists and scientific decisions are taken at the group level. Since a set of universal rules does not exist, the rationality of science must be discovered and reconstructed by describing the activity carried out by scientific, which poses the problem of how to describe best scientific communities. Understanding the activity carried out by the members of a community, in fact, requires an 'entry point' into the culture of that community.

Kuhn decides to consider scientific communities as analogous to linguistic communities. By studying the language of a community, it is possible to begin to understand the constellation of meanings, values, and beliefs shared by its members. The analysis of the linguistic and conceptual elements of science, therefore, must consider that: scientific language is always the language shared by the members of a scientific community; the speakers of a scientific language do not need to agree about the meanings of scientific terms, since all they need to do to maintain successful communication is not disagreeing openly about them; despite these individual differences, that allow language flexibility, language permanence through time must also be allowed. As seen in the previous chapter, with his theory of differentiae, grounded on the results of developmental psychology, Kuhn took numerous steps forward in explaining how speakers who assign different meanings to the same words are nevertheless part of the same linguistic community. In this chapter, I explore how the *lexicon* of a scientific language expresses the belief system of a community and, at the same time, permits language stability.

In Sect. 7.2, I discuss Kuhn's distinction between terms referring to taxonomic kinds, kinds of material, and singletons. The analysis of these different kinds of scientific terms provides a richer but perhaps more cumbersome representation of the conceptual structure of science than that provided by the concept of taxonomy alone. In Sect. 7.3, I propose to consider Kuhn's lexicons as the ground for the inferential patterns followed by a scientific community, and I explain how Kuhn himself pointed towards this direction. In Sect. 7.4, I explain how a seemingly overly linguistic concept such as that of a lexicon relates to Kuhn's early concept of a disciplinary matrix and to a practice-oriented philosophy of science. Bringing the discussion in these last two chapters to a close, in Sect. 7.5 I reject the interpretation for which the mature Kuhn underwent a sort of 'linguistic turn'.

7.2 Kinds of Kinds

In *PW*, Kuhn speaks about *two* different sorts of scientific taxonomies. The first resembles the taxonomies used for zoological classifications. These taxonomies have many ranks, which are (from the more general to the more specific): domain, kingdom, phylum, class, order, family, genus, and species. Some scientific conceptual structures, such as that of celestial bodies, are articulated in a similar way. Kuhn, however, also speaks about another type of taxonomies, that classify not entities, but *materials* (*PW*, in Kuhn 2022: 253–256). In Kuhn's view, kinds of materials and kinds of objects share three characteristics: they are identified through differentiae, they possess a hierarchical structure, and they can be observed (therefore, they can be learned through ostension). At the same time, he also points at some important differences between the two kinds of kinds.

To begin with, the taxonomies that classify materials do not have too many levels of generality: "materials are, for example, divided between solids, liquids, and 'airs' or gases; each of these may be further subdivided (liquids, for example, into oils, acids, and alkalis). But, insofar as these divisions are based entirely on properties observable without human intervention, the hierarchy has few levels, and there are only a few categories in each" (*PW*, in Kuhn 2022: 253).

Second, "materials are not objects: they do not trace lifelines through space over time" (*PW*, in Kuhn 2022: 253). This means that, even though they are learned through ostension, the mechanism for learning kinds of materials may be different from that for learning kinds of entities. Moreover, unlike entities, which may 'transit' across taxonomies (i.e., a 'living being' will be eventually classified together with the 'non-living being'), materials remain constant through time. What Kuhn means here is something specific about the *classification* of materials. Of course, a piece of metal may change through time (for example, it can become rusted), but that does not mean that it will cease to belong to that kind of metal. This observation leads to the question of whether the no-overlap principle plays any role at all in a taxonomy of materials. Kuhn may be tempted to say it does, since there can be no object that, for example, is made of both gold and lead. Here, however, we are not talking about objects made up of certain materials, but of the materials themselves. In the vocabulary of materials, we would just say that 'gold cannot be non-gold'. Rather than the no-overlap principle, what governs the taxonomy of materials seems just the old and revered principle of noncontradiction.

There is a third difference. The infima speciae of the taxonomy for the living beings is constituted by the group of individual beings belonging to that taxonomy and grouped at its lowest rank. An individual living being, such as an animal, may be further 'subdivided' into its components (limbs, bones, entrails, and so on). These, however, are individual entities that do belong to the taxonomy of the living beings. By contrast, as Kuhn explains, "The hierarchy for materials [...] bottoms out, not in individuals, which belong to kinds, but in kinds themselves—iron, water, wood, and so on. These too have parts, but unlike the parts of creatures, they are not directly observable until separated by further intervention (chemical or physical separation) from the material of which they are parts" (*PW*, in Kuhn 2022: 254).

The other kind of kinds Kuhn speaks about are *singletons*, which are very different from taxonomic kinds of entities. Reflecting on the three historical examples that dominate his mature work—Aristotle's physics, Volta's explanation of the battery, and Plank's reinterpretation of his own constant—Kuhn realises that "Of the kind terms drawn illustratively from [these] examples [...]—*motion*, *body*, *property*, and *place*; *cell*, *current*, *liquid* and *resistance*; *energy*, *quantum*, *radiation*, and *resonator*—all but *liquid* (a descendent of Aristotelian taxonomic kind, the element water) are singletons, not members of any contrast set" (*PW*, in Kuhn 2022: 177).

It appears like Kuhn needed some time to see the difference between kinds of entities and singletons, and to admit the crucial importance of the latter for the formulation of scientific laws. Kuhn (1993) proposes a tentative distinction. As already discussed, kind terms are projectible, in the sense that to know the meaning of a kind term means to know how to generalise from its referents. There are two different types of generalisations: *normic generalisations*, admitting exceptions (i.e., 'liquids expand when heated' does not hold for water between 0 and 4 grades centigrade); and *nomic generalisations*, exceptionless and usually identifiable with the so-called laws of nature. Taxonomic kinds support normic generalisations, while singletons support nomic generalisations.

Andersen and Nersessian (2000, from now on AN) interpret Kuhn's late normic/nomic distinction from a cognitivist perspective. AN suggest that concepts supporting normic generalisations and those supporting nomic generalisations are learned in different ways. The former are learned through an ostensive process, the latter cannot.

> With concepts like 'duck', 'goose' and 'swan', their instances can be ostended individually and many similarities and dissimilarities between them discovered through that process. [...] A scientific concept such as 'planet' can be ostended individually and in contrast to concepts like 'star' and 'satellite', but concepts involved in scientific laws usually cannot be ostended individually. Instances of the concept 'mass', e.g., are not pointed out in isolation and no contrasting concepts exist whose instances could reveal how the instances of 'mass' differ from other concepts. It is this difference which Kuhn's late normic/nomic distinction was intended to catch. (AN 2000: S226).

In their attempt to offer a cognitivist interpretation based on dynamic frames, AN propose a two-layered approach to the learning of nomic concepts, claiming that similarity/dissimilarity relations play an important role at the first of the two layers. Nomic concepts, that is singletons, are not learned through definitions but rather through their applications to similar problems. In other words, the student of science will not learn what $F = ma$ is by learning the definitions of force, mass and acceleration, but rather by recognizing that the equations for solving a diverse range of physical problems all look similar to it. The 'first layer' for learning nomic concepts consists in the recognition of similarity/dissimilarity relations not among concepts but among *problem situations*. In some of his post-*SSR* writings, Kuhn had already considered this first layer, by explaining how practice with exemplars provides the acquisition of the meaning of the laws of nature expressed by symbolic generalisations (see Chap. 5). What Kuhn did not explain, for AN, is the 'second layer', which consists in 'extracting' the meaning of the nomic concepts from the first layer constituted by similar problem situations.

For AN, the two layers must be distinguished: recognizing that some problems are similar to one another does not imply that, by solving them, the meanings of *F*, *m* and *a* are acquired. This point had already been made by Nersessian (1984), who argued that it is impossible to 'extract' a representation of, for example, the concept 'electromagnetic field' from the problem situations in which it appears. She then proposes a frame-like structure for representing concepts which can be defined as nomic. Such a frame does not connect superordinate and subordinate concepts through a net of attributes (as in the case of the dynamic frames for taxonomic kinds). Rather, it provides a 'meaning schema' which explicates different components of nomic concepts. Such a 'meaning schema' for nomic concepts explicates causal powers, mathematical structures, functions and ontological status. For example, the meaning schema for 'electromagnetic field' would include, among other things, all the electromagnetic effects (causal power); Maxwell's equations (mathematical structure); the continuous transmission of electrodynamic actions through space (function); and the state of space and matter (ontological status).

To sum up, AN characterise the difference between normic and nomic concepts in the following way:

I. normic concepts (or taxonomic kinds) are learned ostensively, by learning the similarity/dissimilarity relations governing the contrast set they belong to
II. nomic concepts (or singletons) are learned *first* by learning the similarity/dissimilarity relations of a series of problem situations which form a contrast set—a 'first layer' that Kuhn addressed through his theory of exemplars
III. and *second*, by constructing a meaning schema—a 'second layer', that Kuhn did not address but that Nersessian did

There are some problems with (I)–(III). First, (II) recapitulates Kuhn's view on how exemplars work in the learning of the laws of nature. In Chap. 5, however, I highlighted the difficulties of isolating exemplars from other components of the DM. It appears that AN could face similar difficulties in disentangling the first and the second layer since, following their jargon, the 'meaning schema' determines what are the relevant problem situations for learning some nomic concepts instead of others. AN do not consider this problem and simply maintain that scientists learn nomic concepts (or singletons) based on problem situations (or exemplars) that are similar to one another and to the nomic generalisation (or law of nature) they give meaning to. However, to understand that such similarities are indeed relevant, one must *already* know that the problem situations are instances of laws and, therefore, one must already understand what laws mean. Otherwise, going back to the example in Chap. 5, they would end up saying that the problem situation which requires the use of the Ehrenfest theorem is a route to learn the Second Law of classical mechanics. In short, it seems that scientists may already have learned something from the 'second layer' before being capable of constructing the 'first layer'. Therefore, the very relationship between the two layers is not so unidirectional as AN describes.

Furthermore, what is the connection between (II) and (III)? What is the actual cognitive mechanism that allows scientists to extract meaning from the first layer through the schema of the second layer? It is not entirely clear whether AN can answer these questions. As put by Kuukkanen (2008),

> [AN's] application is interesting, but they do not sufficiently prove that the learning of abstract scientific terms actually occurs by the recognition of similar problem situations. [...] More specifically, I think Nersessian's application of the frame schema in the history of science is better described as a suggestion of how to represent historical concepts, i.e. more as an issue in knowledge representation than as a theory of learning concepts. (Kuukkanen 2008: 108).

However, AN do make a valid point: namely, that singletons are learned not through direct ostension, but through the 'mediated' ostension of human-made exemplars. Unfortunately, Kuhn only had time to draft the beginning of the sixth chapter of *PW*, where he was going to talk about singletons, artefacts, and practices. Therefore, one can only speculate about the direction of his thoughts about these topics. Based on Kuhn incomplete writings, it is possible to say the following.

First, for Kuhn, there are two families of singletons: those that are learned and those that are innate.

> Few singletons appear in everyday vocabulary, and the concepts for which those few supply names are the common possession of man and at least the higher animals. In societies with language, they are fundamental categories of thought: space, time, and individuated physical body; perhaps also a root concept of cause and of such basic social categories as self and other. They appear, collectively, to be innate, and [I would] suggest that their likely evolutionary source is the neural processes developed to track moving objects and to match different situations to a repertoire of behavioural responses. (*PW*, in Kuhn 2022: 177)

The 'evolutionary source' of innate singletons is what 'generate' taxonomic kinds. As seen in the previous chapter, Kuhn discusses how the ability of distinguishing kinds is grounded on the ability of grouping individuals in classes based on differentiae, which requires the ability of individuating objects. The ability to distinguish 'bounded regions moving through space over time' requires the awareness of space, time, permanence and, in some instances, causality. These general concepts, which are singletons, are the conditions of possibility for perceiving individuals and, therefore, for constructing classes of individuals.

It is not clear how Kuhn would have developed his line of investigation about 'innate singletons'. Nor is it clear whether, in his view, the pre-linguistic concepts of space and time, necessary for learning about individuals and classes of individuals, are the same of, or at least continuous with, the space and time of scientific theories. Perhaps, one may speculate, Kuhn would have maintained that the modification that science has operated on our innate singletons of space, time, and causality has led to introduction of new scientific kinds, and perhaps new singletons as well.

The unfinished manuscript of *PW* contains a slightly more developed discussion about 'learned' singletons. Kuhn associates them with *artefacts*.

> [It] is an important characteristic of most artefacts that they be widely interchangeable; [...] many members of a given kind of artefact must ordinarily be available to replace each other without loss to the user. Those differences are paralleled by differences in the way the two

> sorts of objects are divided into kinds. In both cases, members of a kind must be identifiable by their observable properties; if they were not, it would be impossible to pick out an artefact of the kind one wanted. But those properties are not what groups them into kinds; rather, they are grouped together by their functions. (All can openers open cans, but they do not all display the same properties to someone unaware of the function they are to serve.) (*PW*, in Kuhn 2022: 260–261)

Artefacts are human-made objects, grouped together not based on their properties, but on their functions. Groups of artefacts must be learned together. For Kuhn, in fact, "one cannot learn *cup* without *bowl* or *fork* without *spoon*" (*PW*, in Kuhn 2022: 262). This process, however, is not like that of 'carving Nature at (some of) its joints'. To distinguish groups of artefacts based on their functions means to learn a *constellation of practices*.

> Artefacts and their functions are [...] nodes in a practice, and the nodes are differentiated by relating their functions (usually through language) to those of other nodes which serve other functions. For artefactual kinds, unlike natural [kinds], there can be no talk of empty perceptual space or of nature's joints. (*PW*, in Kuhn 2022: 262).

How does this discussion relate to the singletons of science? This is how I interpret Kuhn on this point. In his view, singletons are 'abstractions'. They are however learned through human-made artefacts which, in the case of science, are exemplars. AN's 'first layer' consists in the individuation of similarity/dissimilarity relations among exemplars. By contrast, I adopt Kuhn's late view on differentiae, and I apply it to how exemplars are grouped together. Forks and spoons have different functions. Different forks may have different characteristics. There can be forks with two, three, or four tines. There can also be forks with very short tines, which almost look like spoons. Yet, forks and spoons have different functions. What is important, in this context, is not finding the necessary and sufficient conditions that every fork must satisfy but, rather, perceiving forks and spoons as different, based on their functions. Similarly, the Eqs. (5.2)–(5.4) of Chap. 5 are not grouped together based on their notational similarities, but on the function they play within classical mechanics. This, in turn, allows us to distinguish them from Eq. (5.5), that is from the Ehrenfest theorem.

When it comes to differentiate exemplars based on their functions, perception is not enough. To recognise an object as a can opener, one must be already aware of the function of can openers, possibly in contrast with the function of bottle openers. Similarly, to know that Eqs. (5.2)–(5.4) have a different function than Eq. (5.5), one must *already* know that (5.2)–(5.4) define $F = ma$ in classical mechanics. For this reason, I would add a fourth step to AN's account:

IV. While exemplar provide a 'meaning schema' for singletons and the laws in which they appear, singleton and laws *also* provide a meaning schema for individuating exemplars.

In this way, the relation between exemplars and singletons is not 'unidirectional' (from exemplars to the meaning of laws of nature and of the singletons appearing in them), but interactive.

A point that must be stressed is that, like forks and spoons are 'nodes in a practice', so are singletons.

> Sometimes [experimental and instrumental] practices bring singletons and their accompanying laws into being, and sometimes, conversely, those practices are brought into being by the introduction of new singletons. But once in being, those practices can and usually do have a life of their own [...] They can, that is, be carried forward by people who do not control the singletons, laws and theories which scientific development has associated with them. And conversely, the research reports through which science advances can be silent about instrumental practices, describing an instrument only when it is of a new sort and itself still a subject of study. [...] The established, quasi-independent instrumental practices which underlie all research are taken for granted in those reports, together with eyes, ears, and hands, the primitive instruments which these more developed instruments extend. Taken for granted also is the language in which the results of instrumental practices are recorded. (*PW*, in Kuhn 2022: 178–179)

I do not interpret the above passage as defending a strict distinction between 'theory' and 'practice'. The two are evidently intertwined, since a new practice may lead to new singletons, whereas new singletons may open new practices. In my view, in this and similar passages Kuhn is defending is non-extreme form of holism. In his dynamic perspective, systems of beliefs can be compared without falling prey to confirmational holism (see Chap. 3). This is possible because even two incommensurable systems of beliefs share some presuppositions, that are simply taken for granted and, therefore, not tested. Two conflicting paradigms or systems of beliefs can be compared based on measurement or experimental reports, because they both presuppose their validity. But this does *not* make measurements or experiments any less theory laden and, therefore, a 'neutral ground' for inter-paradigm comparison. What happens, rather, is that the supporters of the competing paradigms take for granted the validity of the theories in the light of which measurements or experiments are performed. Measurements and experiments are indeed 'quasi-independent' practices, but by no means are they theory free.

There is no way to know how Kuhn would have pursued further this line of reasoning. Nevertheless, from what I have said in this section, it appears that, in his last writings, Kuhn presented a much richer picture of scientific languages, and that he was interested in the links between the concepts expressed by scientific terms and practice.

7.3 Lexicons and Reasoning

Kuhn's analysis of the terms used to express different families of scientific concepts, the different ways those terms are learned, and the different forms of generalisations their concepts support provides several interesting insights. At the same time, however, one may fear that all these distinctions may obscure the primary reason for studying scientific language. As already explained, Kuhn needs to ground his theory of collective rationality on the description of the activity carried out by a scientific community. To do so, he treats scientific communities as linguistic communities. While studying the scientific/linguistic communities of the past, he encounters the phenomenon of meaning change, which pushes him to develop a theory of meaning. From there, he ends up talking about kinds, singletons, and language acquisition, often getting caught up in an increasing number of distinctions and specifications. What remains, then, of the idea of analysing the activity carried out by scientific communities through the analysis of their language?

Besides, even Kuhn's distinction between kinds and singletons can be questioned. Pirozelli (2024), for example, casts doubt on the viability of distinguishing between terms that are learned through ostension (taxonomic kind terms) and those that are learned through law-like generalisations (singletons). When it comes to scientific concepts, the meaning of the terms expressing them can hardly be learned through ostension. In science, in fact, *both* singletons *and* taxonomic kinds must be learned as part of theoretical generalisations. Pirozelli makes the example of the terms expressing the central concepts of phlogiston theory—such as 'principle', 'element' and, indeed, 'phlogiston'—the referents of which cannot be directly observed, meaning that those terms cannot be learned through ostension. But examples in this sense abound.

For instance, one may look at the elementary particles of the Standard Model, which are divided into fermions (quarks and leptons) and bosons (Higgs boson, photons, and so on). All in all, the conceptual structure of the Standard Model resembles a taxonomy of elementary particles. (Whether the combinations of elementary particles end up forming another taxonomic hierarchy or, rather, a 'tapestry' of entities like that of the chemical periodic table is an issue that will not be discussed here.) Clearly, the terms referring to elementary particles cannot be learned through ostension. Furthermore, the acquisition of their meaning requires some knowledge of the Standard Model. Kuhn could reply that his description of the acquisition of kind terms through ostension was only the first step for a more general theory kind terms acquisition. Perhaps his aim was to show that the cognitive mechanisms for discriminating between ostensible kinds are similar to those for discriminating between theoretical kinds. Even so, however, when it comes to scientific concepts, a distinction between concepts that are learned through ostension and concepts that require low-like generalisations is not only unnecessary but also misleading. In the end, one could even blame Kuhn for spending too much time talking about how we distinguish ducks from goose, instead of talking about quarks and bosons.

Pirozelli suggests that a better theory of meaning is provided by *inferentialism*, according to which

> Sentences are the fundamental units used in the practice of giving and asking for reasons and, for this reason, have an explanatory priority over individual concepts. Moreover, inferentialism asserts that the meaning of a sentence is given by its inferential role in language: what other sentences it presupposes and what sentences can be inferred from it. (Pirozelli 2024: 220)

In a sense, Kuhn himself may have considered a similar route. This appears clear when considering that, for Kuhn, learning a lexicon does not consist only in the acquisition of the meaning of its words but also in the acquisition of a way of reasoning (which, in turn, reflects a way of seeing the world).

In several post-*SSR* writings, Kuhn tells the story of his personal struggle with Aristotle's *Physics*. As he recounts, after the difficulties in trying to make sense of some seemingly unintelligible passages, suddenly "the fragments in my head sorted themselves out in a new way, and fell into places together" (Kuhn 1987, in *RSS*: 16). After this epiphany, Kuhn realises that, in Aristotelian physics, 'motion' and 'change' belong to the same cluster.

> When the term 'motion' occurs in Aristotelian physics, it refers to change in general, not just to the change of position of a physical body. Change of position, the exclusive subject of mechanics for Galileo and Newton, is one of a number of subcategories of motion for Aristotle. Others include growth (the transformation of an acorn to an oak), alteration of intensity (the heating of an iron bar), and a number of more general qualitative changes (the transition from sickness to health). As a result, though Aristotle recognizes that the various subcategories are not alike in all respects, the basic characteristics relevant to the recognition and analysis of motion must apply to changes of all sorts. In some sense that is not merely metaphorical, all varieties of change are seen as like each other, as constituting a single natural family. (Kuhn 1987, in *RSS*: 17)

For Aristotle, 'change', 'motion', 'growth', 'alteration of intensity', and so on, are somehow interconnected, or even co-defined. The reason why Aristotelian and Newtonian physics are incommensurable, which is also the reason why understanding Aristotle is so difficult for Kuhn, is because they group these terms in different clusters. But one cannot stop at this level of analysis. To reason like an Aristotelian scientist would, in fact, it is necessary to understand *why* Aristotle's physics grouped some terms in some clusters instead of others. To 'really get' Aristotle's theory, in other words, it is not enough to learn the Aristotelian meaning of 'motion' by realising that, for Aristotle, motion is a kind of change. Understanding Aristotle, in fact, means understanding why he thought that motion is a kind of change. This means that understanding Aristotle's physics requires an understanding of "the centrality of qualities to its conceptual structure".

> [...] Aristotelian physics inverts the ontological hierarchy of matter and quality that has been standard since the middle of the seventeenth century. In Newtonian physics a body is constituted of particles of matter, and its qualities are a consequence of the way those particles are arranged, move, and interact. In Aristotle's physics, on the other hand, matter is very nearly dispensable. It is a neutral substrate, present wherever a body could be—which means whenever there's space or place. A particular body, a substance, exists in whatever place this neutral substrate, a sort of sponge, is sufficiently impregnated with qualities like heat, wetness, colour and so on to give it individual identity. Change occurs by changing qualities, not matter, by removing some qualities from some given matter and replacing them with others. (Kuhn 1987, in *RSS*: 17–18)

The differences between the Aristotelian and the Newtonian frameworks do not depend only on the differences between the meanings of some terms. The whole frameworks are, in a sense, *structured* around different values and methods: Aristotelian physics is 'qualitative', Newtonian physics is 'quantitative'. Only after understanding this difference, it is possible to understand why some terms in Aristotle's physics have the meaning they have rather than another. For example, in the Aristotelian framework, 'position' is understood as the quality of an object and, since 'motion' is a change of position, then 'motion' is a quality: the reason why, for Aristotle, 'motion' is a kind of 'change' becomes therefore clear. Aristotle puts 'position', 'motion' and 'change' in the same cluster because his framework is qualitative.

The difference between the qualitative and the quantitative character of Aristotelian and Newtonian physics cannot be reduced to questions surrounding the meaning of a small group of words. Rather, Aristotelian and Newtonian physics define small groups of terms differently because of some deeper differences between the two conceptual systems as whole. The semantic aspects of incommensurability are the consequence, if not the superficial manifestation, of such deeper differences.

That Aristotle's physics is qualitative 'makes it possible' to put position, motion and change in the same cluster. The qualitative sense of Aristotle's physics is the precondition which allows him to define motion as a type of change. What is important, here, is not the meaning of the single terms—on which much of the literature on scientific change and incommensurability has been focused—but rather the condition of possibilities which allows the Aristotelian to 'reason' in a certain way rather than in another. Put it in another way: learning the qualitative character of Aristotelian physics has less to do with learning the meanings of the words in a language and more to do with what can be defined in terms of "presuppositions internal to a research programme" (McDonough 2003: 352).

Consistent with the inferentialism suggested by Pirozelli, Kuhn's experience with Aristotle's *Physics* can be described as follows. First, Kuhn comes across some seemingly senseless sentences, which are the fundamental units carrying meaning in the Aristotelian lexicon. From those sentences, he can interpret not only the meaning of single worlds, but also how the Aristotelian lexicon confers those meanings, by grouping words in some clusters rather than others. This allows Kuhn to 'reason like Aristotle', that is to draw the same inferences that an Aristotelian scientist would have drawn. Only when he has reached this point, he can say that he has really learned Aristotle's lexicon.

That his theory of scientific lexicons is compatible with inferentialism becomes even clearer when considering that Kuhn established a parallel between lexicons and Ian Hacking's *styles of reasoning*. Inspired by the monumental work of Crombie (1994) on the styles of Western science, Hacking (1982) argues that styles of reasoning do not serve to determine the truth-value of 'bivalent propositions'; they do not, in other words, determine whether a proposition is either true or false. What styles do, instead, is introduce 'positive propositions', that is propositions that can be evaluated as either true or false, but do not determine their truth value; they only define which propositions can be rationally formulated.

As an example, Hacking speaks of the defunct Renaissance hermetic style, based on doctrines of resemblance and similitude. The contemporary scholar who reads the writings of someone like Paracelsus does not find them to be false, but rather incomprehensible. The reason for such an incomprehensibility is not to be found in the meanings of single words, but in how Paracelsus formulates propositions, hypotheses, and conjectures; in short, in the way he reasons. For the hermetic style, 'mercury salve might be good for syphilis because mercury is signed by the planet Mercury which signs the marketplace, where syphilis is contracted' is a positive proposition: although it cannot be said in advance whether it is true or false, from the perspective of Renaissance scientists who reason following the hermetic style, such a proposition is a good candidate for truth-or-falsity; as such, it is rationally utterable and demands scientific attention.

Styles are not propositions, but ways of conceiving and formulating some propositions rather than others. They introduce possible candidates for truth-or-falsity, but they are neither true nor false in themselves. For this reason, Hacking (1992) characterises them as 'self-authenticating'. He also maintains that each style introduces novelties, such as "new types of: objects, evidence, sentences (new ways of being a candidate for truth or falsehood), laws (or at any rate modalities), possibilities" (Hacking 1992: 11). Styles are not static entities: they emerge, change, and evolve through time. Therefore, the introduction of novelties is not something happening once and for all in the same moment a style emerges; rather, styles' introduction of novelties is an 'open-ended' process.

> [E]ach style should introduce novelties of most or all of the listed types, and should do so in an open-textured, ongoing and creative way. Mathematicians do not just introduce a few sorts of abstract objects, numbers and shapes, and then stop; the type 'abstract object' is open-ended once we begin reasoning in a certain way. Note that on this criterion, logic, be it deductive, inductive or abductive, does not count as a style of reasoning. This is as it should be. Crombie did not list the branches of logic, and no wonder. People everywhere make inductions, draw inferences to the best explanation, make deductions; those are not peculiarly scientific styles of thinking, nor are they Mediterranean in origin. (Hacking 1992: 12).

The introduction of new candidates for truth-or-falsity and the introduction of novelties are intertwined: the new positive propositions are propositions *about* new objects, laws, and so on. In other words, styles introduce new positive propositions which reflect new ways of finding out things, new ways of practising. This is perhaps why, in his more recent writings, Hacking (2007, 2012) does not define them as 'styles of reasoning' anymore but, rather, as 'styles of thinking and doing'.

Kuhn explicitly recognizes important similarities between his late position and Hacking's idea of styles introducing the candidates for truth-or-falsity.

> Ian Hacking [...] spoke of the way in which new 'styles' introduce into science new candidates for true/false. Since that time, I've been gradually realising (the reformulation is still in process) that some of my own central points are far better made without speaking of statements as themselves being true or as being false. Instead, the evaluation of a putatively scientific statement should be conceived as comprising two seldom-separated parts. First, determine the status of the statement: is it a candidate for true/false? To that question [...] the answer is lexicon-dependent. And second, supposing a positive answer to the first, is the statement rationally assertable? To that question, given a lexicon, the answer is properly found by something like the normal rules of evidence. (Kuhn 1991, in *RSS*: 99)

Like Hacking's styles, Kuhn's lexicons (as conditions of possibility) do not tell which propositions are true or false but determine which propositions must be rationally considered to be good candidates for truth or falsity. And, like Hacking's styles, Kuhn's lexicons are not in themselves true or false; rather, they represent the conditions of possibility for formulating true-or-false theories, that is theories which are rational candidates for the truth or the falsity within a lexicon.[1]

In concluding this section, it is necessary to talk not only about the similarities, but also about the differences between Kuhn's lexicons and Hacking's styles. Many disciplines may adopt the same style of reasoning. For instance, the style of statistical reasoning is employed in both the natural and the social sciences. By contrast, a lexicon always pertains to one scientific/linguistic community. Some philosophers have looked at possible connections between the general cross-disciplinary styles and the much more local lexicons. Each on their own ways, both Winther (2012) and Sciortino (2021) propose to consider styles as a more 'fundamental' entity than paradigms or lexicons. In this perspective, changes at the level of styles of reasoning will be reflected by changes at the level of lexicons. The analysis of the evolution of styles of reasoning, therefore, could aid the study of the *long durée*, of the changes occurring at the deeper and more fundamental layers of science. In my view, the relationship between styles and lexicons works in the opposite direction. For example, when looking at the history of the style of statistical reasoning, it turns out that such a style emerged out of the need of solving some problems in the social sciences. After it established itself as part of the dominant lexicon in sociology, it begun to enter the lexicons of other scientific communities, both in the social and in the natural sciences (see Hacking 1990). A good metaphor for the different levels of scientific change is not that of tectonic movements but, rather, that of catastrophic avalanches.

7.4 Communities, Languages, and Practices

The items included in a lexicon (the sentences it produces, the meaning of its terms, and so on) may be interpreted in different ways by the members of the same scientific/linguistic community. What allows communication is not an explicit consensus on the meanings of all scientific concepts or laws, but a lack of explicit disagreement. As already discussed in Chap. 4, individual variations are feature, not a bug, of rational communities. Furthermore, the lexicon of a community evolves through time. While the evolving and flexible nature of a lexicon is rather clear, it is not entirely clear how such an everchanging lexicon can remain stable through time.

[1] An interesting point which unfortunately I am unable to develop in detail here is that both Hacking's styles and Kuhn's lexicon can be interpreted from a 'Kantian perspective'. In Kant's transcendental philosophy, problems surrounding the meaning of terms are, in a sense, dissolved. Kant, in fact, is not interested in the meaning of single words, but in judgements (analytic a priori, synthetic a posteriori, and synthetic a priori). Similarly, both Hacking's styles and Kuhn's lexicons are not conditions of possibility for attaching meanings to single words, but rather conditions of possibility for the rational formulation and utterability of sentences (or judgements). In this interpretation, Hacking's and Kuhn's unit of philosophical analysis is not the meaning of words, but propositions. See also Hacking (1975, Chaps. 11–13).

The stability of a lexicon seems a necessary requirement for the stability of its scientific/linguistic community. Despite its flexibility and all the changes, in fact, a lexicon is passed through generations, and the new members of a community recognize themselves as members of the *same* community of the previous generation. The problem, then, is to harmonize the flexible nature of a lexicon with its stability.

When he attacked the analytic/synthetic distinction, and claimed that the whole web of beliefs faces the tribunal of experience, Quine did not dispense with *any* distinction. In fact, he still considered the web of beliefs as possessing a 'centre' and a 'periphery'. Every belief is revisable, including those contained in the centre or core, though the revision of a core is a less frequent event. Kuhn follows Quine, but only up to a point. In fact, he was against the strong version of Quinean holism. In his view, it is *not* the whole system of beliefs to be tested, but only some of its parts at a time. When comparing two different theories or hypotheses, it is possible to see whether they share some fundamental presuppositions, which will not be tested. Aristotelian and classical mechanics are incommensurable because they do not share such fundamental presuppositions. Galileo's and Torricelli's hypotheses on the cause of vacuum (see Sect. 3.2) are not, because they both presuppose the classical laws of hydrostatic, as well as various beliefs about measurement and instrumentation. At the same time, however, he seems to inherit from Quine the idea of a 'core' less frequently subjected to revisions.

Sometimes, Kuhn defines scientific laws, expressed by a lexicon and containing singletons, as 'synthetic a priori' statements (Kuhn 1989, in *RSS*: 71, 72). These statements are synthetic because, contrary to analytic statements, add something to our knowledge. When it comes to their 'a priori' characteristics, one should ask: *for whom* are scientific laws a priori? Unlike the Kantian definition of a priori judgements, whose validity does not depend on experience, Kuhn is aware that scientific laws are revisable. (Indeed, he builds his model of the development of science on the idea that scientific concepts change through history.) Yet, there are some laws that are a priori for the community holding a particular belief system, in the sense that such laws are *presupposed* by the members of the community. For instance, the lexicon, as well as the laws it contains, is the presupposition to the theories developed by the members of the same community. In this sense, and as explained in the previous section, the lexicon provides the 'conditions of possibility' for other, more 'peripheral' parts of a growing and developing belief system.

Michael Friedman (2001) examines a similar idea, though he thinks that, by doing so, he also develops a 'post-Kuhnian historiography of science'. Like Kuhn, Friedman believes that periods of normal science get their stability from a set of 'contingent synthetic a priori' statements, which often correspond to fundamental scientific laws. Such laws are indeed synthetic a priori, but also 'contingent' on the current state of knowledge and revisable at any moment in the future. Like Kuhn, Friedman regards the changes in the contingent synthetic a priori as revolutionary. He believes, however, that there is one important difference between his view and Kuhn's. Such a difference concerns the *rationality* of scientific revolutions. For Friedman, a change of paradigm is made possible by a philosophical 'meta-paradigm', where old concepts are modified and new concepts created. Thanks to the meta-paradigm, new contingent synthetic a priori statements can be formulated rationally. The meta-paradigm, therefore, guarantees the rationality of scientific revolutions, contrary to Kuhn's (alleged) 'irrationalist' view. Friedman may have misread Kuhn, who never wanted to argue for the irrationality of science. Quite the opposite: Kuhn aimed at showing how a revolution is made possible by the *collective rationality* of the scientific community. Furthermore, Kuhn's model does not need to be amended by a meta-paradigm. New concepts, laws, and theories, for Kuhn, are not developed at the philosophical and meta-scientific level, but from *within* the scientific community itself, which is pervaded by an 'essential tension' (see Chap. 4).

Having clarified that the lexicon is the condition of possibility for the formulation of different theories, it is important to stress the difference between lexicons and scientific theories. Different theories may share the same lexicon (that is, the same 'core' of beliefs), yet the lexicon may not be reduced to any of them in particular. This, in turn, allows one to solve one of the most pressing problems of Kuhn's model, namely the problem of distinguishing 'normal' and 'revolutionary' developments. For example, Barberousse (2007) discusses the problem of 'intra-theoretic incommensurability', that is the problem that there can be incommensurability of concepts between different 'versions' of the same theory (or between different theories belonging to the same paradigm).

Barberousse explains this point by reconstructing the history of the principle of *action* in Newtonian mechanics. Such a concept does not belong to Newton's original theory but was introduced by Maupertuis in 1744. In the same year, Euler proposes a different and mathematically more precise version of what he called the *principle of least action*. In 1760, Lagrange proved the validity of the principle of least action for cases more general than the ones considered by Euler. Decades later, in the 1830s, Hamilton published a series of works in which, among other things, the principle of least action not only is re-defined once again, but it is also deduced from the *law of varying action*. All these intra-theoretical changes cannot be defined as mere logical or aesthetic adjustments, nor as the simple adoption of alternative formal tools with no consequence for theoretical contents. Quite the opposite: Euler's, Lagrange's, and Hamilton's reformulations of mechanics altered the meaning of some pre-existing concepts and introduced new ones, even though they all belonged to the same paradigm.[2] Clearly the existence of intra-paradigm incom-

[2] By insisting on this point, Barberousse dismantles Kuhn's distinction between formal generalisations and their actual exemplary application, through which the meaning of the otherwise uninterpreted generalisations can be learned. For Barberousse, such a distinction is untenable. In fact, "the development of mechanics shows that even generalisation sketches are investigated *qua* meaningful generalisations, leading to new methods which in turn can be used as routines - this was Lagrange's aim. The very meaning of the symbolic generalisations may be a serious domain of scientific investigation within a given disciplinary matrix, contrary to Kuhn's slightly oversimplified conception" (Barberousse 2007: 98). Kuhn clarified that symbolic generalisations may be distinguished from the other components of a DM only for the philosopher who observes a scientific community (see Sect. 5.1). However, it is true that the study of symbolic generalisation *qua* meaningful units of scientific knowledge is an issue that Kuhn did not consider with enough depth.

mensurability challenges the idea that incommensurability arises only in case of conflicting paradigms.

Barberousse is surely right in pointing out that different theories developed within the same normal tradition may attach different meanings to the same terms or expression. This is only a problem if one thinks that the stability of a lexicon is given by the stability of the meaning of its terms, especially if by 'meaning' one means 'definitions'. Kuhn explains that speakers belonging to the same linguistic community may have different ideas about what the correct definitions of some words are. Indeed, the internal variations of a community is beneficial for its own survival, especially when the meaning of some terms will have to be modified. If one keeps in mind that Kuhn's 'theory of meaning' is not much about 'meanings' per se, but about 'sorting things out' (see Sect. 5.2), then the problem disappears. As long as its *structure* is preserved by its speakers, a lexicon remains stable. In the case discussed by Barberousse, then, even though Euler, Lagrange, and Hamilton defined action and the relative laws in different ways, they were still respecting the space of *differentiae* of the Newtonian lexicon, which therefore remained relatively stable despite the different definitions attached to the same terms and laws.

It is also important to stress that the lexicon is not a condition of possibility *only* for theories. Kuhn is interested in the presuppositions at the basis of a belief system, including the presupposed beliefs about practices and tools. For instance, Kuhn (1991) describes the interaction between the group of scientists and the special field they investigate through the biological metaphor of adaptation (see also Sect. 9.6). In doing so, he puts a particular emphasis on the *practice* and *active interaction with the natural environment* necessary for adaptation.

> Creatures born into [an ecological niche] must take it as they find it. They can, of course, interact with it, altering both it and themselves in the process, and the populated world thus altered is the one that will be found in place by the generation which follows. [...] What actually evolves, therefore, are creatures and niches together [...]. It is groups and group practices that constitute the worlds (and are constituted by them). And *the* practice-in-the-world of some of this group is science. (Kuhn 1991, in *RSS*: 101–103)

Even after his (alleged) 'linguistic turn' (see next section), Kuhn speaks of science in terms of practice, as an interactive way of 'being in the world'. Under this aspect, lexicon as a condition of possibility for thinking, theorising and experiencing the world, is also a condition of possibility for practising in the world.

Kuhn's interest in the practice of science is intensified by his rejection of a linear conception of science, for which scientific theories are always prior to technology and experimentation, the latter being a mere applicative extension of the former. Under this respect, Kuhn agrees with the picture of the 'network of science' suggested by Wise.

> The geometry of the network [...] might be depicted as a cylinder, with two gaping holes [namely, 'Theory and Ideology' and 'Reality'] at either end [and with 'Technology' in the middle]. What seems to be missing is any direct connection between ideology and reality. [...] An obvious way to fill that hole is to wrap the cylinder in itself, forming a donut and a direct route between theories and things. (Wise 1993: 244–245)

Speaking about Wise's image of the network of science, Kuhn comments that

> [a] geometry that requires two ends is replaced by one that calls for three symmetrically placed slices. Technology continues to provide a two-way route between theories and reality, but reality provides the same sort of route between theory and technology, and theories provide a third route between reality and technology. Scientific practice requires all three sorts of mediation, and none of them has priority. Each of these three slices—technology, theory, and reality—is constitutive for the other two. And all three are required for the practice whose product is knowledge. (Kuhn 1993, in *RSS*: 246)

For Kuhn, the conceptual structure of scientific theories is constitutive of technology and technology is constitutive of scientific concepts: these two aspects of the scientific practice cannot be fully understood if examined in isolation.[3] But since theories and technology cannot be examined in isolation, the two being co-constitutive of one another, the notion of lexicon cannot be reduced to either the conceptual structure of scientific theories or to the use of experimental devices and scientific instrumentation: theories are acquired and tested through the practical and experiential dimension of science, which in turn is guided by theories. As already said, Kuhn's lexicon "makes possible a corresponding form of life" (Kuhn 1993, in *RSS*: 244).

The fact that Kuhn is interested not only in theoretical or conceptual presuppositions, but also in the beliefs presupposed in practice and instrumentations, has several interesting implications. For instance, the study of the lexicon of a past scientific community does not lead to uncovering just the meanings of some terms and discovering that, indeed, the language of the past is semantically incommensurable with our own. A study of the 'core' of a belief system gives us insights also in the practice of the community of the past and may lead us to understand the degree of methodological incommensurability across different periods of science. As seen in the previous section, reconstructing past lexicons means to understand, for example, the difference between the 'qualitative' and 'quantitative' character of Aristotle's and Newton's views.

Finally, once again, it is important to stress the role of language and language acquisition. Scientists do not begin their research from a 'blank state'. They do not have to re-discover all the scientific laws from scratch, nor do they need to learn the laws of optics to use a microscope. To become a scientist, one must assimilate the presuppositions held by a scientific community and contained in the lexicon. This means that the lexicon must be *transmitted* from one generation of scientists to the next. The intergenerational transmission of a lexicon is possible through guided practice with exemplars as well as ostension of kinds. Through guided practice, the

[3] Kuhn also says, and quite clearly, that theories and technology are constitutive of the world and that the world is constitutive of theories and technology. How to make sense of such a world that constitutes theories and technology, and that is constituted by them? Following the interpretation by Hoyningen-Huene (1993), such a world would be the 'phenomenal world', not the 'world in itself'. Following Kuhn's remarks, however, such a phenomenal world is not just a the perceptual counterpart of the world in itselt, its representation; rather, it is the locus of scientists' activity, their 'niche' (see Sect. 10.5).

junior scientists acquire the lexicon of their community, which eventually will allow them to practise independently. This process is akin to how infants learn the language as well as the cultural codes of their community: they are guided by adults until they reach the point where they can use the language creatively to formulate their own sentences, and to behave and make decisions independently yet respecting the community norms.

The relation between practice, necessary for language acquisition, and indeed language, through which the shared beliefs of a community can be expressed, cannot be underestimated.

> I am, of course, now tracing a circle, from practice to constitutive beliefs and back again. But the circle is by no means vicious. It encompasses only the core of the beliefs and practices of a given tribe at a given time, and they, once grasped, determine a great deal else. Enter the circle, as the historian must, at some selected time; reconstruct the beliefs that then determine the practice and the practices that then determine belief; and, with that temporal starting point established, watch the way in which beliefs and practise together develop from it. The starting point, just because it is the starting point, must remain a merely contingent set of historical states of affairs. But each step forward in time from it can be seen to be the product of choices which, the tribe's circumstances permitting, were made simply because they promised solutions to the problems which the practice in its current state was intended to resolve. Like [the static perspective], the developmental approach [or dynamic perspective] to philosophy of science posits rational criteria for theory choice. But the criteria are now those of a tribe; assimilating them is part of what makes an individual a member of that tribe; and the criteria are therefore non-vacuously applicable only to an explanation of the further development of tribal practice, not to an understanding of its entire current state. (*SKHP*, in Kuhn 2022: 12)

Through practice, the scientists acquire a lexicon, that is a set of 'constitutive beliefs', which allows them to practise following the norms of rationality of their community. The historian and the philosopher of science, as 'ethnographers' of the scientific community, must reconstruct the 'constellation of group commitments' of a scientific community; that is, its DM (see Sects. 4.2 and 4.6). They can do so by acquiring the community's lexicon, which will allow them to understand and reconstruct its practices.

7.5 Did Kuhn Undergo a 'Linguistic Turn'?

In concluding this chapter, I want to examine a particular interpretation of Kuhn's philosophy that has emerged in the past couple of decades. Namely, the interpretation for which, sometimes after the publication of *SSR*, Kuhn underwent a 'linguistic turn'.

To my knowledge, Irzik and Grünberg (1998) were the first to use the expression 'Kuhn's linguistic turn', and they did so in a positive way. For Irzik and Grünberg, Kuhn's post-*SSR* focus on the linguistic aspects of scientific theories allowed him to refine his incommensurability thesis thanks to a fuller neo-Kantian approach. There is, however, another version of the 'linguistic turn' interpretation, which reads Kuhn's mature philosophy in a negative way.

As already mentioned in various parts of this book, Bird (2002) claims that, in his post-*SSR* works, Kuhn turned his back to his early naturalistic approach, in favour of an approach closer to the philosophy of language. He did so to appeal to the philosophy community of his time. In this way, however, he ended up undermining his own legacy.

Bird's 'linguistic turn' interpretation can be summarised as follows:

- the mature Kuhn underwent a 'turn', which was also 'linguistic'
- the mature Kuhn abandoned the naturalistic approach and the use of empirical findings from experimental psychology to draw philosophical conclusions, in favour of philosophy of language and a priori philosophical analyses
- similarly, he abandoned the history of science as a source for the study of scientific practice, again in favour of philosophy of language and a priori philosophical analyses
- Kuhn did so because he wanted to be accepted by the other philosophers of his time
- Kuhn's turn was however a 'wrong' turn, because it undermined Khun's own philosophical legacy

A similar interpretation has been proposed, among the others, by Gattei (2008) and Shan (2020).

At a superficial level, it is true that Kuhn's mature concepts seem overly linguistic. Surely, 'kinds' and 'lexicons' do not sound like 'paradigms' and 'exemplars'. I think, however, that the 'linguistic turn' interpretation is mistaken under several respects.

First, if there was a 'turn', *when* did it happen? In 'Second thoughts on paradigms', the talk he delivered at the Urbana conference in 1969, Kuhn explains how exemplars define symbolic generalisations, but *also*, through the Johnny example, how *kinds* are learned through ostension (see Kuhn 1974; see Sects. 4.2 and 4.3). Already in 1969, that is, he was talking about categorization and language acquisition, that are topics that will dominate his work in the 1980s and in the 1990s. But Kuhn's preoccupation with language even predates *SSR*. In his 1951 *Lowell Lectures*, Kuhn criticised the attempted philosophical project of reducing scientific language to sets of logically related propositions and defended the fruitfulness of 'vague' concepts (*QPT*: Lectures VII-VIII). Indeed, George Reisch (2021: xx) claims that, rather than a 'turn', the mature Kuhn made a "linguistic return" to issues and topics that had interested him since the beginning of his career. Of course, one may argue that a 'return' is *still* a turn, as when one makes a U-turn to go in the opposite direction. But if one considers that Kuhn had *always* been concerned with the relationship between theories and experiments, concepts and practices, one can see that there was never a real 'change of direction' in his philosophy.

In my view, Kuhn's turn, or return, or simple development, was not even 'linguistic' to begin with. Kuhn formulates his views in the wake of Quine's attack on the analytic/synthetic distinction. The demise of such a distinction had several implications on how the stability (or lack thereof) of belief systems is conceived. Kuhn agrees with Quine on many things, but he has his own view about how belief systems are structured and how they change. Beliefs are expressed through language. Changes of belief are reflected on changes of language. Meanings, propositions, languages, and so on, are means to the end of understanding the core of beliefs held by scientific communities, and how they change through time.

Furthermore, in making his turn—even though we do not know when he made it, and maybe it was not even a turn—Kuhn did not abandon the use of empirical sciences for philosophical purposes. The fourth chapter of his (sadly incomplete) *PW* contains such an abundance of discussions about the empirical findings of developmental psychology the surface which I could barely begin to scratch (see Sect. 5.3), and which will surely be analysed for years to come. These are not simple references made for expository reasons. Kuhn finds instances of meaning change in the history of science and he therefore needs a theory of meaning. Furthermore, to explain how it is possible to acquire a 'foreign language' (i.e., a past scientific language), or even a first language, he needs a theory of language (and meaning) acquisition. Such a theory is *grounded* on developmental psychology, which explains how we learn to sort things into group, that is how we learn to subdivide the world into kinds and how we learn the meaning of kind-terms. Apart from developmental psychology, furthermore, in *PW* Kuhn also relies on evolutionary theory, linguistics, the psychology of categorization, and even theories of art as a basis for his philosophical claims. Insights from all these disciplines are all integrated in a way that it would be difficult not to define as naturalistic.

Returning to Quine, he argued in favour of a naturalised epistemology, *and* he also criticised the concept of a priori truth, conveyed by tautological sentences, through a more analytical style of philosophical argumentation. Kuhn did what Quine professed but, in a sense, he did more, and he did it better. He was not only interested in the structure of belief systems 'in abstract', but also on how they are transmitted through generations, and how some beliefs must be acquired and presupposed to be able to practise in the 'world' of a community. To accomplish such tasks, Kuhn employed linguistic analyses and theories of meaning *within* a naturalistic framework.

Nor is it true that he turned his back to the history of science in favour of an a priori philosophy. First, because Kuhn had always been interested in developing philosophical arguments through 'Socratic juxtaposition' (see Sects. 2.3 and 2.4), not from empirical data. Second, because even in his mature writings he relies on historical case studies. Three historical examples dominate his mature production, namely: his personal encounter with Aristotle's *Physics*, Volta's discovery of the battery, and the difference between Planck's first and second version of his derivation of the law of black-body radiation. These three examples appear as the starting point of Kuhn's exposition in his *TL*, delivered in 1984, in "What are scientific revolutions" (Kuhn 1987), in his *SML*, delivered in 1987, and in *PW*. One could contend that these three examples are not discussed with the rigour and richness of details expected by history of science. The obvious reply is that the works in which Kuhn uses them as case studies are not historical, but philosophical; and that they are discussed with the same level of details and historical accuracy that the Chemical Revolution or of the transition from classical mechanics to Einstein's relativity had in *SSR*. (If anything, Kuhn's late three historical examples are discussed with *more* details and depth than those in *SSR*).

Ultimately, Kuhn is a philosopher of scientific practice through and through (Rouse 2003a, b, 2013). He is interested in how a scientific community, through practice, constitutes its own epistemic niche. But practices need to be described. Kuhn believes that historians are like ethnographers. However, while an ethnographer studying a contemporary alien culture can rely on numerous sources, such as artefacts, ceremonies, and so on, the analysis of old scientific texts, which is a form of linguistic analysis, is the primary tool of the historian. This does not mean that it is the *only* tool available. Many times, the historian may also look at past scientific instrumentation. Sometimes, however, past instruments simply do not survive, except in how old scientific texts describe them. For example, when Kuhn speaks about the Volta battery, he is not looking directly at the battery constructed by Alessandro Volta. Rather, he is analysing Volta's *description* of his own battery, which consists of illustrations and written texts. Volta's description of his own battery may look bizarre to contemporary readers, until one understands that Volta attached a different meaning to 'battery' and was working without some of the beliefs that would have become established common knowledge in the physics community only later. Though the analysis of the 'practical' and 'material' dimension of science does not have to be exclusively linguistic, the analysis of past language, which incorporates past belief systems, plays a major role in the study of past scientific activity.

I must also stress that, in this book, I resist any *psychological* interpretation of Kuhn's philosophy. Maybe he really started to use 'linguistic concepts' because he wanted to be accepted by the Putnams and the Kripkes of his time. Maybe he really felt frustrated as a philosopher and he wanted to make his views more palatable to his contemporary peers. Or maybe not. In any case, there are genuine, *philosophical* reasons for dealing with problems of meanings and categorization, as I attempted to show in these past two chapters.

Finally, there can be many reasons for why Kuhn did not leave a 'distinctively Kuhnian school'. The claim for which '*if* Kuhn hadn't undergone a linguistic turn (whatever that was), *then* he would have not undermined his own legacy' may be unwarranted after all. Perhaps a 'distinctively Kuhnian school' is yet to be established, or perhaps it cannot even be established in principle (see Chap. 11). In any case, his (alleged) linguistic turn may not have played a major role in Kuhn's legacy, or lack thereof.

7.6 Conclusions

The dynamic perspective demands a description of the scientific activity carried out within a scientific community. In *SSR*, Kuhn 'discovered' that the members of a community share a paradigm. The concept of a paradigm, however, was met with a disproportionate number of criticisms. For this reason, in his post-*SSR* writings, Kuhn attempted to refine the concept, and, in the end, he decided to consider scientific communities akin to language communities. However, he soon got caught up in subtle distinctions between different types of scientific terms, different types of generalisations, and different ways to learn scientific concepts, risking losing sight of his original intent.

The concept of lexicon may rescue the old aim of studying scientific communities. A lexicon contains all the different types of scientific terms, and it allows for the formulation of different theories. A lexicon possessed a 'core', made up of synthetic a priori sentences in the form of laws, which consent its stability through time. Finally, by studying the lexicon of a community, one also gains insight into the community practices and way of reasoning. A lexicon, in other words, appears to be the linguistic version of a DM, that is of the broader sense of a paradigm, whose function and importance is somehow rescued.

References

Andersen, Hanne, and Nancy Nersessian. 2000. Nomic concepts, frames and conceptual change. *Philosophy of Science* 67:S224–S241.

Barberousse, Anouk. 2007. From one version to the other: Intra-theoretical change. In *Rethinking Scientific Change and Theory Comparison*, ed. Lena Soler, Howard Sankey, and Paul Hoyningen-Huene, 87–101. Dordrecht: Springer.

Bird, Alexander. 2002. Kuhn's wrong turning. *Studies in History and Philosophy of Science* 33:443–463.

Crombie, Alistair. 1994. *Styles of Scientific Thinking in the European Tradition*, Volume I, II and III. London: Duckworth.

Friedman, Michael. 2001. *Dynamics of Reason*. Stanford: CSLI Publications.

Gattei, Stefano. 2008. *Thomas Kuhn's 'Linguistic Turn' and the Legacy of Logical Empiricism*. Hampshire: Ashgate Publishing.

Hacking, Ian. 1975. Why Does Language Matter to Philosophy? Cambridge: Cambridge University Press.

Hacking, Ian. 1982. Language, truth and reason. In *Rationality and Relativism*, ed. Martin Hollis and Stephen Lukes, 48–66. Oxford: Blackwell.

Hacking, Ian. 1990. *The Taming of Chance*. Cambridge: Cambridge University Press.

Hacking, Ian. 1992. 'Style' for Historians and Philosophers. *Studies in History and Philosophy of Science* 23:1–20.

Hacking, Ian. 2012. Language, truth and reason - 30 years later. *Studies in History and Philosophy of Science* 43:599–609.

Hoyningen-Huene, Paul. 1993. *Reconstructing Scientific Revolutions. Thomas Kuhn's philosophy of science*. Chicago: Chicago University Press.

Irzik, Gürol, and Teo Grünberg. 1998. Whorfian variations on Kantian themes: Kuhn's linguistic turn. *Studies in History and Philosophy of Science* 29:207–221.

Kuhn, Thomas S. 1974. Second thoughts on paradigms. In *The Structure of Scientific Theories*, ed. Frederik Suppe, 459–482, Urbana: University of Illinois Press. Reprinted in *ET*: 293–319.

Kuhn, Thomas S. 1987. What are scientific revolutions? In *The Probabilistic Revolution*, ed. Lorenz Krüger, Lorraine Daston, and Michael Heidelberger, 7–22. Cambridge: MIT Press. Reprinted in RSS, 13–32.

Kuhn, Thomas S. 1989. Possible worlds in the history of science. In *Possible Worlds in Humanities, Arts and Sciences*, ed. Allén Sture, 9–32, Berlin: Walter de Gruyter. Reprinted in *RSS*: 58–89.

Kuhn, Thomas S. 1991. The Road since Structure. *PSA: Proceedings of the Biennial Meeting of the Philosophy of Science Association 1990* (2): 2–13. (Reprinted in RSS, 90–104).

Kuhn, Thomas S. 1993. Afterwords. In *World Changes: Thomas Kuhn and the Nature of Science*, ed. Paul Horwich, 311–341. Cambridge: MIT Press. Reprinted in RSS, 224–252.

Kuhn, Thomas S. 2022. *The Last Writings of Thomas S. Kuhn*, ed. B. Mladenovic. Chicago: University of Chicago Press.

Kuukkanen, Jouni-Matti. 2008. *Meaning Changes: A Study of Thomas Kuhn's Philosophy*. Saarbrucken: VDM - Verlag Dr. Mueller e.K.

McDonough, Jeffrey. 2003. A *Rosa Multiflora* by any other name: Taxonomic incommensurability and scientific kinds. *Synthese* 136:337–358.

Nersessian, Nancy. 1984. *Faraday to Einstein: Constructing Meaning in Scientific Theories*. Dordrecht: Martinus Nijho.

Pirozelli, Paolo. 2024. Reappraising kind concepts. In *Rethinking Thomas Kuhn's Legacy*, ed. Yafeng Shan, 209–228. Chem: Springer.

Reisch, G. 2021. Thomas Kuhn's *Quest for Physical Theory*: Editor's Introduction. QPT: xi–xxxvi.

Rouse, Joseph. 2003a. *How scientific practices matter: Reclaiming philosophical naturalism*. Chicago: University of Chicago Press.

Rouse, Joseph. 2003b. Kuhn's philosophy of scientific practice. In *Thomas Kuhn*, ed. Thomas Nickles, 101–121. Cambridge: Cambridge University Press.

Rouse, Joseph. 2013. Recovering Thomas Kuhn. *Topoi* 32:59–64.

Shan, Yafeng. 2020. Kuhn's "wrong turning" and legacy today. *Synthese* 197:381–406.

Sciortino, Luca. 2021. The emergence of objectivity: Fleck, Foucault, Kuhn, and Hacking. *Studies in History and Philosophy of Science* 88:128–137.

Wray, Brad K. 2007. Kuhnian revolutions revisited. *Synthese* 158:61–73.

Wray, Brad K. 2011. *Kuhn's Evolutionary Social Epistemology*. Cambridge: Cambridge University Press.

Winther, Rasmus. 2012. Interweaving categories: Styles, paradigms, and models. *Studies in History and Philosophy of Science* 43:628–639.

Wise, Norton. 1993. Mediations: Enlightenment balancing acts, or the technologies of rationalism. In *World Changes: Thomas Kuhn and the Nature of Science*, ed. Paul Horwich, 207–256. Cambridge: MIT Press.

Part III
Kuhnian Issues

Chapter 8
Specialisation and Interdisciplinarity

Abstract Kuhn's mature writings contain some interesting but rather underdeveloped insights about the process of specialisation, which is the topic of this chapter. I begin by discussing the formation of what Kuhn calls 'scientific specialties', the kind of incommensurability that drives such a phenomenon, and the relation between specialisation and scientific revolutions. My take on these topics draws from Kuhn's philosophy but leads to conclusions that Kuhn himself could not see or failed to appreciate. I also I discuss interdisciplinarity, a phenomenon that Kuhn did not address, even though it threatens his ideas about the existence of an incommensurable divide among scientific specialties. Finally, I consider how the theory of specialisation enriches our understanding of Kuhn's views on scientific change and of his remarks about the 'plurality of worlds'.

Keywords Specialisation · Interdisciplinarity · Scientific revolutions · Incommensurability · Scientific community · Plurality of worlds

8.1 Introduction

Late in his career, Kuhn realised that scientific change, rather than always being a 'substitutive' process involving the replacement of a paradigm with a new one, often leads to the proliferation of "cognitive specialties or fields of knowledge" (Kuhn 1991, in *RSS*: 97). This may occur in two ways: either the new specialty splits from one 'parent-discipline', or it emerges from the area of partial overlap between two or more disciplines. Although this second type of specialty formation looks like an instance of unification, Kuhn specifies that the new branch of science "seldom or never gets assimilated to either of its parents. Instead, it becomes one more separate specialty, gradually acquiring its own new specialists' journals, a new professional society, and often also new university chairs, laboratories, and even departments" (Kuhn 1991, in *RSS*: 97).

V. Politi, *Thomas Kuhn's Philosophical Perspective*, Boston Studies in the Philosophy and History of Science 348,
https://doi.org/10.1007/978-3-032-13540-7_8

It is important to notice, to begin with, that Kuhn uses the terms 'specialty' and 'specialisation' in a loose sense. As they are commonly intended, the specialties of a discipline investigate *different parts of the same ontological domain*. For example, astrodynamics and astrophysics are both specialties of astronomy, with which they share the ontological domain of the astronomical entities. The branching off from pre-existing disciplines Kuhn describes, however, may lead to the *opening of new domains*. For example, in his interpretation of Kuhn, Wray (2011) refers to the creation of virology from bacteriology, and of the creation of endocrinology from physiology as examples of specialisation. Clearly, virology is *not* a specialty of bacteriology, nor is endocrinology a specialty of physiology. In such cases, it would be more appropriate to talk about the creation of new disciplines, rather than of new specialties. Or, if anything, it would be more correct to classify virology as a specialty of microbiology (alongside bacteriology) and endocrinology as a specialty of medicine (alongside physiology). Leaving aside the issue of the correct definition and classification of scientific disciplines and subdisciplines, as it will become clear in this chapter, what Kuhn describes would be better defined as the *creation of new domains of research*, possessing their own ontology. Having clarified this point, in the remainder of this chapter I will keep talking about 'specialisation' to conform with the terminology used by Kuhn and by some of his commentators.

Although suggestive, Kuhn's view on specialisation is rather underdeveloped. The aim of this chapter is to develop a 'Kuhnian model' of specialisation that does not consists only in the recapitulation of what Kuhn wrote about the formation of specialties. The model that I will develop uses Kuhn's *dynamic perspective* (see Sect. 3.2) and his *collective approach* (see Chap. 4) to explain the kind of scientific change brought about by specialisation. In this way, I may end up reaching conclusions that the same Kuhn failed to see.

In Sects. 8.2 and 8.3, I discuss the formation of scientific specialties and the kind of incommensurability that drives such a phenomenon. I develop my account my first analysing the limits of Wray's interpretation of Kuhn's view on specialisation. As I will show, Wray's interpretation is not consistent with the dynamic perspective and must therefore be corrected. In Sect. 8.4, I discuss interdisciplinarity that Kuhn did not address. This is surprising for at least two reasons. First, one of the two modes of specialty-formation he describes consists in the convergence of two (or more) 'parent disciplines' towards an overlapping area of research, which will eventually grow up autonomously. It is difficult not to see such a process as involving interdisciplinarity. Second, and more importantly, interdisciplinarity threatens Kuhn's ideas about the existence of an incommensurable divide among scientific specialties. If specialties are kept separate by an incommensurable divide, how is interdisciplinary possible? The fourth section will answer this question; or, rather, it will reformulate it from the dynamic perspective. In Sect. 8.5, I defend the claim that specialisation and scientific revolutions are not different kinds of scientific change. Rather, in my view, the creation of a new specialty is a special case of scientific revolution, especially if looked from the perspective of the scientific community involved in such a change. Finally, in Sect. 8.6, I consider how the theory of specialisation enriches our understanding of Kuhn's late remarks about the 'plurality of worlds'.

8.2 Scientific Specialties

Kuhn managed to sketch a theory of scientific specialisation that is both *descriptive* and *prescriptive*. It is descriptive because, as a matter of fact, scientific specialties do proliferate: "The point is empirical and the evidence, once faced, is overwhelming: the development of human culture, including that of the sciences, has been characterised by a vast and still accelerating proliferation of specialties" (Kuhn 1993, in *RSS*: 250). Such an account is also normative insofar as the proliferation of scientific specialties is necessary to comprehend and solve a higher number of specific problems. Specialisation is, in fact,

> prerequisite to the continuing development of scientific knowledge. The transition to a new lexical structure, to a revised set of kinds, permits the resolution of problems with which the previous structure was unable to deal. But the domain of the new structure is regularly narrower than that of the old, sometimes a great deal narrower. What lies outside [the domain of a specialty] becomes the domain of another scientific specialty [...]. Proliferation of structures, practices, and worlds is what preserves the breadth of scientific knowledge; intense practice at the horizons of individual worlds is what increases its depth. (Kuhn 1993, in *RSS*: 250)

Kuhn welcomes the irreducible character of the newly formed specialties. The fact that science does not get increasingly unified through time but, rather, gets fragmented among an ever-growing number of sub-disciplines is what allows the increase of its problem-solving power. In virtue of their being exclusively focussed on a narrow ontological domain, specialties can refine their research without being distracted by problems arising in neighbouring fields. The increased focus and precision of scientific specialties, in turn, guarantees the resolution of more problems.

> To anyone who values the unity of knowledge, this aspect of specialisation [...] is a condition to be deplored. But such unity may be in principle an unattainable goal, and its energetic pursuit might well place the growth of knowledge at risk. [...] Very likely it is the [process] of specialisation [...] that permits the sciences, viewed collectively, to solve the puzzles posed by a wider range of natural phenomena than a [...] homogeneous science could achieve. (Kuhn 1991, in *RSS*: 98–99)

Unfortunately, Kuhn did not really develop a full account of specialisation, nor is it clear whether he would have done so in the chapters of *PW* he could not finish to write. Surprisingly, most Kuhnian scholars have overlooked Kuhn's view on specialisation, with at least one notable exception, namely Karl B. Wray (2011).

Wray rightly observes that Kuhn has the merit of considering a type of scientific change that, despite being so important and ubiquitous, has been discussed mainly by sociologists and historians, but neglected by most philosophers. The problem with the sociologists' and historians' explanations of the process of specialisation, however, is that they tend to be 'mono-causal': they tend, that is, to take only one factor as the cause for the creation of a new specialty. Such mono-causal explanations are purely sociological, based on the assumption that scientists just create new sub-fields toward which they can 'migrate'. The migration allows them to leave older fields that, being already overcrowded, offer less chances for career development and recognition. These analyses fail to explain how new research fields come into being to begin with. That a handful of scientists decide to invent a new special field and create it 'out of nowhere' just to have some career opportunities, is whimsical to say the least.

Wray contends that, although sociological factors surely play some role in accelerating it, the process of specialty formation must be initiated by a conceptual change and for epistemic reasons. First, it is found that the conceptual structure of the pre-existing discipline is incapable of solving some problems arising in some areas of its research domain; new concepts are therefore introduced to deal with such esoteric problems. After a while, scientists may reclaim as their own the part of the domain of the pre-existing discipline where the problems requiring conceptual adjustment arises. It is only at this point that the new specialty is created and scientists can migrate towards it. Following Wray, therefore, Kuhn's late work answers the question: "how can conceptual or epistemic developments in science contribute to the development of a new scientific specialty?" (Wray 2011: 135), that sociologists and historians leave unanswered, and that most philosophers of science neglect.[1]

For Wray, the problems that will eventually lead to the creation of a new speciality are often associated with the *discovery of a new kind of entity* that cannot be accommodated within the conceptual structures of pre-existing disciplines (Wray 2011: 118–122). Wray's chief examples are the creation of *endocrinology* and *virology*. In both cases, the new specialty was created as a response to the discovery of a new kind of entity that conflicted with the pre-existing classification systems. In the first case, the discovery of 'hormones' led some physiologists to re-conceptualize the co-ordination of certain body functions in terms of chemical transmission, rather than nervous mechanisms. In the second case, sub-groups of bacteriologists and biochemists realised that some micro-organisms were relevantly different from bacteria and toxins and, as a consequence, they converged toward an independent specialty to study the properties of the newly discovered 'virus' (Wray 2011: 127–130).[2]

[1] Although Kuhn is indeed one of the few philosophers who pays attention to the process of specialisation, he is by no means the only one. In works that predate Kuh's mature reflections on specialisation of almost two decades, Shapere (1974) and Darden (1978) provide invaluable insights for the study of the formation of new disciplines, even though neither Kuhn nor Wray ever mention them. It seems therefore more correct to characterise Kuhn's underdeveloped insights on specialisation not as opening a new field of philosophical investigation, but rather as the late contribution to a debate that never gained momentum among philosophers; or, perhaps, as an attempt to revitalise it. Incidentally, and as Wray notices, even sociologists and historians seem to have prematurely abandoned the topic.

[2] As already said in Sect. 8.1, virology is *not* a 'specialty' of bacteriology. Even though it came into existence when a group of scientists branched off from the pre-existing community of bacteriologists, virology is just a different discipline than bacteriology. This is why it is important to stress that what Kuhn and Wray talk about is not just specialisation, but the creation of new domains of research with their own ontology.

With these examples, Wray shows how discoveries and conceptual changes, and not just sociological factors, drive the process of specialisation.

I have a problem with Wray's identification of the discovery of a new kind of entity as the main epistemic cause of the creation of a new specialty. To begin with, several new specialties were created around a newly discovered kind of entity that, however, did *not* conflict with pre-existing taxonomies and that did *not* require radical conceptual changes. One example is the creation of *biochemistry*, the discipline studying the chemical processes within living organisms. Biochemistry was created to study 'enzymes', a kind of macromolecular biological catalysts responsible for some chemical reactions, such as fermentation and metabolism.

Although the term was coined by Kühne in 1877, the discovery of the entity itself is traditionally backdated to 1833, when Anselme Payen and Jean-François Persoz, while working as chemists at a French sugar factory, discovered the first enzyme, the 'diastase', responsible for the breakdown of starch into maltose. The newly discovered entity, however, did not conflict with any of the pre-existing conceptual taxonomies. Surely, diastase could explain some previously unclear phenomenon, such as fermentation. However, it did not require a conceptual re-systematization and a change in the very criteria of pre-existing classifications.

The real conflict with the scientific credo happened with the discovery of 'zymase', made by Eduard Buchner in 1897. Zymase is an enzyme that is naturally present in yeasts, and that catalyses the fermentation of sugar into ethanol and carbon dioxide. Buchner's discovery challenged the established idea that cells were the sole carrier of all the vital functions. Zymase, in fact, demonstrated the existence of 'cell-free fermentation'. This generated a fierce controversy within the scientific community of the time (for a historical analysis of the debate over the impact of zymase, see the classical works of Kohler 1971, 1972, 1973, 1975). Like every historical process, determining the exact point in which biochemistry as a discipline was created and established is not easy. Unarguably, Buchner's discovery of zymase played a crucial role in such a process. It must be stressed, however, that this entity was not directly in conflict with the pre-existing classification: after all, other sub-kinds of enzymes had *already* been discovered and conceptualised. Rather than between different classification systems, Buchner's discovery of zymase sparked a competition among incompatible *explanations* of some chemical mechanisms underlying some living functions. In short, neither enzymes (in general) nor zymase (in particular) required a relevant change in the taxonomy of the pre-existing fields: such entities were simply added to the ontology of the time. What required an adjustment was the explanation of some biological processes.

Furthermore, although the discovery of enzymes and, in particular, of zymase, was a necessary ingredient for the creation of biochemistry, it was by no means sufficient. The subject matter of biochemistry is the set of so-called 'intermediary metabolic processes' (i.e., glycolysis and biological oxidation). These processes were discovered and conceptualised by pioneering works on metabolism made by physiologists and chemists. Such cross-disciplinary research opened a new research field. Intermediary metabolic processes involve more than simple chemical reactions which, however, do not occur at the level of physiologically distinct units: as such, they cannot be reduced to either the chemical or to the physiological level. Lately, it was theorised that enzymes play a pivotal role in these intermediary metabolic processes (Bechtel 1986). While some enzymes, including the crucially important zymases, were discovered in the nineteenth century, intermediary metabolic processes were discovered around the 1930s. The discovery of a new ontological field, at the intersection of the chemical and the physiological levels of organisation, with all its specific and irreducible processes and mechanisms, is what pressed scientists to create and establish a new, independent discipline.[3] This type of discovery, however, can be hardly accommodated within a philosophical conception that defines the main driver of scientific change mainly in terms of violation of the hierarchical-taxonomic structure of kinds of entities.

Not only the discovery of a new kind of entity is not sufficient for the creation of a new scientific specialty but, as already explained by Darden (1978), it is not even necessary. Following Darden, the emergence of several scientific specialties

> do[es] not result from the discovery of a new kind of entity but from the discovery of repeated patterns in data; new kinds of entities may be postulated to account for the data. For example, Dalton's atomic theory accounted for the laws of chemical combination. In the emergence of genetics, genes were postulated to account for patterns of inheritance of characteristics. In the atom and gene cases, the entities were postulated, not discovered experimentally [...]. Darwin's theory of the origin of the species postulated, not a new kind of entity, but a new kind of process—selection. The claim that selection is the general cause of the origin of the species generated a promising line of research which continues today. (Darden 1978: 157)

Sometimes new kinds of entities can be simply postulated rather than discovered; incidentally, Shapere (1974) also discussed how 'hypothetical entities' become 'domain items'. Other times, what is postulated is not even an entity, but a process. There are indeed many examples of new specialties that emerged without the prior discovery of a new kind of entity.

Statistical mechanics is a branch of physics that investigates the microstates of gases as the cause of their macro-states. One of its chief aims is to find a mechanical explanation of the Second Law of thermodynamics, which roughly states that every isolated system always and irreversibly evolves toward a state of thermodynamic equilibrium. In the study of the properties and behaviours of gases, what the Second Law says is that the gas contained in an isolated system will inevitably end up in a state of equilibrium—that is, every part of the closed system will be occupied by the same amount of gas and such a state of equilibrium is irreversible. Such a law is 'phenomenological', that is, based on the observation of the behaviour of gaseous phenomena. What statistical mechanics does, then, is to study the mechanics

---

[3] In fact, the first journals of biochemistry appeared in the beginning of the 1900, while the discipline as such was academically established even later than that, long after the discovery of enzymes.

governing the micro-state of the gases—i.e., how the tiny particles of gases move around. No discovery of a new kind of entity preceded the creation of statistical mechanics.[4]

So-called *String Theory* was created from the hypothesis about the existence of kinds of entities that, however, have not been discovered yet. It is debatable whether String Theory forms, as a matter of fact, a well-formed and established discipline, but for sure it opened a new, more 'fundamental' domain of investigation. Whether it will ever become a 'mature' field or not, the point is that, so far, String Theory owes its development not to the actual discovery of 'strings' but, rather, to the creation of new mathematical models and 'fuzzy logics'.[5]

So-called *complexity science*, or complex systems science, provides new mathematical models—i.e., chaotic and non-linear models—for the study of systems exhibiting properties which do not seem to be reducible to the properties of their constitutive components. It cannot be said, however, that complexity science was born out of the 'discovery' of complex systems. What complexity science has done is to introduce the concept of 'complex system' to re-conceptualize and re-describe some collections of already known entities and their processes. Nor can it be said that the research on complex systems has a *narrow* domain of research, separated from the domains of the other sciences. A notable aspect of complexity science, in fact, is its broad applicability, encompassing both natural and social sciences. Complex systems are present in disciplines such as physics, chemistry, biology, earth sciences, computer science, economics, anthropology, and many others.[6]

Rather than the discovery of new kinds, an element that deserves more attention for the understanding of specialisation, and which is central in Darden's analysis, is the *problem-solving power* of the new emerging specialty. For Darden, a new

---

[4] One of the most authoritative philosophical introductions to statistical mechanics is Sklar (1993); in its second chapter the relations between statistical mechanics and classical thermodynamics are discussed from a historical perspective. Albert (2000) is an accessible introduction to the philosophical problems that statistical mechanics and the concept of entropy pose to the classical idea of the 'arrow of time'. For a recent discussion of some philosophical issues with the two major approaches in statistical mechanics—namely, Boltzmann's approach and Gibbs's approach—see Frigg (2008).

[5] String theory is assessed in remarkably different ways: the string theorists themselves regard it as a promising and fruitful field of research, whereas a great part of the rest of the community of physicists is rather sceptical; such a conflicting assessment can be explained as (or, even, exemplifies) the 'paradigmatic rift' between two different sides which disagree over theory assessment (Dawid 2009). For a very philosophical discussion of the issues surrounding string theory - its methods as well as its status among and relations with the other physical sciences—see Dawid (2013).

[6] For an introduction to complexity science, see Mitchell (2009). One philosophical problem with complexity science is that the very notion of 'complex system' is not rigorously defined, nor is it clear what are the necessary conditions for a system to be 'complex' and hence to be taken as the object of research of complexity science; for an introduction to these issues, see Ladyman et al. (2013). Philosophical discussions of the main methodological and metaphysical issues surrounding the discipline are found in Hooker (2011), which also devotes several chapters to the study of complex systems in different domains, from biology to anthropology.

special field emerges around a new theory that provides "a new solution to an old problem", that is also "recognized as a promising one" and which leads to "a new line of research using the new instruments or techniques begins" (Darden 1978: 155). Darden's account puts more emphasis on how new entities, patterns, processes, theories, techniques, and new methods interrelate and conspire for the emergence of a new discipline. To be fair, Wray does not deny that the discovery of a new kind of entity is made possible by both conceptual and technological developments (Wray 2011: 132). Nevertheless, most of the time he restricts his attention to the conceptual changes associated with the discovery of new entities.

As have already discussed in Chaps. 6 and 7, kinds of entities are not all there is in the conceptual structure of science. Instead of focusing on the study of the kind-relation within a taxonomy, I have suggested that we focus on both taxonomic and non-taxonomic kinds, as well as on both kinds of entity and kinds of process, and also on methodological assumptions and techniques. Instead of seeing a new specialty as being constructed around a new kind of entity which somehow branches off a pre-existing taxonomy, I propose to consider the emergence of a new specialty as the emergence of a new lexicon, held by a new DM and constituting a new 'world'. After all, as shown in one the quotations at the beginning of this chapter, Kuhn (1993) himself speaks of the proliferation of "structures, practices and worlds".

To sum up: while I agree with Wray about the importance of looking at the epistemic factors driving the process of specialisation, I disagree with his generalising the discovery of a new kind of entity as the most important of such epistemic factors. The discovery of a new kind of entity may lead to the creation of a new specialty, but not every new specialty is created because of the discovery of a new kind of entity. Moreover, even when a discovery is involved in the process, very often the new kind of entity is simply not enough for the establishment of a new specialty. Other types of changes—for instance, in the methods and techniques of research—must be considered.

As already mentioned at the beginning of this section, Kuhn's sketch for a model of scientific specialization, like his more famous model of scientific revolutions, is both descriptive and normative. The development of Kuhn's views, therefore, must be grounded on good descriptions of specialty formation, which involves more than the discovery of new kinds of entities and, sometimes, it does not even involve such discoveries.

8.3 Specialty-Incommensurability

For Kuhn, specialisation is a process of isolation. Only if specialists focus on their narrow special domain, without the distraction of the problems pertaining to the domains of neighbouring fields, can each specialty increase in precision. The members of a scientific specialty "share a lexicon that provides the basis for both the conduct and the evaluation of their research and which simultaneously, by barring full communication with those outside the group, maintains their isolation from practitioners of other specialties" (Kuhn 1991, in *RSS*: 98). Isolation is therefore a prerequisite for the development of specialties, which is in turn a prerequisite for the increasing the problem-solving power of science.

The process of isolation that leads to specialisation, for Kuhn, is driven by a form of incommensurability. He describes specialty-incommensurability as a barrier that makes cross-specialty communication difficult:

> what makes [...] specialties distinct, what keeps them apart and leaves the ground between them as apparently empty space [...] is incommensurability, a growing conceptual disparity between the tools deployed in the two specialties. Once the two specialties have grown apart, that disparity makes it impossible for the practitioners of one to communicate fully with the practitioners of the other. And those communication problems reduce, though they never altogether eliminate, the likelihood that the two will produce fertile offspring (Kuhn 1992, in *RSS*: 120).

Wray (2011) defines specialty-incommensurability as akin to semantic incommensurability. Both Kuhn and Wray argue that:

1. Specialty-incommensurability is a type of semantic incommensurability
2. As such, it poses a linguistic barrier that makes cross-specialty communication difficult
3. *Conclusion*: because of specialty-incommensurability, scientists from different specialties do not communicate (or, at least, have difficulties in communicating) and become more isolated from each other (and such an isolation is beneficial to specialties)

There are several problems with this argument. As I will discuss in the next section, the idea that scientists from different specialties do not communicate because of incommensurability, which corresponds to (3) in the above argument, is contradicted by the very existence of interdisciplinarity. In this section, however, I focus on the issues with (1) and (2).

The problem with (2) is that it is not entirely clear what kind of difficulties specialty-incommensurability poses to cross-specialty communication. For Kuhn, specialty-incommensurability impedes 'full communication'. In principle, a lack of 'full communication' does not imply the impossibility of partial communication, which could still produce some limited yet valid outcome. For Wray, however, "[s]scientists working in neighbouring specialties are often impeded in *effective communication* across specialty lines because *they attach different meanings to the same terms*" (Wray 2011: 75, *my emphasis*).

Wray's strong position about specialty-incommensurability seems to conflict with Kuhn's own view. As mentioned in previous chapters, in a series of post-*SSR* replies to his critics, Kuhn adamantly rejects the idea that incommensurability implies incommunicability. If, following Kuhn's own argument, incommensurability is always local (that is, if meaning variation affects only a small cluster of inter-defined theoretical terms); if the 'communication breakdowns' Kuhn's critics were so repulsed by are very occasional; and if, incommensurability notwithstanding, and in virtue of Kuhn's own argument, communication is always possible across revolutions and between incommensurable scientific cultures, then it is not clear why, in the context of specialisation, incommensurability should become the huge obstacle to cross-disciplinary communication Wray seems to think it is.

Wray does not elaborate a valid argument to support the idea that incommensurability impedes communication across specialties. As an example of such a linguistic/conceptual divide, and following Ereshefsky (1998), he speaks about the existence of several incommensurable concepts of *species*. He uses this example to show how scientists from different specialties may attach different meanings to the same terms. It is still not clear, however, why attaching different meanings to the same term should be an obstacle to communication. There would be incommunicability only if scientists were totally unaware of the fact that their colleagues from the neighbouring specialties use some terms differently. However, if Ereshefsky and Wray know that 'species' may have different meanings, and they also know what these meanings are, it is likely that all the different specialists using that word in different ways may know these things too.

As already said, Kuhn's position is more cautious. In his view, incommensurability is an obstacle only to "full communication". For Kuhn, in other words, incommensurability is an obstacle to cross-specialty communication; but an obstacle is not an impossibility. Although weaker than Wray's, Kuhn's view is equally problematic. If incommensurability is an obstacle that can be overcome—and it *is* indeed overcome in interdisciplinary collaborations, as I will discuss in the next section—then the problems it would pose to cross-specialty communication would be rather innocuous, if not trivial. Specialty-incommensurability would even risk becoming philosophically uninteresting.

My argument is that specialty-incommensurability should not be regarded as posing problems to communication. It is possible that two different scientific communities, or two groups of specialists, can perfectly understand each other's language and can communicate, yet still have some fundamental and irreconcilable disagreements. In other words, I argue against a strictly semantic interpretation of specialty-incommensurability. Before developing this view any further, it remains to examine the first premise of the argument above, which both Kuhn and Wray seem to take for granted. It remains to be seen, that is, *why* Kuhn and Wray characterise specialty-incommensurability as a 'linguistic barrier'.

As explained in the previous section, for Wray the conceptual changes driving specialisation are associated with the discovery of a new kind of entity that cannot be accommodated within a pre-existing conceptual taxonomy. In his view, such 'significant discovery' requires "radical changes to the taxonomy of a field with the result that a new field [is] born and the domain of the original field [is] subsequently truncated" (Wray 2011: 129). However, as already discussed, a closer look at the actual process of specialisation shows that some specialties are not created after the discovery of a kind of entity that conflicts with the pre-existing conceptual taxonomy, while others do not even require a discovery to begin with. In the absence of a new kind that defies current classifications, or even of a new kind altogether, it becomes difficult to regard specialty-incommensurability as a matter of conflicting taxonomies.

In a sense, it seems like both Kuhn and Wray *presuppose* a semantic/taxonomic version of the incommensurability thesis, which they then apply to their view on specialisation. By contrast, I suggest that a better understanding of specialty-incommensurability should *follow*, and not precede, a better understanding of how new scientific specialties are created. In the examples mentioned in the previous section (i.e., biochemistry, statistical mechanics, String Theory, and complex science) the driver for the emergence of the new specialties was the establishment of a new way of framing and solving problems, which was relevantly different from that of the pre-existing disciplines. This relevant difference can be explained in terms of *methodological incommensurability*, that is the way in which different groups of scientists weigh and rank epistemic values for the assessment and choice of theories (and models, methods, and so on; see Sect. 1.4.3). Some philosophers have provided further examples of the role of methodological incommensurability in specialty-formation. De Benedetto and Lucchetti (2023) explain how the process of specialisation is driven by 'value divergence' that, in virtue of the feedback loop between values and theories, helps establish new research niches, as in the case of molecular biology.[7] Haufe (2022) analyses the emergence of evolutionary palaeontology as a process involving much more than semantic incommensurability. Gefaell and Saborido (2022) characterise the divide between the Modern Synthesis and the Extended Evolutionary Synthesis as a mixture of semantic and methodological incommensurability.

I close this section with the following considerations. First, in *SSR*, Kuhn spoke about revolutions and characterised incommensurability as a complex phenomenon, which includes semantic, methodological, and cognitive aspects. In his later works, he spoke about specialisation and characterised specialty-incommensurability as a kind of semantic incommensurability. However, nothing forbids applying Kuhn's 'early' view of incommensurability to his 'late' view on specialisation. Such a move is not only possible, but also legitimised by, on the one hand, an interpretation of

[7] As it will be shown at the beginning of Sect. 8.5, Kuhn develops his theory of specialisation through the analogy with biological speciation (*RSS*: 97–98; 119–120). The latter process requires a physical barrier separating two subpopulations that will eventually become two separate subspecies. Similarly, Kuhn describes specialisation as a process of diversification made possible by a linguistic barrier. A possible criticism to the definition of specialty-incommensurability in terms of methodological rather than semantic incommensurability is that it would represent a disanalogy with biological speciation. Differences in problem evaluation and in the application of scientific values and standards do not represent a 'barrier' that makes cross-disciplinary contact impossible. One may answer that this is not really an issue: Kuhn made use of biological analogies for illustrative purposes, he was not claiming that the processes of scientific change and of biological change unfold in the same way. Others, however, read Kuhn's insight through the more recent developments in evolutionary biology. In their article, De Benedetto and Luchetti (2023) points out that, as it turns out, biological speciation does *not* always require a physical barrier segregating two subgroups. In fact, even within the same environment, specialisation may happen through a process known as 'gene divergence'. They then discuss the methodological version of specialty-incommensurability in analogy with such a phenomenon and in terms of 'values divergence'. A fuller account of how evolutionary biology may illuminate Kuhn's theory of scientific change is developed by Haufe (2022).

Kuhn's late concept of lexicon as the re-description of his early concept of DM; and, on the other hand, by the description of the actual process of specialisation.

Second, and to repeat what I have already mentioned in Sect. 8.1, by now it has become even clearer that what Kuhn calls 'specialisation' is actually the creation of a new sub-field of research, with its own sub-community. On the one hand, the process of creating new domains of research and new sub-communities, which, I believe, was what Kuhn was attempting to explain, cannot be reduced to what is commonly referred to as specialisation. Indeed, 'virology' is not a specialty of 'bacteriology', even if the field was born when a sub-community of bacteriologists split off from the rest. On the other hand, some instances of specialisation per se may not correspond to the creation of new domains and may not be driven by incommensurability at all. In short, they may not be described as instances of scientific change of the sort Kuhn had in mind. For instance, 'organic chemistry' studies compounds that contain carbon, while 'inorganic chemistry' studies all the other compounds. It is questionable whether these two branches are separated by a semantic or a methodological incommensurability, or that the fragmentation of the chemical field in these two sub-branches was nothing more than a re-organisation of the chemical communities for sociological reasons. Therefore, when talking about his view on 'specialisation', one must keep in mind that what Kuhn was trying to explain, whether he was fully aware of this or not, is the creation, driven by incommensurability, of new domains of research.

8.4 Specialisation, Interdisciplinarity, and Incommensurability

Not everybody agrees with the idea that there is an incommensurable divide across different scientific specialties. Hanne Andersen points out that while, during a revolution, incommensurable paradigms compete for the explanation of the same range of problems and phenomena, different specialties are just about different things altogether (see Andersen 2012, 2013a, b, 2016). Consequently, it is not the case that scientists from different specialties *cannot* communicate. Rather, they just *do not* communicate because they do not have much to say to each other to begin with.

One could reply that different specialties are incommensurable simply by virtue of their being different. Such a reply would clearly trivialise the whole concept of incommensurability, which implies more than difference. Theories, methods, and paradigms that are about different things may peacefully and unproblematically coexist within the 'Republic of the Sciences'. The incommensurability thesis, however, acquires its philosophical relevance in the context of the assessment of competing and conflicting views about the same problems; for instance, when an exclusive choice between two incompatible paradigms is demanded. Therefore, if one claims that specialties are incommensurable, one should also be able to clarify in which sense they are not 'just different'.

If specialties were incommensurable, and not just about different things, then their domains should at least partially overlap. Otherwise, there would be no ground for conflict and competition. The problem is that specialisation does not lead to the creation of overlapping domains. Rather, it is a process of diversification that leads to ontological fragmentation, with each newly created discipline being focused on its own special domain.

To understand why it is legitimate to consider incommensurability within a model of specialisation, it is important to remember Kuhn's *dynamic perspective*, that looks at science in its developing dimension, not as a static body of knowledge. Philosophers like Andersen argue against specialty-incommensurability because they look at how different specialties are *now*. By doing so, they dismiss the very dynamic process of specialisation. What I am suggesting here, by contrast, is to look at how scientific specialties *came to be* differentiated. For example, *during* the emergence of virology, but *before* its establishment as a fully formed and independent field of investigation, 'bacteriological' and 'virological' theories of virus were indeed in a state of competition for the explanation of the same range of phenomena and entities—namely, for the explanation of the causes of some diseases. That *now* virology and bacteriology are different disciplines, with their own separate domain, is beside the point. Incommensurability, in fact, should be understood as playing a role during the process of specialisation, not after specialisation has been accomplished.

Granted that incommensurability drives the process of specialisation/differentiation, it remains to be understood whether it also plays any role in interdisciplinary research. In fact, a point that I have raised in the previous section, but that I have not tackled yet, is the tenability of Kuhn's (and Wray's) conclusion, which amounts to say that, because of specialty-incommensurability, scientists from different specialties cannot 'fully communicate' (as Kuhn maintains) or cannot 'effectively communicate' (as Wray contends). This view of specialty-incommensurability as a conceptual/linguistic barrier hampering cross-specialty communication is contradicted by the existence of interdisciplinary research. Scientists involved in interdisciplinary research not only interact, but sometimes they even achieve theoretical and methodological integration. While Andersen may be a bit too hasty in dismissing the idea of specialty-incommensurability across specialties, she surely is right when she highlights that, no matter *how* or *when* they have been differentiated, disciplines which are *now* different are not in a state of competition and, therefore, they are not incommensurable.

For Andersen, rather, the disciplines involved in interdisciplinary research are *complementary*, with the scientists coming from different disciplinary backgrounds possessing *interlocking skills*. This view is further developed by Andersen and Wagenknecht (2013), who apply to the intra-disciplinary model of epistemic dependence of Hardwig (1985, 1991), based on the idea of the *joint integration* of what different agents know, to interdisciplinary cooperation.

My suggestion is, once again, to look at interdisciplinarity from the dynamic perspective. In virtue of their historical trajectories, different disciplines may *become* incommensurable when they end up 'converging' towards the same domain.

After all, the whole idea of interdisciplinarity is founded on the belief that *different disciplines* may have something to say about a *common range of complex problems and phenomena*. Once they have become 'about the same range of phenomena', the different disciplines involved in interdisciplinary research may conflict, for example, about the methods for solving a complex problem. In short, interdisciplinarity is not achieved by fiat. Rather, it is the result of a process in which scientists may experience those kinds of cognitive barriers and fundamental disagreements which may go under the label of 'incommensurability'.

Even without excluding a priori the existence of some incommensurability across the specialties involved in interdisciplinarity, the very existence of interdisciplinary research shows that barriers across disciplines can be and are eventually overcome. A philosopher like Andersen, then, could accommodate her views and accept the existence of an 'innocuous' incommensurability across disciplines. If it is residual, minimal and non-threatening, then incommensurability does not seem to pose any interesting problem to interdisciplinary collaborations and, therefore, can be treated as a non-necessary obstacle for understanding of interdisciplinarity. The advantages of not delving too much into incommensurability when talking about interdisciplinarity would be of economic order. The cost of dismissing incommensurability in the context of interdisciplinarity, however, may be an overly simplistic view of the latter.

Andersen's account seems to presuppose the complementarity of different disciplines. The risk of such an assumption is that the possibility of integration could somehow be taken for granted or reduced to an operation almost as simple as a sum of smaller pieces of knowledge. However, in the same way in which incommensurability should not be uncritically presupposed, the idea that different disciplines are 'complementary' and that scientists' expertise is indeed 'interlocking' should not be taken for granted either. Whether scientists coming from different specialties have skills and knowledge which can be (more or less easily) integrated for the resolution of a complex problem can be determined only within the actual context of interdisciplinary problem-solving practice and, often, only after their interdisciplinary collaboration has revealed to be fruitful. It would be more correct to say that scientists involved in interdisciplinary research must learn *how to make* their skills and expertise interlock. This is a process which may involve the breaking of cognitive and methodological barriers, of the same order of those described by Kuhn's (early formulation of) the incommensurability thesis. In other words, from a Kuhnian perspective, interdisciplinarity itself ought to be regarded as a kind of scientific change.

These ideas can be better explained by looking at some recent ethnographical and philosophical studies of interdisciplinarity, which show that putting different specialists at work for the resolution of some common problems is not a straightforward operation. In interdisciplinary practice, scientists must adapt their skills to a set of problems. Problems *as well as* scientists' skills needing to be re-shaped accordingly. For example, in a study conducted on groups working in Integrative System Biology, an emerging field which seeks the integration of computation and engineering modelling with biological experimentation, MacLeod and Nersessian (2016) observe that

> Modellers are not trained as systems biologists, but as a kind of engineer (e.g., electrical or telecommunications) [...]. The importance of their background is not their knowledge of specific engineering systems, but rather their particular approach to problem solving which they reference as the ability to "think systematically" and "debug problems". This enables them to cope with lack of routines and unstructured task environments. Epistemic values that favour precision and exactness that come with handling engineered systems have to be transformed when dealing with "messy" biological systems. (MacLeod and Nersessian 2016: 407)

As MacLeod (2018) explains, there can be deep cognitive causes for the difficulties in achieving interdisciplinarity. One of the consequences of disciplinarity is the development of domain-specific expertise. Because of their domain-specific preparation, scientists from different disciplines may encounter serious difficulties in interdisciplinary interactions. Some practices may be too opaque to the outsiders of a particular domain. Furthermore, scientists coming from different fields and trained to the resolution of different domain-specific problems may experience conflicts in the application of some epistemic values for choosing the best problem-solving approach. Finally, the very ways in which the problems to be solved are defined and framed in an interdisciplinary environment may remain vague for a long time. Another consequence of disciplinarity is a strong sense of *epistemic identity* (Osbeck and Nersessian 2017). On the one hand, epistemic identities shape and influence the cognitive practices of the members of the same discipline. On the other hand, they may create obstacles for the collaboration of scientists coming from different scientific communities.

For Boon and Van Baalen (2019), cognitive-ethnographic studies on actual interdisciplinary research groups challenge the idea that successful interdisciplinarity is a matter of putting together complementary bits of knowledge. In their view, such a conception is based on an epistemology of science which regards scientific knowledge as consisting of propositional knowledge. One of the implications of this epistemological perspective is the conceptualisation of interdisciplinarity in terms of inter-theoretical relations and reductions. However, Boon and Van Baalen continue, if one looks at the scientific practice in general, and at the practice of interdisciplinarity in particular, it will appear that theories, models, concepts, or data by themselves do not tell (a) how they have been produced, and (b) how they are going to be used and applied to solve problems. For Boon and Van Baalen, (a) and (b) are determined by Kuhn's DM. From this rather Kuhnian perspective, therefore, they regard interdisciplinarity as a complex activity involving different components of different DMs, which requires scientists to understand how different disciplinary perspectives lead to different ways of doing research and to use different devices, protocols, and techniques. Boon and Van Baalen do not talk explicitly about incommensurability in interdisciplinarity. However, from their talking about the differences between different DMs it follows that incommensurability in interdisciplinarity goes beyond semantic differences and involves cognitive and methodological dimensions (as supported by the cognitive-ethnographic studies in interdisciplinary research teams).

Three points deserve to be highlighted. First, I am not just assuming the existence of incommensurability. Rather, incommensurability could be considered as a hypothesis that would explain the empirical evidence collected during several empirical studies—such as the 5-years ethnographic study of two Systems Biology labs, conducted by Prof. Nancy Nersessian and her team, which have produced, among other things, the studies cited in the previous paragraphs (see also Nersessian 2022).

Second, in these studies, integration, when and if achieved, appears to exceed the sum of its integrated elements. Far from being an activity in which different pieces of knowledge somehow interlock with one another, interdisciplinarity looks more like a 'transformative' activity, in which the actors involved must modify some of their beliefs, adapt their knowledge and devise new problem-solving methods.

Third, what emerges from these empirical studies is something closer to Kuhn's early formulation of the incommensurability thesis, which describes the disparity not only between theoretical languages, but also between problem-solving methods, values and research agendas. In its early formulation, that is, the incommensurability thesis seemed to capture more of the actual problems scientists encounter in scientific practice. The problems faced by interdisciplinary researchers is that finding a way to communicate across specialties may be necessary but not sufficient for successful problem-solving. Scientists from different disciplines may understand each other's theoretical language and communicate with one another without too many problems. This does not mean, however, that they will all agree about the methods to employ and the problem-solving strategies to implement. This is all the more evident in those cases in which the knowledge and expertise to be integrated is not only theoretical, and in which interdisciplinary collaborations involve theoreticians, on the one hand, and experimentalists, on the other.

Finally, is incommensurability overcome in interdisciplinary practice? Often, though not necessarily always, it is. This does not mean that incommensurability is a trivial or minor problem. By focussing on the cognitive and practical difficulties scientists must overcome in interdisciplinary settings, it is possible to understand interdisciplinarity as a kind of scientific change, that is as a process that requires modifications of concepts, techniques, and methodologies. Unfortunately, Kuhn himself contributed to some of the confusion surrounding the role of incommensurability in both specialisation and interdisciplinarity. In some of his mature writings, Kuhn distinguishes between processes that involve scientific traditions in a 'diachronic relation', that is tradition in temporal succession, and 'synchronic relations' among coexisting scientific relations (see Kuhn 1991, in *RSS*: 96–97). Scientific revolutions, for Kuhn, involve scientific traditions that have the first kind of relation. Talking about specialisation means talking about scientific traditions that are in the second kind of relation: different specialties do co-exist at the same time. This distinction may obscure the fact that, like revolutions, the creation of a new specialty or the establishment of an interdisciplinary line of research are instances of scientific change, in which scientists must construct a new DM or mixing pre-existing theories and methods. Changes, by definition, are processes unfolding in time. Although, from a contemporary perspective, scientific revolutions look like a succession of paradigms while the different specialties look like a collection of coexisting specialties, the scientists involved in revolutions, specialisations, and interdisciplinarity participate in a process. From a dynamic perspective, the controversy between, for example, the supporters of the phlogiston and the oxygen theory during Chemical Revolution and the controversy between the bacteriologist and virologist theorists appear similar. Both controversies were generated by the conflict of incommensurable conceptual systems, and, at the end of the process, one conceptual system was discarded by the other. In the same way, it is possible to see why different disciplines become incommensurable when participating in interdisciplinary collaboration, and how overcoming such incommensurability requires a transformative process. In short, when applied to specialisation and interdisciplinarity (in ways that Kuhn himself did not do), Kuhn's dynamic perspective clarifies how incommensurability enters these instances of scientific change.

8.5 Specialization and Revolutions

Notoriously, towards the end of *SSR*, Kuhn employs a Darwinian metaphor to explain the view of scientific progress that emerges from his history-centred model. In this view, science does not make progress by getting closer to the ultimate truths of nature, but by moving from periods of normal research to new periods of 'better' normal research (i.e., research with higher problem-solving power). In short, in this Darwinian view of science, scientific progress is a process pushed from behind, with scientific revolutions playing a role analogous to *mutations*. In his more mature writings, he proposed another biological analogy to illustrate his view on specialisation. Science makes progress through the proliferation of specialties, each with their own restricted domain which allow them to increase their precision and problem-solving power, and this process is not akin to mutation but, rather, to *speciation*.

Kuhn's early Darwinian metaphor has been criticised by Bird (2000), while the validity of his later metaphors has been challenged by Renzi (2009). Both Bird and Renzi highlights the disanalogies between the progress of science and biological evolution, or even how Kuhn misinterpreted or mischaracterised the latter. Reydon and Hoyningen-Huene (2010) stress that Kuhn was not trying to justify his model by referring to biological analogies: Kuhn never claimed that the process of scientific development and the process of biological evolution are the same process; rather, he used the biological analogies in a didactical and rhetorical fashion, for expository reasons. Therefore, in their view, the various attempts to 'debunk' Kuhn's model by revealing the inadequacies of his biological analogies, are simply misconceived. Other philosophers defend Kuhn's analogy directly from a biological perspective. During the past few decades, biology has relied on new evidence and has formulated new hypotheses. If one takes Kuhn's collective approach and his focus on the scientific community seriously, newer biological theories actually vindicate his model of science (Haufe 2022).

In this section, however, I do not want to discuss whether Kuhn's biological analogies are tenable and to what extent. Rather, I want to discuss how the use of different analogies—the revolutions-as-mutation and the specialisation-as-speciation—has been interpreted as evidence that Kuhn speaks about different kinds of scientific change, and that he shifted from a 'revolutionary' to an 'evolutionary' view on the progress of science.

For Wray (2011), Kuhn's philosophy does indeed describe two distinct kinds of scientific change. On the one hand there are scientific revolutions, which are disruptions with the normal tradition, leading to the abandonment and replacement of an old paradigm. On the other there is specialisation, which is driven by incommensurability but does not seem to possess the same 'destructive' character of revolutions: the creation of a new specialty does not replace its parent-discipline(s) and it does not discard a pre-existing paradigm; it simply adds a new one. The problem, then, is to harmonize the tensions between the revolutionary and the evolutionary conceptions within a general theory of scientific progress. For Wray, these tensions are resolved by looking at scientific development as an evolutionary tree, the branches of which gradually grow (normal science), break (revolutions) and split in sub-branches (specialisation) (Wray 2011: 125, Fig. 3). In short, Wray takes for granted that revolutions and specialization are different in kind, but he does not investigate any further investigation on the potential link between the two types of scientific change. In analysing his theory of specialisation, however, he seems to ignore Kuhn's own hints at a possible connection between revolutions and specialisation.

Kuhn sometimes speaks of revolutions and specialisation as if they were somehow linked.

> After a revolution there are usually (perhaps always) more cognitive specialties or fields of knowledge than there were before. […] [R]evolutions, which produce new divisions between fields in scientific development, are much like episodes of speciation in biological evolution. The biological parallel to revolutionary change is not mutation, as I thought for many years, but speciation. And the problems presented by speciation (e.g., the difficulty in identifying an episode of speciation until sometime after it has occurred, and the impossibility, even then, of dating the time of its occurrence) are very similar to those presented by revolutionary change and by the emergence and individuation of new scientific specialties (Kuhn 1991, in *RSS*: 97–98).

> [T]he episodes that I once described as scientific revolutions are intimately associated with the ones I've […] compared with speciation. […] Though the process of proliferation is often more complex than my reference to speciation suggests, there are regularly more specialties after a revolutionary change than there were before (Kuhn 1992, in *RSS*: 119–120).

My claim is that Kuhnian unified theory of scientific change has always been there all along. To discern it, one should give up the idea that revolutions and specialisations are different kinds of change, related to a revolutionary and an evolutionary view respectively.

As explained in Chap. 4, one of Kuhn's major innovations in philosophy of science is the development of a collective approach to science. This approach takes the scientific communities, rather than individual scientists, to be the loci and agents of science. When it comes to scientific revolutions, the collective approach urges us to ask *for whom* the change is indeed revolutionary. In the same chapter, it was also pointed out that Kuhn speaks about 'high-level' and 'low-level communities'. If revolutions involve communities, the existence of high-level and low-level communities implies the existence of high-level and low-level revolutions; that is, of revolutions that may affect scientific communities of different sizes and levels of generality. In short, when applying Kuhn's collective approach to specialisation—that is, when considering *who* the agents involved in the breaks and splits of the tree of science are—it becomes difficult to distinguish different types of scientific change as Wray does. From a Kuhnian dynamic and collective perspective, in fact, the creation of a new specialty can be described as a low-level scientific revolution, that is, as a revolutionary change driven by and experienced by a low-level scientific community.

Scientists always create a new specialty from within their parent-discipline(s): no new specialty comes into being without the direct involvement of scientists who already practise in pre-existing disciplines. What happens is that, first, these specialists recognize the inability of their discipline to solve some recalcitrant problems. Scientists' dissatisfaction with the methods and proposed solutions of their own discipline looks similar to a perceived sense of *crisis*. It is because they are dissatisfied with some concepts and methods of their parent-discipline(s) that they begin to consider alternative problem-solving approaches. By doing so, they create new concepts and invent new strategies to solve some old problems in new ways, thus entering a period which is not too dissimilar from what Kuhn in *SSR* defines as *extraordinary science*. In the transition toward the emerging specialty, the sub-group of special scientists will focus exclusively on problems arising from a restricted domain, abandoning the concepts and methods of the parent-discipline(s) and replacing them with the new ones. This sort of *loss* is reciprocal: with the emergence of the new specialty, the parent disciplines will also lose a fragment of their old ontology. The sub-group of scientists which has migrated from some pre-existing discipline to the new specialty will inhabit (and will have to adapt to) a new 'niche': they will, in other words, live 'in a different world'. Finally, as discussed in the previous section, the establishment of new specialties is driven by a form of *incommensurability*. Specialisation, in short, follows the same steps and to be characterised by the same elements as a scientific revolution. It must be specified, however, that the fact that a sub-group of specialists is dissatisfied with how their discipline deals with some problems arising from a restricted area does not mean that the *whole* pre-existing discipline is in a state of crisis and must be replaced in its totality.

To use one of Wray's own examples, historians like van Helvoort explicitly use the Kuhnian model of scientific revolutions to describe the emergence of virology (van Helvoort 1991, 1992, 1993, 1994). To begin with, it is not entirely clear *when* or by whom viruses were discovered. Although the term 'virus' was used, in a very loose sense, in the context of the medical sciences of the Nineteenth century, for a long time it was believed that viruses were just special types of bacteria. That viruses could be a different kind of entity was first claimed by Martinus Beijerinck, a scientist researching the causes of the so-called 'mosaic disease' affecting plants and flowers. This rather common plant disease had been studied for centuries, but its pathogen agents were isolated only in 1895 by Beijerinck, who maintained that the cause of the infection was not bacteriological. His claims generated priority disputes. On the one hand, Adolph Mayer's earlier works on infected tobacco leaves influenced Beijerinck's research, to the point that it may be possible to regard Mayer as the first scientist who observed the new entity. On the other hand, a few years after Beijerinck's publication, the Russian bacteriologist Dmitrii Ivanovsky claimed that he had discovered the viruses independently.

More importantly, the establishment of the concept of virus as a separate kind of entity was met with resistance, if not controversies. Not every scientist was persuaded that Beijerinck, Ivanovsky or others had discovered a new natural kind. Quite the opposite: for a long time, the scientific community of bacteriologists tended to regard viruses as a sub-kind of bacteria. Not only the discovery of virus appears to have the same complicated historical structure described by Kuhn (1962), but the establishment of virology can be regarded as the branching from some bacteriologists, who undergo a change of paradigm, while the rest of the bacteriology community 'loses' a part of its domain. For the bacteriologists turned virologists, the establishment of virology as a separate discipline was, indeed, a revolutionary change.

The history of virology deserve a further comment. During the establishment of the new discipline, the emerging group of virological theorists was split between those, like Beijerinck, who believed viruses were liquid entities, and those who thought they had a different structure; and, furthermore, between those who thought viruses belonged to the living realm and those who regarded them as non-living. Even now that virology is an established discipline, disputes about the correct definition and classification of viruses persist, to the point that some argue that the idea of a monistic virological taxonomy should be given up in favour of a pluralistic classification (Morgan 2016). In proposing and endorsing the new 'virological paradigm', virologists did not need to reach a consensus about what a virus is. As explained in Chap. 4, all it takes for scientists to accept and work on a paradigm is the lack of an explicit dissensus, rather than an explicit consensus of every detailed aspect of the paradigm. Second, the history of virology, with scientists deciding to create a new discipline about an entity they don't even know how to unanimously classify, exemplifies Kuhn's late theory of differentiae, explained in Sect. 6.4. No matter what they really are and how they are defined, all virologists needed to establish their new discipline was classifying viruses as *different from* bacteria.

My view of specialisation as a special instance of scientific revolutions is consistent with Kuhn's dynamic perspective and collective approach. If one takes Kuhn's philosophical perspective seriously, then one should look at specialisation from such a perspective. By contrast, Wray oscillates from an appreciation of Kuhn's philosophical perspective to a rather theory-centred and conservative interpretation of his model. On the one hand, Wray is deeply aware of the centrality of the concept of a scientific community in Kuhn's philosophy—which he even describes as a form of 'social epistemology'. On the other hand, however, his account of specialisation, with its focus on the conceptual differences between different kinds of entities, does not seem to pay attention to the social dimension of the creation of a new specialty. More importantly, Wray seems to forget that scientific language is always the language of a scientific community, which is akin to a linguistic community. To analyse changes in the meanings and conceptual structure of such a language without considering who are the 'speakers' who make such changes and are affected by them, is to tell only half of a Kuhnian story. Furthermore, to look at the branches of the tree of scientific knowledge as breaking during a revolution and as splitting through specialisation requires an external bird-eye view—or, if we want, an 'Archimedean platform'—on science. For the scientists involved, however, a splitting of their branch *is* indeed a break.

And finally, is it even true that Kuhn had two different views on scientific progress, one 'revolutionary' and another 'evolutionary'? I cannot see why scientific revolutions must be considered as conflicting with an evolutionary philosophy of science. In the same way in which biological species evolve through mutations, so science evolves and makes progress through revolutions (with specialisation being a special, lower-level case of revolutionary changes).

8.6 The Plurality of Worlds

So far, then, I argued that the process of specialisation is a special instance of the process of scientific revolutions. Could it be that the process of scientific revolutions is actually a special instance of the kind of scientific change that leads to specialisation?

An important issue that Kuhn overlooked is that, sometimes, the creation of a new discipline may have some important consequences for the pre-existing disciplines and for the scientists who do *not* migrate. For example, the concept of 'gene' as developed by old genetics still survives in some scientific contexts and remains crucial for more applied research. It is fair to say, however, that old genetics as such did not survive as a research field in its own right or, at least, that it underwent some profound modification after 'losing' some of its concepts to molecular biology. This point shows that the creation of a new discipline is not a simply additive process in which a new sub-branch is incorporated to some pre-existing trunk. Rather, a new discipline may *supersede* some of the existing branches. Science, therefore, may not develop by simple 'proliferation' of new disciplines; rather, it appears to grow through fragmentation *and* dissolution: in some cases, though not necessarily every time, the creation of a new discipline does not change just the number of the branches of the tree of science, but the very structure of the tree.

If this is the case, then low-level revolutions, which occur with the creation of a new discipline, may actually have some large-scale effects. To understand this point, instead of just considering the creation of a new specialty as akin to scientific revolutions, I propose to also consider scientific revolutions as akin to the process of specialisation, in the loose sense of 'creation of new ontological domains', each studied by different communities.[8]

As already discussed, Wray (2011) believes that revolutions and the creation of new specialties are two different kinds of scientific change because, in the first case, the new paradigm completely supersedes the old one whereas, in the second case, there is not such a total replacement. If we look at the history of physics of the XX century, however, things do not look as neat. Kuhn considers the transition from classical mechanics to Einstein's relativity as a genuine case of revolutionary scientific change. However, many of his critics, starting from Shapere (1980), have contested the idea that Einstein's theory was 'revolutionary', in the sense of causing an epistemic rupture in physics. This line of criticism argues that some of the classical laws can be considered as 'limiting cases' of Einstein's relativity and, therefore, Newton's theory can be 'derived' from Einstein's. This would prove that the transition from the two does not require any break or rupture. This argument can be challenged. The fact that some classical laws can be shown to work in some 'limiting cases' of Einstein's does not mean that they are also some of their 'logical consequences', as argued by Nickles (1973).[9] Furthermore, it is not entirely clear why the fact that some of Newton's laws can be regarded as limiting cases of relativity theory implies that the whole Newtonian view is continuous, reducible, or even reduced to Einstein's.[10]

My main point, however, is not whether there is continuity or discontinuity between concepts, laws and theoretical representations. My point is that, as suggested in the previous chapter, to understand when a scientific revolution occurs, one has to look at the structure of the scientific community first and then reconstruct how many components of its DM have changed. In the case of the classical-mechanics-to-relativity shift, terrestrial and planetary gravitation are regarded as the same problem within the community of classical mechanics but, with the acceptance of Einstein's theory, planetary gravitation became a problem to be solved on its own. In other words, Einstein brought a revolution in (the community of) physics

[8] Indeed, communities and their ontological domains cannot be separated, in the same way as a species is inseparable from its niche. See also Sect. 9.5.

[9] "Reductions in the tradition of Newton and Einstein do not primarily involve the consolidation of concurrent theories, nor are the central issues of these reductions ontological. It is not as if we had both STR [Special Theory of Relativity] and CM [Classical Mechanics] and suddenly realised that we did not need CM anymore, or as if we discovered that we do not need to posit both Einsteinian masses and Newtonian masses. Rather, STR came to be entertained specifically as a replacement for CM, and Einstein's theory could not have succeeded without simultaneously making its predecessor fail" (Nickles 1973: 183).

[10] On this point, I agree with Worral (1989), when he observes that "there is much more to Newton's theory than the laws of motion and the principle of universal gravitation considered simply as mathematical equations. These equations were interpreted within a set of very general theoretical assumptions which involved amongst other things the assumption that space is infinite, that time is absolute so that two events simultaneous for one observer are simultaneous for all, and that the inertial mass of a body is constant. Einstein's theory entails on the contrary that space is finite (though unbounded), that time is not absolute in the Newtonian sense and that the mass of a body increases with its velocity" (Worrall 1989: 109).

because, although it is possible to show some logical continuity between classical mechanics and the theory of relativity, in practice after Einstein the problem of planetary gravitation began to be tackled by a restricted community of scientists, while the rest of the community of physics kept on using the paradigm of classical mechanics. A similar point can be made about the emergence of quantum mechanics.

What happened in the past century, then, is that the old Newtonian paradigm was not superseded by *one* new paradigm. Rather, the ontological domain of classical mechanics got fragmented and 'lost' some of its parts to at least *two* new fields, namely Einstein's relativity and quantum mechanics, investigated by two different sub-communities. This ontological fragmentation corresponds to the creation of a 'plurality of worlds', as Kuhn himself would say. Einstein's lexicon, constitutive of the 'Einsteinian world', is not the same as Newton's and the quantum lexicon, constitutive of their respective worlds. Of course, philosophers inclined to formal and logical analyses may respond that Newtonian mechanics can be derived from Einstein's relativity as a 'limiting case'. Such a reply has a rather logical empiricist flavour that, of course, is incompatible with Kuhn's dynamic perspective.

Furthermore, it is not even the case that classical mechanics was *completely* replaced by relativity and quantum mechanics. The *theory* of classical mechanics is not regarded as providing a true representation of reality any longer. However, as a 'system of practice', classical mechanics indeed persists, *alongside* the Einsteinian and the quantum 'worlds'. All and all, the scientific revolutions of the physics of the xx century resembles the process of specialisation described in this chapter.[11] And yet, at the same time, Einstein's relativity and quantum mechanics are competing for the explanation of the fundamental laws of nature. Future developments in physics may lead one of the two fields to supersede the competitor, or perhaps the emergence of a 'synthesis' between them will lead to the emergence of a new paradigm. The point here is that the fragmentation of the ontological domain of classical mechanics led to the emergence of new worlds, but these new worlds, instead of being simply added to what was practised before, may eventually 'cannibalise' the pre-existing world from which they came from.

Was this what Kuhn had in mind? Apart from the futility of asking what Kuhn 'really' had in mind, one cannot help but notice that, in several of his late writings, Kuhn seems to flirt with similar ideas. He observes, for example, that one of the results of the process of specialisation is that "[t]he older, more encompassing modes of practice simply die off: they are the fossils whose palaeontologists are historians of science" (Kuhn 1992, in *RSS*: 120). From this perspective, therefore, revolutions stop to be exclusively intra-disciplinary one-to-one paradigm replacements to become something affecting (and creating) a plurality of worlds with their respective communities.

---

[11] Obviously, Einstein's relativity and quantum mechanics are not 'specialties' of classical mechanics, but neither is virology a 'specialty' of bacteriology. Specialisation in this context means creation of new ontological domains, therefore akin to scientific revolutions (in the same way in which some revolutions may be akin to the process of specialisation). See above, ft. 2.

8.7 Conclusions

If one accepts Kuhn's dynamic perspective—that is, if one accepts that there is no absolute and neutral 'Archimedean platform' for the evaluation of scientific claims, that such an evaluation involves different and competing belief systems, and that scientific rationality is not provided by universal rules but, rather, it is embedded in the practice of scientific communities—then one must also look at the phenomenon of scientific specialisation from such a perspective. From the dynamic perspective, then, it appears that the creation of a new scientific specialty involves the emergence of a sub-community that, for a time, will be competing with the rest of the community for the explanation of the same phenomena. During this time, and lacking a neutral ground for comparison, the belief systems of the two competing communities will be incommensurable. The establishment of the new specialty will be resolved following a process akin to that of scientific revolutions.

Ultimately, embracing the dynamic perspective allows to look at revolutions, specialization, and interdisciplinarity as changes involving scientific communities experiencing incommensurability. The emergence of a new speciality or of an interdisciplinary group, as well as the ontological fragmentation following some scientific revolutions, result in the 'plurality of scientific worlds' Kuhn hinted at in his last writings.

References

Albert, David. 2000. *Time and Chance*. Cambridge: Harvard University Press.

Andersen, Hanne. 2012. Conceptual development in interdisciplinary research. In *Scientific Concepts and Investigative Practice*, ed. Uljana Feest and Friedrich Steinle, 271–292. Berlin: Walter de Gruyter and Co.

Andersen, Hanne. 2013a. Bridging disciplines. Conceptual development in interdisciplinary groups. In *Classification and Evolution in Biology, Linguistics and the History of Science*, ed. Heiner Frangerau, Hans Geisler, Thorsten Halling, and William Martin, 33–44. Stuttgart: Franz Steiner Verlag.

Andersen, Hanne. 2013b. The Second Essential Tension: On tradition and innovation in interdisciplinary research. *Topoi* 32:3–8.

Andersen, Hanne. 2016. Collaboration, interdisciplinarity and the epistemology of contemporary science. *Studies in History and Philosophy of Science* 56:1–10.

Andersen, Hanne, and Susan Wagenknecht. 2013. Epistemic dependence in interdisciplinary groups. *Synthese* 190:1881–1898.

Bechtel, William. 1986. Biochemistry: A cross-disciplinary endeavour that discovered a distinctive domain. In *Integrating Scientific Disciplines*, ed. William Bechtel, 77–100. Dordrecht: Springer.

Boon, Mieke, and Susan Van Baalen. 2019. Epistemology for interdisciplinary research – shifting philosophical paradigms of science. *European Journal for Philosophy of Science* 9:16. https://doi.org/10.1007/s13194-018-0242-4.

Darden, Lindley. 1978. Discoveries and the emergence of new fields in science. *PSA: Proceedings of the Biennial Meeting of the Philosophy of Science Association* 1:149–160.

Dawid, Richard. 2009. On the conflicting assessments of the current status of string theory. *Philosophy of Science* 76:984–996.

Dawid, Richard. 2013. *String Theory and the Scientific Method.* New York: Cambridge University Press.

De Benedetto, Matteo, and Michele Luchetti. 2023. Specialisation by value divergence: The role of epistemic values in the branching of scientific disciplines. *International Studies in the Philosophy of Science* 36:121–141.

Ereshefsky, Marc. 1998. Species pluralism and anti-realism. *Philosophy of Science* 65:103–120.

Frigg, Roman. 2008. A field guide to recent work on the foundations of statistical mechanics. In *The Ashgate Companion to Contemporary Philosophy of Physics*, ed. Dean Rickles, 99–195. Hampshire: Ashgate Publishing Limited.

Gefaell, Juan, and Cristian Saborido. 2022 Incommensurability and the Extended Evolutionary Synthesis: taking Kuhn seriously. *European Journal for Philosophy of Science* 12: 24, https://doi.org/10.1007/s13194-022-00456-y

Hardwig, John. 1985. Epistemic dependence. *Journal of Philosophy* 82:335–349.

Hardwig, John. 1991. The role of trust in knowledge. *Journal of Philosophy* 88:693–708.

Haufe, Chris. 2022. *How Knowledge Grows: The Evolutionary Development of Scientific Practice.* Cambridge: MIT Press.

Hooker, Cliff, ed. 2011. *Philosophy of Complex Systems.* Oxford: North Holland.

Kohler, Robert. 1971. The background to Eduard Buchner's discovery of cell-free fermentation. *Journal of the History of Biology* 4:35–61.

Kohler, Robert. 1972. The reception of Eduard Buchner's discovery of cell-free fermentation. *Journal of the History of Biology* 5:327–353.

Kohler, Robert. 1973. The enzyme theory and the origin of biochemistry. *Isis* 64:181–196.

Kohler, Robert. 1975. The history of biochemistry: A survey. *Journal of the History of Biology* 8:275–318.

Kuhn, Thomas S. 1962. Historical structure of scientific discovery. Science, 136:760–764. Reprinted in *ET*: 165–179.

Kuhn, Thomas S. 1991. The Road since Structure. *PSA: Proceedings of the Biennial Meeting of the Philosophy of Science Association 1990*, volume 2: 2–13. Reprinted in *RSS*: 90-104.

Kuhn, Thomas S. 1992. The Trouble with the Historical Philosophy of Science. *Occasional Publication of the Department of History of Science*, Cambridge: Harvard University. Reprinted in *RSS*: 105-120.

Kuhn, Thomas S. 1993. Afterwords. In *Word Changes*, ed. Paul Horwich, 311–341. Cambridge: MIT Press. Reprinted in *RSS*: 224-252.

Ladyman, James, James Lambert, and Karoline Wiesner. 2013. What is a complex system? *European Journal for Philosophy of Science* 3:33–67.

MacLeod, Miles. 2018. What makes interdisciplinarity difficult? Some consequences of domain specificity in interdisciplinary practice. *Synthese* 195:697–720.

MacLeod, Miles, and Nancy Nersessian. 2016. Interdisciplinary problem-solving: Emerging modes in integrative systems biology. *European Journal for Philosophy of Science* 6:401–418.

Mitchell, Melanie. 2009. *Complexity. A guided tour.* Oxford: Oxford University Press.

Morgan, Gregory. 2016. What is a virus species? Radical pluralism in viral taxonomy. *Studies in History and Philosophy of Biological and Biomedical Sciences* 59: 64–70.

Nersessian, Nancy. 2022. *Interdisciplinarity in the Making*. Cambridge: MIT Press.

Nickles, Thomas. 1973. Two concepts of intertheoretic reduction. *The Journal of Philosophy* 70:181–201.

Osbeck, Lisa, and Nancy Nersessian. 2017. Epistemic identities in interdisciplinary science. *Perspectives on Science* 25:226–260.

Renzi, Barbara. 2009. Kuhn's evolutionary analogy and its being undermined by inadequate biological concepts. *Philosophy of Science* 76:143–159.

Reydon, Thomas, and Paul Hoyningen-Huene. 2010. Kuhn's evolutionary analogy in *The Structure of Scientific Revolutions* and *The Road since Structure*. *Philosophy of Science* 77: 468–476.

Shapere, Dudley. 1974. Scientific theories and their domains. In *The Structure of Scientific Theories*, ed. Frederik Suppe, 518–565. Urbana: University of Illinois Press.
Shapere, Dudley. 1980. The Structure of Scientific Revolutions. *The Philosophical Review* 73:383–394.
Sklar, Lawrence. 1993. *Physics and Chance. Philosophical Issues in the Foundations of Statistical Mechanics*. Cambridge: Cambridge University Press.
van Helvoort, Ton. 1991. What is a virus? The case of tobacco mosaic disease. *Studies in History and Philosophy of Science* 22: 577–588.
van Helvoort, Ton. 1992. Bacteriological and physiological research styles in the early controversy on the nature of the bacteriophage phenomenon. *Medical History* 36:243–270.
van Helvoort, Ton. 1993. A bacteriological paradigm in influenza research in the first half of the twentieth century. *History and Philosophy of the Life Sciences* 15:3–21.
van Helvoort, Ton. 1994. History of virus research in the twentieth century: the problem of conceptual continuity. *History of Science* 32:185–235.
Worrall, John. 1989. Structural Realism: The Best of Both Worlds? *Dialectica* 43:99–124.
Wray, Brad K. 2011. *Kuhn's Evolutionary Social Epistemology*. Cambridge: Cambridge University Press.

Chapter 9
Kuhn and the Value-Free Ideal

Abstract Now that the Kuhnian scholarship has clarified that Kuhn's philosophy does not support externalist, constructionist, or even irrationalist images of science, a new worry is starting to loom large. Some argue, in fact, that Kuhn may be read as a supporter, or even a promoter, of the so-called Value-Free Ideal (VFI), which many contemporary philosophers of science find untenable. Other philosophers, however, interpret Kuhn in the opposite direction, that is as a supporting the idea that non-epistemic values are necessary to science. This chapter aims at clarifying the place of Kuhn's philosophy within the contemporary debate about values in science.

After defining VFI, I will discuss the 'Value-Free Kuhn' interpretations which, as I will argue, are based on some misunderstandings about some fundamental elements of Kuhn's philosophy. Furthermore, I will argue that the issue of values in science must be disentangled from the issue of the political autonomy of science, and that Kuhn's approach shows how this is so. Finally, I suggest how Kuhn's philosophy may help us reconsider some aspects of the debate on values in science, as well as some issues concerning the science/society relationship.

Keywords Value-free ideal · Value incommensurability · Inductive risk · Autonomy · Scientific community · Science and society

9.1 Introduction

For a long time, Kuhn was regarded as a philosopher who conceived of science as being porous to numerous social and political factors. This reading was largely based on *CR*, where Kuhn included some mystical and religious beliefs, as well as some social and political factors (such as, for example, the reform of the calendar), as driving the transition to Copernican cosmology. This reading led to a constructionist interpretation of his philosophy, championed by sociologists and STS scholars. In reality, Kuhn was repulsed by social constructionism (see Sect. 2.3). He

V. Politi, *Thomas Kuhn's Philosophical Perspective*, Boston Studies in the Philosophy and History of Science 348,
https://doi.org/10.1007/978-3-032-13540-7_9

regarded himself as an internist: he did believe that some social or political factors may 'accelerate' some decisions in science but not determine their outcome or substitute scientific rationality. Ultimately, reading Kuhn as an externalist only based on *CR*, a book published in the 1950s, means to either ignore or dismiss the vastness of his work and the development of his thought. As pointed out by Melogno (2021), already with *SSR* Kuhn abandons the externalist historiography that characterised *CR* to focus on internalist explanations of science only. Apart from some textbook oversimplifications or plain misreadings, by now Kuhn's internalist approach can be considered as a settled issue, at least among Kuhnian scholars.

What is not settled yet, however, is whether Kuhn was a strenuous supporter of the old Value-Free Ideal (VFI) or an early proponent of a value-laden conception. Why? Because he has been interpreted as *both*. On the one hand, some argue that Kuhn may be read as a supporter, or even a promoter, of VFI. Moreover, different philosophers develop different arguments about Kuhn's relations with VFI. Lacey, himself a supporter of VFI, claims that Kuhn supports VFI, but not in the right way. Douglas, a fierce critic of VFI, claims that Kuhn supports VFI, and that it is not right. In the following two sections I will examine their arguments respectively. On the other hand, philosophers like Longino develops their 'value-laden' conception of science by making explicit and deepening some of Kuhn's insights. In this chapter, I argue that the 'Value-Free Kuhn' interpretations misconstrue some crucial aspects of Kuhn's philosophy.

In Sect. 9.2, I define VFI, and I briefly discuss some of the philosophical arguments against it. In Sects. 9.3 and 9.4, I discuss Hugh Lacey's and Heather Douglas' interpretation of Kuhn as a supporter of VFI, criticizing them both. Finally, in Sect. 9.5 I argue that looking into Kuhn's views on values in science may bring a new perspective to contemporary debates.

9.2 What Is VFI?

The idea that scientific inquiry must be free from the influence of social and political values has been held by philosophers and scientists throughout the history of science, although often implicitly. Henry Poincaré, in one of the few explicit formulations of this idea, stated that "Ethics and science have their own domains, which touch but do not interpenetrate. The one shows us to what goal we should aspire, the other, given the goal, teaches us how to attain it. So they never conflict since they never meet" (Poincaré 1920 [1958]: 12).

The separation between science, regarded as the domain of objectivity, and ethics, regarded as a domain populated by subjective tastes and judgements, is at the basis of the so-called Value-Free Ideal (VFI). Even after granting that the two domains are indeed separated, one still needs to explain whether and how they may enter in contact with each other. Or, to use Poincaré' terminology, how they may touch without interpenetrating. In its contemporary formulation, VFI states that *non-epistemic values* must play no role in the *internal stage* of scientific research.

This formulation presupposes two important distinctions, namely that between 'epistemic' and 'non-epistemic values', and that between the 'internal' and the 'external stage' of science.

VFI does not prohibit every kind of value in the internal stage of science. Since evidence alone does not determine the truth of a hypothesis, philosophers of science (including Kuhn, see Sect. 4.2) argue that many decisions in science require the application of so-called epistemic values, such as simplicity, coherence, fruitfulness, and so on (see Kuhn 1977; Laudan 1984; McMullin 1993). These values drive science towards its epistemic ends and are accepted as supporting scientists in drawing objective conclusions. By contrast, so-called non-epistemic values, such as moral, social, or political values, are regarded as compromising or distorting the methods and results of science in favour of subjective preferences or even wishful thinking. VFI forbids these values from the internal stage of science. Strictly speaking, VFI amounts to a 'Non-epistemic Value-Free Ideal', although it is customary to use a simpler formulation.[1]

Moving on to the second distinction, what philosophers mean by the 'internal stage' of science includes those phases in which scientists make inferences and draw conclusions; for instance, when they must accept or reject a hypothesis on the face of the available evidence. The so-called 'external stage' of science, by contrast, includes all those phases which are not related to the epistemic content of science. For instance, the setting of research priorities tells scientists what they have to investigate, but it does not determine which hypothesis of explanation they will end up accepting. Similarly, it is possible to organise in different ways the communication and deployment of the results of scientific research, yet communication and deployment of results does not alter scientists' conclusions.

Having clarified the two distinctions presupposed by VFI, it is possible to refine its definition. VFI does not deny that epistemic values play a role, and sometimes a crucial role, in the internal stages of science. In fact, it does not forbid scientists to recur to values such as simplicity, fruitfulness, and so on, when they must choose between two empirically equivalent theories. Moreover, VFI does not deny that non-epistemic values play a role, and sometimes even a crucial role, in the external stages of science, such as during the setting up research priorities and agenda. All VFI states is that non-epistemic values must be kept out of the epistemic phases of science, such as when scientists have to appraise the appraise the strength of a hypothesis in the face of the available evidence.

[1] Talking about 'epistemic' and 'non-epistemic values' is also a simplification. Different philosophers have defined in different ways the kinds of values that may or may not enter the various stages of a scientific enquiry. Some of them differentiate between 'cognitive' and 'non-cognitive values' (Longino 1996), or between 'constitutive' and 'contextual values' (Longino 1990, 2002), or between 'broad', 'cognitive', and 'probative schemes of values' (Kitcher 2011, 2024). For the sake of brevity, here I characterise VFI as concerned with 'epistemic' and 'non-epistemic values', with the previous being defined as those traditionally accepted as good scientific standards and believed to be conducive to objective knowledge, and the latter being traditionally associated with moral and social standpoints, as well as with rather subjective stances.

Defenders of VFI regard freedom from non-epistemic values as the mark of the epistemic integrity of science. It is in virtue of such value-freedom that science is capable of providing objective knowledge and to represent the main remedy to irrational beliefs and wishful thinking. As such, value-freedom also grants science its epistemic authority: scientists, being free from the potential distortions and manipulations of social, political or religious ideologies, are the ultimate authorities in the explanation of the facts of the world. The epistemic integrity and authority of science, in turn, are the source of the public trust in science. Finally, it should not be forgotten that VFI is a normative ideal. Even if it was the case that many or even most scientists were indeed not free at all from non-epistemic values, that would be no ground for rejecting VFI, which tells how science ought to proceed, not how it actually proceeds.

There are different ways for criticising VFI. For instance, some philosophers have questioned the very distinction between epistemic and non-epistemic values (see, for example, Rooney 1992). If the two kinds of values are not so easily distinguishable, then VFI is put on rather shaky grounds. Other philosophers have questioned the distinction between the internal and external stages of science, by arguing that decisions made in the agenda setting stage may have an impact on various methodological choices, thus impacting the internal stage (Elliott and McKaughan 2014). However, most philosophers of science still find some value in maintaining the epistemic/non-epistemic distinction and the internal/external distinction, even if it should turn out that they are not always as sharp or easy to draw. Here I will follow this trend, and I will simply presuppose the validity of these distinctions without further specifications.

There are some arguments that challenge VFI without questioning the distinctions it relies on. These arguments, in other words, aim at showing why non-epistemic values are necessary to the internal stages of the scientific inquiry, and why they would not compromise scientific objectivity. For Longino (1990), theory choice does not happen in a vacuum, but it is always made in the light of 'background assumptions', which include both epistemic and non-epistemic values. Epistemic values are regarded as the necessary (or 'constitutive') features of science. Kuhn would agree with this characterisation: indeed, he himself maintained that theories, to be considered scientific, must possess some values, and that such values remain pretty much unaltered throughout the history of science. However, as explained in previous chapters, Kuhn also maintained that, though agreeing on what such values are, different individual scientists may weigh and apply them in different ways during theory choice. This is why Longino considers the role of context- and problem-sensitive non-epistemic values: they help scientists apply the constitutive/epistemic values, and therefore make decisions, in specific contexts and towards the resolution of specific problems. Therefore, contextual/non-epistemic values play a central role in scientific rationality. Furthermore, for Longino, such values do not put the idea of scientific objectivity in peril. In her view, objectivity is not achieved through a progressive detachment that would eventually lead to an ideal 'view from no-where'. Rather, objectivity is the result of intersubjective debate, which allows the exploration and mutual criticism of scientists' individual stances.

The so-called 'argument from inductive risk' (AIR) states that epistemic values are necessary to counterbalance the risk of error in scientific appraisal (Churchman 1948; Rudner 1953; Douglas 2009). It may happen, for example, that during theory choice scientists end up accepting a false theory or rejecting a valid one. Of course, one does not expect scientists to be infallible. After all, many scientific results are achieved through trial and error. What one should expect, however, is that scientists act responsibly. Errors in hypothesis appraisal may have undesirable or even harmful consequences. Scientists, like everybody else, must prevent harm; hence, they must mitigate the potential risk of errors. This cannot be done in a purely quantitative and algorithmic way. A 5% chance that the accepted theory is wrong can be either an acceptable or a worrying margin of error, depending on what the potential consequences of such an error would be. For example, the 5% probability that a new drug will cause a mild headache would not be evaluated in the same way as the 5% probability that a new drug will cause irreversible coma. Non-epistemic values, therefore, help scientists in setting the criteria for scientific appraisal in the face of the potential risks of error.

Both Longino and AIR show how non-epistemic values play a necessary role in the internal stage of science. Supporters of VFI may respond in different ways. For instance, they may argue that scientists do not have the duty of making value-judgements to mitigate the risk of error, but only to communicate uncertainties to regulators and policy makers, who will then make the choice (Levi 1960; Betz 2013; Henschen 2021; Carrier 2022). To this argument, however, one may respond by saying that, especially in innovative research, the distinction between 'scientists', 'advisers', and 'policy makers' is blurred.

The debate about VFI is one of the most lively in contemporary philosophy of science. On the one hand, supporters of a more 'value-laden' conception of science face the 'new demarcation problem' (Holman and Wilholt 2022), that is the challenge of finding a principled way to distinguish the non-epistemic values that play a legitimate role in science from those that, by contrast, would have a corrupting effect. On the other hand, recently some philosophers have revitalised VFI. Some argue that, although legitimate, the role of non-epistemic values in science ought to be minimised (Stamenkovic 2024). Others maintain that, even considering the arguments about its untenability and undesirability, VFI is still a worthy 'regulative' ideal (Menon and Stegenga 2023). Counterarguments to these and similar arguments have already been developed (Brown 2024). All and all, though most philosophers seem to endorse a value-laden conception of science, the debate between supporters and critics of VFI is not entirely settled yet.

9.3 Hugh Lacey on Kuhn

9.3.1 Lacey's Formulation of VFI

Hugh Lacey's book *Is Science Value Free?* (1999) develops the contemporary formulation of VFI. As explained in the previous section, the general idea that science and ethics belong to separate domains has been held by philosophers and scientists throughout history. Following Lacey, this idea relies on three views, namely: *neutrality*, *impartiality*, and *autonomy*. Neutrality is the product of Galileo's metaphysical divorce between 'facts of Nature' and human wishes and desires. Holding the view of science as neutral corresponds to accepting that its aim is to represent and explain pure facts, independently of how such knowledge will bear on human values and activities. More specifically, Lacey distinguishes three dimensions of neutrality. The first dimension tells that scientific theories do not have value judgements among their logical implications. The second dimension, which follows on the first one, tells that accepting a theory has no cognitive consequences on one's own values. Finally, the third dimension tells that neutral scientific knowledge is available to everybody to be applied and used in projects and activities that may be value laden. This third dimension corresponds to the traditional separation between the production of pure scientific knowledge and its applications.

The source of impartiality is Bacon's view on scientific method, and it has an epistemological and methodological character. Sound scientific knowledge is impartial if it is obtained through observation, experimentation, replication, and agreement, independently of our wishes, desires, and perspectives. The scientific method is trans-cultural, available to everyone and everywhere; this means that scientific claims may be produced and tested in different social contexts. For a long time now, philosophers of science have contested the naive conception of an absolute, trans-historical, and trans-cultural scientific method shared by all the proper sciences. The contemporary version of the idea of impartiality, therefore, holds that scientific inferences are made in accordance with the available evidence, some formal rules, and the shared epistemic values, with no space for non-epistemic values.

Autonomy is a political thesis about the relationship between science and society. Such a relationship is regulated by a 'social contract', that establishes that the scientific community is free to pursue the truths of nature, of which it remains the sole and uncontested authority, while the responsibility for the application and deployment of its results is left to other societal actors, being them politicians, policymakers, and so on. At the same time, being autonomous, the scientific community dictates the rules for regulating its own internal dynamics and which may take the form of a special 'ethos', like that theorised by Merton (1973).

Neutrality and impartiality are about the content of scientific knowledge. The Galilean idea of neutrality is at the basis of the conception of scientific objectivity as detachment, which in turn needs Bacon's methodological impartiality to be achieved. Autonomy, from its part, protects scientists from social, political, and ideological intrusions and manipulations; as such, it guarantees impartiality.

Having clarified the core of the traditional version of VFI implicitly supported by many throughout history, Lacey proposes his own version, that consists in rejecting autonomy and redefining neutrality, while defending impartiality. His conclusion pretty much corresponds to the contemporary formulation of VFI, for which the internal, epistemic, and inferential phases of science are value-free—that is, impartial—while non-epistemic values may play important roles in the external stage. This means that, one the one hand, Lacey recognises that scientific knowledge may have an impact on, and ought to respect, moral and social values (hence his modification of neutrality) and, on the other hand, that the social and political context may play a role in the setting up of research priorities (hence his rejection of autonomy). To clarify his position, Lacey uses Kuhn—who, in his view, also supports VFI—as a foil (see Lacey 1999: Chap. 7).

9.3.2 Lacey on Kuhn

Lacey begins to examine Kuhn's relation with VFI by discussing the notoriously controversial claim for which "though the world does not change with a change of paradigms, the scientist afterwards works in a different world" (*SSR*: 121). For Lacey, the controversy dissolves once one understands the difference between 'the world' and 'worlds'. The (unchangeable) world is the totality of fundamental causal principles and ultimate constituents of reality; or, to follow Hoyningen-Huene's (1993) Kantian interpretation, the 'world-in-itself'. The (changeable) worlds are also self-contained totalities, but they do not exist independently of the perceptions, practices, meanings, and self-understanding of its inhabitants: they are Kantian phenomenal worlds. The sole and only (unchangeable) world is accessible only through 'a world'.

From the distinction between 'the world' and 'worlds', Lacey concludes that Kuhn was right when he defined scientific revolutions as instances of world change. For example, the emergence of the Aristotelian and the Galilean paradigm—or, to use Lacey's jargon, of 'research strategy' A and G—depended on the different worlds of practices, meanings, and articulations within which they were formulated. A and G were not only different, but also incommensurable, insofar as they competed for the explanation of a sub-set of phenomena that overlapped between the two. As it is well known, Kuhn has the burden to explain how the transition between two incommensurable worlds is not only possible but also rational.

In Lacey's account, for Kuhn the A-to-G transition was driven by one epistemic value, namely 'fruitfulness'. Scientists of the time realised that G was more fruitful than A in the sense that it solved more problems and better, and it generated more new hypotheses and theories. It is however crucial to stress that G displayed a higher degree of fruitfulness than A *only* in relation to a sub-set of shared problems.

> Let D*a* and D*g* be the domains of phenomena of which A and G respectively manifest the cognitive values to a reasonably high degree. They only overlap; neither is contained in the other (so "comparative comprehensiveness" does not help to settle the dispute between A

> and G). Let D be the domain of overlap. The movements of the planets, moon and sun belong to D. Furthermore, data concerning these phenomena, observational reports of the varying angular positions, brightness and other appearances of these bodies as obtained with naked-eye vision are selected to fit theories developed under both strategies. Attending to significant areas of incommensurability between A and G should not obscure that these theories are expected to manifest the cognitive values highly with respect to *some* of the same data. (Lacey 1999: 166–167)

A and G are articulated, interpreted, and applied in incommensurable worlds, so that a direct comparison between the two is impossible. When it comes to planetary motion, however, it is possible to see which one provided better explanations and solved more problems. From its success in a restricted sub-set of shared problems, scientists' confidence in G increased while their confidence A decreased. They therefore choose to adopt G, which in the long run proved its efficacy in solving more problems even in areas outside D.

That a paradigm displays a high degree of fruitfulness in one subdomain does not imply that it will be overall better than its competitor in all the other areas as well. In other words, when scientists choose G based on its success in explaining planetary motion, they had no guarantee that it would have also been better than A when applied to other problems. In principle, it cannot be excluded that A was actually better than G in solving some problems outside of the subdomain of planetary motion. Considerations like this could justify the kind of pluralism advocated by Chang (2012). *Contra* Kuhn, Chang maintains that, after a revolution, *both* competing paradigms should be kept as live research options, since they may be both good at getting something right in different domains, or in different parts of the same domain. Lacey, however, defends the normativity of Kuhn's model.

> As Kuhn made us all aware, there were numerous anomalies confronting A, addressing some of which led to its late medieval modifications, including projectile and circular motions, free fall, quantitative treatment of motions and forces, the discord between Ptolemaic astronomy and Aristotelian physics. The anomalies were well known, pertinent to phenomena encountered in the realm of daily life and experience, not ignored, but typically dealt with by either introducing auxiliary (including ad hoc) hypotheses or by instrumentalist interpretations. Since G came to shed light on these phenomena, it is tempting to say, counterfactually, that G could have solved problems faced in the earlier social "worlds". But G could not have done so without creating discord in those "worlds" and their self-understanding (as it later did). [...] It is not a matter of ignoring the evidence or shutting off alternative modes of inquiry, but of focusing inquiry on key objects of the realm of daily life and experience prevailing in the "worlds". Theories, even *our* best ones, cannot be grafted onto any social order. (Lacey 1999: 164)

Someone supporting Chang's pluralism would argue for the need of keeping different paradigms alive because, by doing so, we increase our changes to solve as many problems as possible 'in general'. For Lacey, however, the aim of science is not to get things right but getting what is *relevant* in a specific situation. In his view, in fact, theory, paradigm or research strategy choice do not happen in a social vacuum (hence his rejection of autonomy). Each theory is formulated within a scientific world, which is also part of a wider social world. This is the reason why it is impossible to work on paradigms belonging to different social words at the same time. In other words, “One cannot enter the ‘world’ of G without exiting the ‘world’ of medieval Christendom, and so without entering a new social ‘world’ that cannot coexist with that ‘world’” (Lacey 1999: 164). From this perspective, it makes sense to accept one paradigm at a time, in so far as it conforms to the values and needs of the wider social world.

All and all, it appears that Lacey appreciates Kuhn’s views on incommensurability, revolutions, and ‘world changes’. The quotes about the impossibility of living in two different worlds, however, also contains the seed of his criticism of Kuhn. For Lacey, in fact, Kuhn provides a *far too* internalist account of the dynamic and rationality of scientific chance, therefore going against the once popular reading of Kuhn as an externalist philosopher.

Lacey argues the following. It is possible to evaluate which paradigm, between the A and G, displayed a higher degree of fertility in a subdomain overlapping between the two. This comparison not only tells us that G is more effective than A in solving some known problems, but also that G provides more *significant* problem solutions and novel hypotheses than A. For Lacey, the ‘significance’ of a paradigm amounts to its desirability and applicability outside of the scientific world and is therefore determined by the wider social world (of which ‘scientific worlds’ are part of). In the case of the transition to G, during a certain period of history the social world came to value technical and experimental control of nature over intellectual understanding and speculation. This shift allowed the construction of new tools, such as lenses and telescopes, which were welcomed by scientists and often became objects of scientific investigation themselves. It was this new socio-cultural context that finally allowed the emergence of the Galilean paradigm, which would have been otherwise inconceivable.

In Lacey’s view, Kuhn focuses exclusively on the degree with which paradigms and theories display epistemic values, without understanding the crucial role of the social context in determining their significance. To explain the differences between his own view and Kuhn’s, he summarises the transition between A and G in the following way:

1. G’s strategy became more fruitful than A
2. G’s theory came to provide a better account of certain phenomena (and to fit shared data about them) that are of considerable importance to A
3. G’s theory came to apply better to certain objects of increasing importance in the realm of daily life and experience (Lacey 1999: 167–168)

For Lacey, Kuhn considers (1) and (2) sufficient reasons for accepting G, and disregards (3); whereas Lacey regards (3) as a necessary condition for the formulation and then adoption of G.

To sum up the bulk of Lacey’s argument, ‘scientific worlds’ are part of larger ‘social worlds’, which contain social, political, and pragmatic values. The transition from one scientific world to the other is made possible by changes in the social world; and, consequently, by changes in the shared system of social, political, and pragmatic values, which therefore play a role in science, albeit one that does not impair its impartiality. In other words, the non-epistemic values of the wide social world (i.e., 'control', 'applicability', and so on) assess the significance of paradigms and research strategies and may even lead to the emergence of new ones, not the epistemic content of the theories scientists formulate or the way they solve problems. At the same time, the comparison between theories formulated within incommensurable scientific worlds may still rely on the use of epistemic values alone (i.e., 'fertility'), so that the impartiality of scientific reasoning, and therefore VFI, is not compromised.

9.3.3 *Problems with Lacey's Interpretation*

Following Lacey's interpretation, what sets Kuhn and him apart is the *version* of VFI they hold. Lacey allows non-epistemic values to play a role in the external phases of science, while Kuhn's exclusive focus on epistemic values alone would lead him not only towards a stronger but ultimately untenable version of VFI, but also towards a rather unrealistic view of scientific world(s) as avulsed from the social world(s). This interpretation presents some problems.

First, it mischaracterizes Kuhn's theory of values and choice in science. In Lacey's account, it looks like the proponents of both A and G agreed that their dispute had to be resolved by appealing to fertility. Furthermore, it appears like both groups agreed on how to understand 'fertility' and on how to use it to assess their research strategies. This is clearly far from the picture Kuhn proposed (see Chap. 4). In a Kuhnian perspective, one would look at the disagreement between Aristotelians and Galileans about the fertility of each approach, or even about the centrality of fertility over other aspects, without deeming one of the two parties as mistaken or irrational. Describing a revolutionary transition as being driven by *one* epistemic value (in this case, fertility), which is interpreted and applied in *the same way* by both competing groups, really misses the point of Kuhn's view on scientific rationality.

Moreover, what differentiates Kuhn and Lacey may not be how strong their versions of VFI are. It may turn out that Kuhn did not even hold VFI to begin with. For Kuhn, evidence alone cannot determine the superiority of one paradigm over another. Epistemic values, therefore, assist scientists in the process of theory or paradigm choice. The fact that such values assist in the choice, does not mean that they determine it. In the same way in which evidence is not a neutral means for inter-paradigm comparison, so epistemic values are interpreted and applied in the light of a paradigm. This work of interpretation, definition, application, and ranking of epistemic values may be driven by scientists' own philosophical or ideological affiliation, or by their own moral, social, and political stances. It may be driven, in other words, by non-epistemic values.

Of course, it would be a stretch to claim that Kuhn was a critic of VFI and a supporter of a value-laden conception of scientific reasoning. What is not a stretch, however, is to notice that his theory of values and choice leaves the door open for a conception of scientific reasoning as involving both epistemic and non-epistemic values. Not surprisingly, Kuhn's work had a clear influence on Helen Longino, one of the first and most prominent supporters of a value-laden conception of science.

For Longino, the problem with Kuhn is not a too strong version of VFI, but incommensurability. She agrees with Kuhn in regarding scientific reasoning as requiring not only hypotheses and data, but also background assumptions that fill the gap between them by specifying the 'context of application' of the previous on the latter and by determining evidential relevance. Such background beliefs consist of both epistemic and non-epistemic values (Longino 1990: Chap. 3). Longino, then, characterises this Kuhn-inspired position as 'wholism', not to be confused with 'meaning wholism'.

> Wholism is the view that the meaning of any statement can only be understood in the context of the entire theory to which it belongs, that no part of a theory can stand independently of the whole. [...] For their part the scientists who invoke Kuhn to explain how particular social and cultural values have affected a body of research seem to make use of a different form of wholism. They seem to think that the political valence of scientific theorising is inevitably a function of observational data laden with cultural assumptions and values. [...] In this interpretation of wholism the ensemble of data, theory, and assumptions must be understood as a whole. Each element, including value-based assumptions, can only be understood in the context of the others. In this case value claims that are a part of a theory would be validated by the factual data of the theory as much or as little as any other claim is. Given the incommensurability of theories, however, values conflicts could not be settled independently of any theory (Longino 1990: 181–182)

Because of wholism, incommensurability cannot be settled by appealing to values, which are never neutral with respect to paradigms. Longino realises, in a way that Lacey fails to appreciate, how incommensurability makes it difficult not only the comparison of theories and theory-laden evidence, but also of different systems of (epistemic and non-epistemic) values, that are also paradigm-dependent. At the same time, however, she does not buy into this strong version of incommensurability and maintains that some of the elements of the scientific activity may be specified independently from a paradigm. Her 'contextual empiricism' concedes to the empiricist tradition that some facts may function as the neutral arbiter between incommensurable paradigms. Without this conception, for Longino, incommensurable paradigms would be like 'monads', self-contained totalities that cannot be compared with others.

Two things must be noticed. First, Longino accepts the view of the neutrality of data because she needs a remedy to a strong version of incommensurability, which would make inter-paradigm communication impossible. However, Kuhn himself did not hold such an extreme view, which stems from a strong holism (or 'wholism') that he criticised. Second, for Longino, while incommensurability does not pose a threat to the process of choice within the scientific community, things become more complicated when scientific theories are discussed outside, in the wider social world. Because of incommensurability,

> There would be no independent way of choosing between a theory that claims that some relationship is natural and one denying this, or between a theory prohibiting interference in natural relationships and one permitting it. If we are appraising theories in a scientific context and using Kuhnian criteria of evaluation such as problem-generating and

> problem-solving capacity, this is not a problem. If, however, we move outside the research context to that of social action and public policy, the resulting circularity becomes vicious and invalidating. On neither interpretation of wholism, then, can scientific inquiry be understood as independently relevant to the support of values or ideals. (Longino 1990: 182)

In short, while Lacey thinks that the social world permits to overcome incommensurability, Longino regards the same social world as the arena where the problems of incommensurability are exacerbated. In her view, while scientists may agree on some facts as a basis for theory evaluation, in the social world theories, values, and political preferences are even more intertwined, so that comparison between conflicting views becomes extremely difficult.

The difference between Lacey's and Longino's readings of Kuhn is therefore the following. Lacey accepts a strong interpretation of incommensurability, so that incommensurable scientific traditions correspond to separate worlds; the transition between such incommensurable worlds cannot be made only *via* epistemic values alone, as Kuhn (in Lacey's view) claims, but it is resolved into the wider social world. Longino reads Kuhn as defending the view for which scientific reasoning requires evidence, theories, but also background beliefs, which include epistemic and non-epistemic values; but she rejects the idea that, in virtue of incommensurability, different scientific communities are isolated monads. Or, to be more synthetic: Lacey reads Kuhn as holding both a strong version of incommensurability and a strong version of VFI, and he rejects the latter; Longino reads Kuhn as holding a strong version of incommensurability and a value-laden conception of scientific reasoning, and she rejects the previous. Since, as I argued in previous chapters, Kuhn actually did not hold a strong version of incommensurability, and since his accounts of values is more sophisticated than Lacey's interpretation, I am inclined to think that his philosophy could be consistent with Longino's contextual empiricism, despite the grain of neo-positivism it contains, and, therefore, with a value-laden conception of science.

That Lacey misinterpreted Kuhn, of course, does not mean that Kuhn (and Longino) are right. However, other aspects of Lacey's view that appear problematic. His argument risks putting the cart before the horse. In his view, the wide social world contains the scientific worlds and arbitrates between their rivalries by determining the significance of research strategies and problem solutions. For example, values such as 'manipulability' and 'control over nature' allowed for the emergence of the Galilean paradigm and then for its victory over the Aristotelian one. What Lacey does not consider is that things could be read in the opposite way. It was the emergence of the quantitative and experimental method that changed the system of values held by the wide social world and contributed to question long held philosophical and religious beliefs. While it is not too controversial to claim that the external social world has an influence on science (for instance, on how funds are allocated), Lacey seems to be oblivious to the impact that science has on the external social world. But if that is the case—that is, if scientific worlds modify and change the non-epistemic values of the wider social world—then the non-epistemic values of the wider social world cannot function as the available neutral means for inter-paradigm comparison. A more realistic picture would have scientific worlds and social worlds interact and influence each other. The degree of such influences and interactions, as well as the permeability of these worlds to each other's values, would determine the tenability of VFI.

Ultimately, the differences between Kuhn and Lacey run even deeper and involve their philosophical perspectives. Lacey accepts that there exists a neutral ground for inter-paradigm comparison (even though he may develop his own definition of neutrality) as well as the idea that there is something as an impartial methodology to be followed, while he rejects the autonomy of science from society because of the embeddedness of the previous in the latter. The dynamic perspective consists in the rejection of both neutrality and impartiality. It is therefore unclear how Lacey could defend his interpretation of Kuhn as a supporter of VFI. As I will discuss in the next section, Kuhn accepts the only thing Lacey rejects, namely the autonomy of science. Ironically, he does so in a way that does not imply acceptance of VFI.

9.4 Heather Douglas on Kuhn

9.4.1 Douglas' Formulation of AIR

Heather Dougals supports an expanded version of AIR. In its original formulation, credited to Churchman (1948) and Rudner (1953), the argument started that non-epistemic values are required in the face of risk of error during theory choice. For Douglas, theory choice is only the last step of the epistemic phase of science. Through the discussion of several case studies, she persuasively argues that scientists face inductive risk also at several intermediate steps, such as during the collection and interpretation of data, when they may come across some 'ambiguous' evidence.

For example, Douglas (2000) compares three different interpretations of the same data collected to assess the correlation between dioxins and liver cancer. In the study, dioxins were administered to lab mice throughout their whole life. Specifically, three groups of mice received three different doses of dioxins, while a control group remained dioxin-free. All the mice completed their natural life cycle, meaning that they all died of old age, and then autopsies were performed to check for damages in their livers. Livers with few and small lesions, however, cannot be said with certainty to be affected by cancer. In fact, they could be caused by other factors, such as old age. Interpreting the ambiguous cases as signalling cancer increases the evidence for the harmfulness of dioxins, while interpreting them as non-carcinogenic increases the evidence for the harmlessness of at least the low doses of dioxins. These two different interpretations may then inform completely different policies, leading to either—an over or an under-regulation of the use of dioxins. It turns out that three different analyses of the same data, performed at different times and by different researchers, gave three different results, since the different groups of scientists considered the implications of their choices based on different sets of values.

This and similar examples, for Douglas, show how value judgements do not play a role only during theory choice, but they also play a role when making methodological choices (for example, in setting the threshold of statistical significance) and in collecting and interpreting evidence.

Two aspects of Douglas' version of AIR need to be highlighted. First, Douglas defends the idea that non-epistemic values play an important role in several steps of the internal stage of science, but such a role is *indirect*. In other words, she does not claim that non-epistemic values may trump evidence and epistemic values. In her view, non-epistemic values do play a legitimate role in science only in mitigating the risk of error associated with uncertainty.

Second, Douglas supports a rather individualistic perspective. In the previous section, I have explained how philosophers like Longino regard the scientific community as the arena where different perspectives and stances are expressed and discussed, and decisions are taken. Indeed, for Longino, objectivity is achieved through subjectivity. Douglas, however, states: "while I appreciate and utilise Longino's emphasis on social aspects of science, I think we also need clear norms for *individual* reasoning in science [...]. Thus, while my views on objectivity [...] draw insights from Longino, I do not rest the nature of objectivity on social processes alone" (Douglas 2009: 19). Whether she accomplishes the task of providing "clear norms" for scientific reasoning is debatable. Apart from this, one may interpret the difference between Longino and Douglas as a simple difference of focus: Longino emphasises the kind of collective reasoning taking place within a scientific community, while Douglas emphasises individual reasoning. However, emphasising individual reasoning implies the possibility of distinguishing it from collective reasoning, and that such a distinction is significant. Saying that collective reasoning requires the existence of individual reasoning is as obvious as saying that the existence of groups requires the existence of individuals. However, philosophers who endorse a collective approach to reasoning and rationality, like Kuhn and Longino, would argue that group dynamics position and constraints individuals. This would motivate the need of taking the group, rather than the individual, as the collective agent engaged in scientific reasoning.

9.4.2 Douglas on Kuhn

Douglas argues that there are two ways for rejecting AIR. The first, as already mentioned in Sect. 9.2, is to simply deny that scientists have to make decisions in the face of uncertainty. Levi (1960), in replying to Churchman (1948) and Rudner (1953), claimed that scientists must only communicate uncertainties to other societal and political actors. In the contemporary debate, Levi's argument has been revitalised by, among the others, Betz (2013).

The second strategy, in Dougals' view, is related to Kuhn's view on the scientific community. In *SSR*, Kuhn explained that a paradigm may isolate a scientific community, so that researchers will be entirely focussed on the set of problems and puzzles it dictates. In some passages, he even claims that the isolation of the scientific community from the external context contributes to scientific progress.

> Some [aspects of scientific progress] are consequences of the unparalleled insulation of mature scientific communities from the demands of the laity and of everyday life. That insulation has never been complete — we are now discussing matters of degree. Nevertheless, there are no other professional communities in which individual creative work is so exclusively addressed to and evaluated by other members of the profession [...]. Even more important, the insulation of the scientific community from society permits the individual scientist to concentrate his attention upon problems that he (sic) has good reason to believe he will be able to solve (*SSR*: 164).

For Douglas, Kuhn's 'isolationist view' implies an endorsement of VFI, but not only. In her view, Kuhn had such a great impact in the philosophy of science of the second half of the XX century that, though many of his ideas were challenged and critically discussed, the validity of many others was taken for granted.

> It is hard to underestimate the impact of Kuhn's book on the philosophy of science. Many philosophers developed alternative views of scientific change, the history of science, and scientific progress to counter Kuhn's account. The issue of scientific realism picked up steam in part in reaction to Kuhn's seemingly antirealist view of the history of science. The need for more robust historical accounts in the philosophy of science became a central aspect of the field, leading to programs and departments in the history and philosophy of science. Kuhn challenged the prevailing views in philosophy of science in many ways, and his work shaped the field for years to come. As philosophers grappled with some aspects of Kuhn's work, others were left unchallenged, including Kuhn's isolated view of science, which may have been widely accepted in part because philosophers were already leaning toward that idea. The field of philosophy of science was greatly simplified and streamlined if science was thought of as cut off from the society in which it operated. (Douglas 2009: 61)

Contrary to Lacey, Kuhn rejected both neutrality and impartiality. The only way to define him as a supporter of VFI, then, would be to show how his endorsement of the autonomy of science corresponded to an endorsement of value-freedom. This is what Douglas does. In her view, it is like Kuhn's isolationist view of the scientific community provides a *justification* to Levi's argument: since science makes progress when the scientific community is isolated from the rest of society, scientists must not worry about making decisions that may have social implications; therefore, they must keep their reasoning free from non-epistemic values. Following Douglas' reconstruction, this view, that she even defines as "the Kuhn-Levi view" (Douglas 2009: 64), provides the philosophical ground for the *contemporary* support to VFI.

To sum up, Douglas' view about the relation between Kuhn and VFI contains the following two theses:

1. Kuhn's isolationist view of the scientific community corresponds to an endorsement of VFI
2. Kuhn's endorsement of VFI is at the basis of the contemporary support to a value-free image of science

Both these theses can be challenged.

9.4.3 Problems with Douglas' Interpretation

Kuhn proposed an isolationist view of the scientific community. In other words, he endorsed the 'autonomy thesis', which states that scientists should work without political interference. This is a political thesis regulating the 'social contract' between science and society. VFI is a normative thesis about how scientists ought to reason. The autonomy thesis does not imply VFI. On the one hand, in the previous section I discussed how Lacey, the champion of the contemporary version of VFI, defends the value-freedom of science based on neutrality and impartiality, while rejecting autonomy. On the other hand, to say that science enjoys political autonomy tells nothing about whether scientists ought to refrain from using non-epistemic values. Yet, VFI and autonomy often confused, and even conflated, even though the relationships between these two normative theses are (Politi 2024).

When it comes to Kuhn, it appears that he held the autonomy thesis, but *not* VFI. As already explained numerous times, he admits that personal idiosyncrasies may drive scientists to define, rank, and apply the same epistemic values in different ways. Such idiosyncrasies may include social and political values. Even though it is insulated from the rest of society, Kuhn does not exclude that the scientific community may be driven by non-epistemic values *from within.*

Even in his more 'linguistic' phase, Kuhn refers to the role of non-epistemic values in determining kind membership. As explained in Chap. 6, Kuhn distinguishes between individuals and kinds, which are collections of individuals. Since individuals, for Kuhn, are 'tracers' of bounded regions of space moving through time, it may happen that they may transform during their lifetime, to the point of losing their membership to a kind and becoming members of another one.

> There are cases in which an object traversing a spatially and temporally continuous lifeline does become, slowly or quickly, a member of another kind. [...] How such lifelines are to be analysed is problematic, for it raises controversial issues about human intentionality. [...] [T]he choice of the point in time at which the corresponding changes take place introduces an element of arbitrariness that sometimes has vast consequences for human behaviour. Think of the passion invested in current debates about abortion and euthanasia: both involve the stipulation of the point during an observationally continuous change at which a change of kind takes place, *a change that requires alteration in the acceptable responses to and responsibilities toward the altered object. Such choices are not avoidable, but they are also not determined by matters of fact.* (*PW*, in Kuhn 2022: 234–235, *my emphases*)

In this quote, Kuhn explicitly addresses issues concerning human intentionality and responsibilities in establishing kind membership when individuals undergo a continuous process of change. He also says that "the resulting difficulties [are] intrinsic to the very concept of a kind" (*PW*, in Kuhn 2022: 235, *my emphasis*).

Even though they do not have much to do with 'continuous processes', in a sense many of Douglas' examples could also be read as difficulties in establishing kind membership, that is difficulties in classification. After all, her example about the interpretation of the lesions in the rat livers point out at the problem of establishing membership to the groups of 'livers with cancer' and 'sane livers', and of deciding how many individuals in the fuzzy 'space of differentiae' in between the two belong to either group. Kuhn would agree with Douglas that deciding how many of the ambiguous livers ought to be categorised as affected by cancer requires a reflection on "responsibilities" and on what constitutes an "acceptable interpretation" of data. These choices "are not determined by matters of fact", meaning that they are determined by values.

What about Douglas' second thesis? Perhaps Kuhn has been misunderstood: he did not endorse VFI and he may have regarded science as a quite value-laden enterprise. Or maybe he did not make his mind up on this point. Regardless, he may have unintentionally inspired the contemporary support for VFI. It is undeniable that *SSR* was a hugely influential book. To claim that a couple of paragraphs of *SSR* about the insularity of the scientific community led philosophers of the second half of the XX to take VFI for granted, however, might be a bit of a stretch.

In the US, despite the popularity of American Pragmatism, influential thinkers like Polanyi, Hayek and Merton had *already* argued in favour of the political autonomy of science long before the publication of Kuhn's best-seller. Furthermore, when the logical empiricists moved to the US in the wake of WWII, they displaced American Pragmatism from its position of intellectual dominance. Although the early phase of the logical empiricism movement may have been concerned, among other things, with social change through scientific progress, its 'American' phase was dominated by Reichenbach's distinction between the context of discovery and the context of justification, which relegated philosophy of science to the latter as well as to a rather detached form of logical and conceptual analysis. Instead of being the result of his influence, this intellectual milieu was the pre-existing stage in which Kuhn made his entrance, and which was prepared long before the publication of *SSR*.

Ironically, Longino (2003) suggests that Kuhn may have inspired the rise of feminist philosophies of science. Longino refrains from making the exaggerated claim that Kuhn had a direct influence on the emergence and establishment of feminist philosophies, or even that he was a feminist philosopher *avant la lettre*. Rather, in her view, Kuhn's inspiration to feminist scholars was like an 'unintended consequence' of his work. His non-algorithmic theory of rationality and his view about individual variations in the applications of scientific values could not be overlooked by feminist philosophers, who aimed precisely at challenging the idea of a universal method of scientific rationality and at exploring the social dimension of science. The influence of Kuhn's views is visible (and recognized) in Longino, who takes Kuhn's theory of collective rationality as a starting point for her own investigation, which expands towards a rich and explicit account of the interplay between epistemic and non-epistemic values in scientific reasoning. The collective approach, which is one of the central elements of Kuhn's philosophical perspective, is the main difference between Longino's (and Kuhn's) account and Douglas' individualistic view.

For his part, Rouse (1987) credits Kuhn with an even deeper transformation in philosophy of science. In his view, Kuhn brought attention to the practical dimensions of scientific research. As he rightly points out, what goes on during periods of normal science is not only the refinement of scientific representations, but an intense activity of paradigm articulation and problem resolution. The decision of measuring one thing instead of another, or to performing an experiment and not another one, however, requires a decision about what instruments and tools to use, or even to build, and which implications of the experimental activity are acceptable. This opens the way to a 'political philosophy of science', which must assess the relationship between knowledge production and (political and scientific) power.

9.5 Kuhn, Values, and Society

The idea that Kuhn supported VFI has been recently questioned by George Reisch (2019). Through an extensive archival work, Reisch has concluded that Kuhn himself was deeply concerned with social and political issues. At the time, Reisch speculates that the political milieu in which Kuhn found himself at the beginning of his academic career may have had an impact on him. When Kuhn was moving his first steps in the academic world, in fact, McCarthyism and the anti-Communists hunt were rampant in the US. Academics were particularly targeted by MacCarthy's political repression. David Bohm, Albert Einstein, and Gian Carlo Wick are only some of the notable scientists who suffered some form of political repercussions at that time. This political climate influenced the young Kuhn who, following Reisch, decided to provide a philosophical account of science that could not be perceived politically problematic. Reisch bases this claim on Kuhn's modifications of the original texts he has prepared for *QPT*, through which he edited out 'controversial' references to the social impacts and functions of science. Later, to make it clear that he rejected the sort of 'central planning' adopted in the Soviet Union, Kuhn established the political autonomy (or 'insularity') of the scientific community as a philosophical norm for the progress of science. At the same time, however, he kept describing how different individual perspectives are endemic to the scientific community. It is of course impossible to know whether this is what Kuhn 'really' thought. The point is that it is not even necessary to examine Kuhn's biography to reflect on how his philosophy may be fruitful for the contemporary debate as well.

One of the most debated issues concerns the source of values that may play a legitimate role in science. As a response to the 'new demarcation problem', several philosophers of science argue that science ought to endorse democratically held values. Such values would enter science through scientists' active engagement with representative members of the public. The criterion for discriminating between legitimate and illegitimate values in science, therefore, would be that of democratic legitimacy (for variation of this argument see, among the others, Elliott 2017; Elliott and Resnick 2014; Intemann 2015; Intemann and de Melo-Martín 2010; Kitcher 2001, 2011; Schroeder 2017). This philosophical argument coheres with the increasingly popular discourse about the need for a more 'open' and 'democratised' science. Democratisation, however, is not the only possible way to insert values into science. As shown in previous sections, Kuhn argues for a scientific community that is autonomous with respect to its external socio-political context yet internally governed by both epistemic and non-epistemic values. Or, to say it differently, Kuhn rejects the democratisation of science but endorses the value-ladenness of scientific reasoning. Of course, supporters of democratised science would protest about the lack of democratic legitimisation in scientists' decisions and in the choice of values driving such decisions. Whether democratic legitimation is the only possible form of value legitimation, or at least the best one, is however open to discussion. Unfortunately, Kuhn himself cannot participate in this debate. Moreover, it would be anachronistic to hold that his writings contain the answer to contemporary questions. Nevertheless, examining the possible justifications for why Kuhn endorsed his view may still shed a new light on current debates.

To begin with, Kuhn's isolationist view does not imply 'indifference' towards social problems, let alone a-morality or irresponsibility on scientists' side. Kuhn says that the isolation of the scientific community from society may be one of the drivers of *scientific progress*. In his view, science makes progress when scientists solve paradigm-dependent problems and puzzles (more on this in the next chapter). Scientific problems do not exhaust the list of all the problems humanity faces. In other words, while scientific problems are problems that (some) humans attempt to solve scientifically, not every problem that humans attempt to solve are scientific problems. Contemporary science is more and more expected to solve several 'big problems' and 'global challenges'. While there is much rhetoric surrounding the current discourse on science's challenges, there is not enough reflection about whether such challenges are 'scientific' (meaning that they can be framed and eventually solved by scientists) or not.

Kuhn's position does not even preclude that the scientific community could work on some of the challenges that society deems are pressing and relevant. The point he implicitly makes is that, before scientists can address them, such socially relevant problems must be transformed into scientific problems. In other words, problems from the external social-political context must first be 'internalised' to be then solved scientifically. How such problems get modified during the process of internalisation and how the scientific community adjusts itself to these problems are important issues that risk to be marginalised by the dominant and somehow triumphant rhetoric that conceives science as capable of solving every challenge.

It may even be the case that some of such challenges will never be solved scientifically, or at all. This rather pessimistic conclusion is a reminder of the limits of science and of scientists' capabilities. Reflecting upon such limits is part of a realistic philosophy of science. Instead, holding the conviction that science can solve every human problem, thus equating human progress with scientific progress, is the mark of a rather unwarranted scientism.

Not only Kuhn does not exclude in principle the possibility of scientists working on socially relevant problems, once such problems have been 'scientificised'. His isolationist conception of the scientific community is not even too strong to begin with. Even though, in his view, the insularity of the scientific community is 'unparalleled', Kuhn understands science is part of a larger context, without which it could not exist. Yet, once again, he invites us not to simply accept the existence of a relationship between science and society, but to problematize it.

> External environment—intellectual, ideological, or economic—must systematically affect the value system of much larger groups, and the consequences can include difficulties in introducing the scientific enterprise to societies with inimical values or perhaps even the end of that enterprise within societies where it had once flourished. In this area, however, *great caution is required.* Changes in the environment where science is practised can also have fruitful effects on research. Historians often resort, for example, to differences between national environments to explain why particular innovations were initiated and at first disproportionately pursued in particular countries, e.g., Darwinism in Britain, energy conservation in Germany. At present we know substantially nothing about the minimum requisites of the social milieu within which a science like enterprise might flourish. (Kuhn 1977, in *ET*: 333, ft.8)

Kuhn does not have a black-or-white position on this matter. Some socio-political contexts may repress science. Kuhn is not necessarily thinking about the direct repression of scientific thought in, for example, societies ruled by religious dogmas. Other systems of political, economical, or social values may also affect science (for instance, a society driven by economic interests would systematically question the scientific knowledge about environmental risk, to the point of spreading doubt or fabricating 'alternative truths'). Some other socio-political contexts, however, may allow science to bloom. Kuhn admits that it is not clear which kind of social structure affects the development of science in a positive way. The fact that he is reflecting upon this issue sets him apart from the most extreme and naive versions of the current discourse, which assumes democratised science to be just better science, without considering whether a 'wide open' science could risk being dissolved into politics or propaganda.

9.6 Conclusions

By moving away from a *Structure*-centred interpretation, one may end up finding Kuhn where one least expects him. By moving away from the old discussions about the tenability of his history-based model of science and about the concepts of revolution, paradigm, or incommensurability, for example, one may notice that, outside of the restricted circle of the Kuhnian scholars, Kuhn's philosophy has been used within, and at times it has even been dragged into, the debate about the role of non-epistemic values in science. And one may even find out that Kuhn has been interpreted both as a supporter of VFI (by Lacey and Douglas) and as an early proponent of a value-laden conception of science (by Longino), or even as inspiring a political philosophy of science (by Rouse).

In this chapter, I have shown how the 'Value-Free Kuhn' interpretation rests on some misunderstanding about his non-algorithmic theory of collective rationality, which is one of the main implications of his dynamic perspective. I have also shown how his view about the scientific community, which is insulated yet internally driven by values, may bring a new perspective to the debate about the science/society relationship.

References

Betz, Gregor. 2013. In defence of the value free ideal. *European Journal for Philosophy of Science* 3:207–220.

Brown, Matthew. 2024. For values in science: assessing recent arguments for the ideal of value-free science. *Synthese* 204:112. https://doi.org/10.1007/s11229-024-04762-1.

Carrier, Martin. 2022. What does good science-based advice to politics look like? *Journal for General Philosophy of Science 53*:5–21.

Chang, Hasok. 2012. *Is Water H2O? Evidence, Realism and Pluralism.* Dordrecht: Springer.

Churchman, C. West. 1948. Statistics, pragmatics, induction. *Philosophy of Science* 15:249–268.

Douglas, Heather. 2000. Inductive risk and values in science. *Philosophy of Science* 67:559–579.

Douglas, Heather. 2009. *Science, Policy, and the Value-Free Ideal.* Pittsburgh: Pittsburgh University Press.

Elliott, Kevin. 2017. *A Tapestry of Values*. Oxford: Oxford University Press.

Elliott, Kevin, and Daniel McKaughan. 2014. Non-epistemic values and the multiple goals of science. *Philosophy of Science* 81:1–21.

Elliott, Kevin, and David Resnick. 2014. Science, policy and the transparency of values. *Environmental Health Perspectives 122*:647–650.

Henschen, Tobias. 2021. How strong is the argument from inductive risk? *European Journal for Philosophy of Science* 11:92. https://doi.org/10.1007/s13194-021-00409-x.

Holman, Bennett, and Torsten Wilholt. 2022. The new demarcation problem. *Studies in History and Philosophy of Science* 91:211–220.

Hoyningen-Huene, Paul. 1993. *Reconstructing Scientific Revolutions.* Chicago: University of Chicago Press.

Intemann, Kristen. 2015. Distinguishing between legitimate and illegitimate values in climate modelling. *European Journal for Philosophy of Science* 5:217–232.

Intemann, Kristen, and Inmaculada de Melo-Martín. 2010. Social values as evidentiary standards: The case of the HPV vaccine. *Biology and Philosophy* 25:203–213.

Kuhn, Thomas S. 1977. Objectivity, value judgement and theory choice. In *The Essential Tension,* Thomas S. Kuhn, 320–339, Chicago: Chicago University Press.

Kuhn, Thomas S. 2022. In *The Last Writings of Thomas Kuhn*, ed. Bojana Mladenović. Chicago: Chicago University Press.

Kitcher, Philip. 2001. *Science, Truth, and Democracy*. Oxford: Oxford University Press.

Kitcher, Philip. 2011. *Science in a Democratic Society*. New York: Prometheus.

Kitcher, Philip. 2024. The scientist, *qua* scientist, is an ethical agent. *Filozofia* 79:231–243.

Lacey, Hugh. 1999. *Is Science Value Free? Values and Scientific Understanding*. London: Routledge.

Laudan, Larry. 1984. *Science and Values*. Los Angeles: University of California Press.

Levi, Isaac. 1960. Must scientists make value judgments? *Journal of Philosophy* 57:345–357.

Longino, Helen. 1990. *Science as Social Knowledge: Values and Objectivity in Scientific Inquiry.* Princeton: Princeton University Press.

Longino, Helen. 1996. Cognitive and non-cognitive values in science: Rethinking the dichotomy. In *Feminism, Science, and the Philosophy of Science*, ed. Lynn Hankinson Nelson and Jack Nelson, 39–58. Dordrecht: Springer.

Longino, Helen. 2002. *The Fate of Knowledge*. Princeton: Princeton University Press.

Longino, Helen. 2003. Does *The Structure of Scientific Revolutions* permit a feminist revolution in science? In *Thomas Kuhn*, ed. Thomas Nickles, 261–281. Cambridge: Cambridge University Press.

McMullin, Ernan. 1993. Rationality and paradigm change in science. In *World Changes: Thomas Kuhn and the Nature of Science*, ed. Paul Horwich, 55–78. Cambridge: MIT Press.

Melogno, Pablo. 2021. From externalism to internalism: The historiographical development of Thomas Kuhn. *Foundations of Science* 27:371–385.

Merton, Robert. 1973. *The Sociology of Science: Theoretical and Empirical Investigations*. Chicago: University of Chicago Press.

Menon, Tarun, and Jacob Stegenga. 2023. Sisyphean science: Why value freedom is worth pursuing. *European Journal for Philosophy of Science* 13:48. https://doi.org/10.1007/s13194-023-00552-7.

Poincaré, Henry. 1920 [1958]. *The Value of Science*. (English translation by G. Halsted). New York: Dover Publications.

Politi, Vincenzo. 2024. The value-free ideal, the autonomy thesis, and cognitive diversity. *Synthese* 204:24. https://doi.org/10.1007/s11229-024-04673-1.

Reisch, George. 2019. *The Politics of Paradigms*. New York: SUNY Press.

Rooney, Phyllis. 1992. On values in science: Is the epistemic/non-epistemic distinction useful? *PSA: Proceedings of the Biennial Meeting of the Philosophy of Science Association 1992* 1:13–22.

Rouse, Josep. 1987. *Knowledge and Power. Toward a political philosophy of science*. Ithaca: Cornell University Press.

Rudner, Richard. 1953. The scientist *qua* scientist makes value judgments. *Philosophy of Science* 20:1–6.

Schroeder, Andrew. 2017. Using democratic values in science: An objection and (partial) response. *Philosophy of Science 84*:1044–1054.

Stamenkovic, Philippe. 2024. Straightening the 'value-laden turn': Minimising the influence of extra-scientific values in science. *Synthese* 203:20. https://doi.org/10.1007/s11229-023-04446-2.

Chapter 10
Truth and Progress

Abstract Kuhn was unable to fully develop a comprehensive theory of truth and scientific progress, which continue to be unresolved topics in his philosophy. This chapter discusses the reasons these issues were significant for Kuhn's work and expands on his observations concerning both of them. Although he did not support traditional theories of truth, Kuhn also chose not to eliminate the idea entirely. Some of his later writings provide arguments for interpreting him as aligned with coherentism; however, it remains uncertain how he might have formulated his own theory of truth. In addressing the concept of progress, I challenge two prevalent assumptions: first, that Kuhn espoused a purely instrumentalist view of progress; and second, that he believed postulating a general aim for science is unnecessary. Drawing on my interpretation of Kuhn's dynamic perspective, I describe his conception of scientific progress as an epistemic theory. Furthermore, I elucidate how this framework posits a constitutive, albeit not ultimate, aim for science.

Keywords Truth · Correspondence theory of truth · Pragmatic theory of truth · Scientific progress · Aim of science · Evolutionary philosophy · Science and society

10.1 Introduction

Due to his untimely death, Kuhn could not satisfactorily address the concepts of truth and scientific progress, which remain open issues in his philosophy. Even though it is impossible to know which direction he would have followed had he had the time to complete his last book, in this chapter I discuss Kuhn's last thoughts about truth and progress.

In Sect. 10.2, I analyse Kuhn's proposal to overcome two theories of truth, namely the correspondence and the pragmatic theory. However, it is not clear what his own theory of truth would have amounted to. In Sect. 10.3, I discuss Kuhn's problem-solving conception of science from the perspective of the contemporary

V. Politi, *Thomas Kuhn's Philosophical Perspective*, Boston Studies in the Philosophy and History of Science 348,
https://doi.org/10.1007/978-3-032-13540-7_10

debate about scientific progress. As I explain in Sect. 10.4, if one considers the implications of his dynamic perspective on the nature of knowledge and on the comparison of different systems of beliefs, it appears like, in his last writing, Kuhn's view on scientific progress cannot be characterised uniquely in instrumental terms. In Sect. 10.5, I address a relevant issue for the debate on scientific progress, namely the problem of whether science has an overall aim, and I examine whether Kuhn needs to postulate such an aim or not. In Sect. 10.6, I suggest science does have an aim, but that it is one that does not conflict with an anti-teleological and evolutionary perspective.

10.2 Kuhn's Predicament

Owing to his premature passing, Kuhn did not have time to complete the manuscript of *PW*, managing to write only two chapters of the second part, about the biological pre-conditions of language acquisition and the theory of kinds, and to barely begin a third chapter, which would have contained a theory of how singletons are learned through practice. He could not write the whole third part of the book, in which the philosophical perspective of the first part and the theory of language acquisition of the second would have been synthesised within a theory which would have also touched upon truth and scientific progress. Even though it is impossible to reconstruct what Kuhn 'really' thought about these issues, it is still possible to explain which positions he did *not* accept and why, and to assess his arguments in this regard.

As mentioned in Sect. 3.2, one of the main implications of the dynamic perspective is the rejection of the traditional view of truth-as-correspondence. Such a view held 'truth' as the factor pulling science forward: the world itself tells which theories best correspond to it. From this perspective, the truths about the world provide the neutral ground for the comparison between rival scientific explanations and representations. The replacement of truth-as-correspondence with probabilism does not challenge the traditional perspective, since it simply replaces the 'truth' with the 'most likely proposition' as the neutral ground for scientific comparison and choice. Kuhn's dynamic perspective, by contrast, dispenses with these concepts.

In the same way Kuhn's challenge to the received view of rationality does not amount to a claim about the irrationality of science, so he does not aim at dispensing with the idea of truth altogether, but rather at showing the inadequacies of the traditional conception. In aiming to do so, he marks his distance from the social constructionists of science, who explained the success of science by substituting the concepts of truth and reality with the concept of social construction and power relations. For Kuhn, "the world is not invented or constructed [...] it is entirely solid: not in the least respectful of an observer's wishes and desires; quite capable of providing decisive evidence against invented hypotheses which fail to match its behaviour" (Kuhn 1991, in *RSS*: 101).

Kuhn, therefore, finds himself between the Scylla of the correspondence theory of truth and the Charybdis of social constructionism, for which there are no such things as truths, facts, and reality, but only discourses imposed by social and political power. Kuhn does not want to be read in this way.

> Some formulations of the developmental perspective, particularly my own, have been taken to imply that not simply the rationality of a scientific conclusion but also its truth is relative to time, place, and culture. [...] But, however tempting that reading, it was not the one I intended. What I took to be at risk was not so much the permanence of truth as its nature. I could not [...] reconcile the correspondence theory of truth, on which I had been raised, with what I had come to see as the way in which scientific knowledge developed. But neither could I see a way to account for scientific progress without recourse to it. What resulted was for me a deep predicament. (*PW*, in Kuhn 2022: 129)

> This is the fundamental form of the dilemma of the developmental philosophy of science. Properly understood, its evolutionary viewpoint is in irreconcilable conflict with the correspondence theory of truth. But it nevertheless needs some cognitive principle which requires scientists to choose between incompatibles, and none is supplied by the available substitutes for the correspondence theory. (*PW*, in Kuhn 2022: 131)

Let's examine the nature of 'Kuhn's predicament'. Prima facie, from a developmental perspective, the concept of truth could appear as an unnecessary obstacle for understanding the progress of science. Scientists choosing among different and conflicting paradigms (or lexicons, or systems of beliefs) do it without knowing which of the two is 'truer' than the other. With hindsight, of course, one can be tempted to say that scientists chose the paradigm that was 'closer to the truth' than the other. However, often the new paradigm was not even proven yet to be empirically better than the one it replaced. When operating a choice, scientists often appraise the 'heuristic potential' of the new paradigm, rather than its correspondence to the truth (Nickles 1989, 2006). And Kuhn is mainly interested in scientists' 'prospective rationality' (Kindi 2011).

Nevertheless, for Kuhn, a concept of truth is inevitable to avoid relativism. Conflicting paradigms can be compared, since incommensurability does not imply incomparability; and, as a matter of fact, scientists do compare paradigms and eventually choose between them. Without the possibility of a comparison even talking about a 'choice' between paradigms would not make much sense, since in that case scientists would simply keep living in their segregated words, in their 'monads'. The comparison must be rational, grounded on some cognitive principle, and not driven by power and politics. The problem is that it is difficult to see how to make sense of cross-paradigm comparison without the concept of truth.

One way out of this predicament is to avoid both the correspondence theory and constructionism by looking for a different theory of truth. Kuhn considers so-called 'pragmatic' or 'instrumentalist' theories of truth.

> [T]here is a long-standing family of philosophical viewpoints which holds that a far weaker notion of scientific truth is sufficient to permit understanding. Under such names as pragmatism or instrumentalism, these viewpoints suggest that scientific laws and theories are to be evaluated not for their correspondence to some hypothetical reality but for their ability to realise some other, more immediate objective: predicting or explaining natural phenomena are two goals that have been suggested; controlling nature is a third. On this viewpoint truth as correspondence continue to govern the evaluation of the descriptive statements current in everyday affairs as well as the observation statements which provide evidence for

> scientists. But, for the evaluation of theories and of laws whose subject transcends brute observation, one or another looser conception of truth is taken to be sufficient. (*PW*, in Kuhn 2022: 130)

In this quote, Kuhn does not address any theory of truth in particular. Instead, he refers to a "family of philosophical viewpoints". In the following page, he does (briefly) mention the theory of truth as the *ideal limit of inquiry* and that of truth as *warranted assertability*, perhaps having in mind the works of Charles S. Pierce and John Dewey respectively. However, the above quote could also refer to the logical empiricists' instrumentalism. As explained in Sect. 2.1, logical empiricists divided language into 'observational' and 'theoretical' and maintained that, except for the analytic statements of logic, only the observational statements possess a truth value (in the sense of correspondence), while the theoretical statements of a linguistic frameworks are neither correct or incorrect, but have only an instrumental value for the formulation of scientific theories (which can be later proved or disproved by empirical data). As it is well known, Kuhn rejected the observational/theoretical divide. Furthermore, for Kuhn, the theories of truth included in the "family of philosophical viewpoints"—the theory of truth as the ideal limit of inquiry, the theory of truth as warranted assertability and, perhaps, instrumentalism—all fail to explain how a *change* from a system of belief to another is possible. At the same time, he has still not come up with a theory of truth of his own.

Some have argued that Kuhn's rejects the correspondence theory of truth because he mischaracterized it as an epistemological thesis. As Bird (2000: 225–227) argues, one thing is to discuss the inability to assess the degree of correspondence of a set of beliefs with an independent reality, another thing is to define truth as the correspondence with an independent reality. There can be truth as correspondence, whether we are able to recognize it or not. It remains the fact that, through his 'epistemological' interpretation of the correspondence theory, Kuhn remarks that truth, in its traditional sense, is not available as a *criterion* for choice in science.

Elaborating on this point, Kuukkanen (2007) suggests reading Kuhn through the perspective of *coherentism*. First, he rightly reads Kuhn's dynamic perspective as shifting the focus from the evaluation of single beliefs to the evaluation of whole systems of beliefs.[1] Then he argues that changes of systems of beliefs can be justified based on 'increased coherence', with the problem-solving practice being one of the coherence-increasing activity. Lastly, Kuukkanen argues that since coherentism is compatible with convergent realism, reading Kuhn as a coherentist would make his position compatible with convergent realism. Kuukkanen maintains that Kuhn's rejected convergent realism based on an 'empirical argument' coming from the

[1] "The focus is not on the evaluation of beliefs as such, because the whole inherited system of belief is more or less just taken for granted, as if presumptively justified, making it pointless or even impossible to ask justification of individual beliefs. If our beliefs are holistically justified, that is, if the beliefs, in a set, mutually justify each other, then we have to understand the whole set as given and try to improve it, rather than to try to find a justification for individual beliefs on a one-by-one basis or (even less) to start the construction of a new system from scratch" (Kuukkanen 2007: 559).

history of science. In principle, therefore, his position is not incompatible with convergent realism.[2] Moreover, if convergent realism will receive some empirical support in the future, even a philosopher like Kuhn would change his mind.

Kuukkanen's interpretation has been criticised by Šešelja and Straßer (2009), on the grounds that it runs against Kuhn's anti-realism. As Šešelja and Straßer point out, Kuhn's rejection of convergent realism is not motivated by the study of empirical data about science. His argument, rather, is a priori. Ultimately, if one accepts the Kantian reading of Kuhn, one must also accept his anti-realism. Kuukkanen (2009) replies by minimising the role of a priori arguments in Kuhn, claiming that they were just an 'alternative' to his early naturalistic and empirical oriented philosophy. He also questions whether Kuhn was ever committed to a "full-blooded idealism that discounts the existence of the mind-independent world" (Kuukkanen 2009: 328). I think that both parties of the debate are onto something.

On the one hand, Kuukkanen reading of Kuhn as a coherentist makes a lot of sense in several respects. Especially in his last writings, Kuhn increasingly talks about his dynamic perspective as focussing on systems of beliefs, which may need to be revised or substituted when facing the risk of incoherence. Kuhn would say that that would happen when a new kind of entity appears to violate the no-overlap principle of a conceptual taxonomy. Even if one does not buy this restrictive taxonomic reading (see Chap. 6), it appears legitimate to say that, for Kuhn, a system of belief enters in a state of 'crisis' when its coherence is put into question. A change is justified, then, when a new system results capable to deal with the anomaly more coherently. On this point, I agree with Kuukkanen.

On the other hand, however, the fact that coherentism is in principle compatible with convergent realism does not mean that Kuhn's specific view, should it really amount to a form of coherentism, is also compatible with convergent realism. Kuukkanen dismisses Kuhn's late remarks about a priori reasoning, but the point is that, as I have argued in Chap. 3, his dynamic perspective was never meant to be supported by empirical data. The rejection of the correspondence theory of truth and convergent realism are just some of the implications of a philosophical perspective developed through 'Socratic juxtaposition' with the static perspective, which can be challenged by other philosophical perspectives, not by empirical data. Furthermore, I suspect that Kuukkanen misunderstands Kant's own brand of idealism, namely 'transcendental idealism', often attributed to the mature Kuhn, that is indeed very different from an extreme form of idealism "that discounts the existence of the mind-independent world". On these points, therefore, I agree with Šešelja and Straßer.

At the same time, Šešelja and Straßer do not consider that the rejection of truth as correspondence and convergent realism does not imply the rejection of other theories of truth and even of other kinds of realism. One of the aims of *PW* was to develop a new theory of truth as a way out of a 'predicament' Kuhn felt with

[2] To say that Kuhn's position is in principle compatible with convergent realism is different from claiming that, to be coherent, Kuhn's position must give up anti-realism, as argued by Dimitrakos (2023).

increasing discomfort. Moreover, Kuhn sometimes defined himself as a realist, even though he did not know exactly what kind of realist he was.

> There is a great deal about which Boyd and I agree. And in the remaining area, where we clearly differ, I have great difficulty articulating just what we disagree about. Both of us are unregenerate realists. Our differences have to do with the commitments that adherence to a realist's position implies. But neither of us has yet developed an account of those commitments. (Kuhn 1979, in *RSS*: 203)

Kuhn's last writings clarify his position with respect to many issues. He clearly explains that his dynamic perspective is supported by philosophical arguments (which eventually led him to look at the history of science and to other empirical sources for the study of the scientific activity), not by empirical data. He rejects the theory of truth as correspondence, the pragmatic theories of truth, and instrumentalism, and claims that he is looking for a different theory of truth. He rejects convergent realism, but he nevertheless defines himself as some sort of realist. What these writings do not clarify are Kuhn's 'positive' position with respect to truth and realism. What kind of theory of truth did he have in mind? What kind of realism was he defending? These questions remain without any precise answer, and open to future interpretations.

10.3 Progresses

Traditionally, Kuhn's conception of scientific progress is regarded as a reaction against the received view of progress as an approximation to the truth. Before him, Popper (1972) already challenged such a view. In virtue of his falsificationism, Popper maintained that it is not possible to be sure when a theory or conjecture has been verified. Empirical tests can determine with certainty only if a theory has been falsified. When a theory is not falsified, then, it is not automatically verified; rather, it is corroborated and maintained until further tests. Even though one cannot predict how or when, every corroborated theory may end up in the Procrustean bed of falsification. Therefore, for Popper, the idea that we can know when science got closer to the unfathomable truth so that we can say that it made a progress is untenable. In its place, he suggested that progress consists in *increasing truthlikeness,* or increased *verisimilitude.*[3] In this view, the transition for a theory to another is progressive

[3] Popper consistently employed the term 'verisimilitude', whereas subsequent formal developments and contemporary philosophical discourse predominantly adopt the term 'truthlikeness'. While some distinctions can be made between the two concepts—where *verisimilitude* might connote the closeness of a theory to the truth in terms of the truth value of the propositions derived from it, and *truthlikeness* a more formalized notion that, for instance, defines distance to the truth in terms similarity between logically possible worlds—the terms are often treated as interchangeable, or at least closely aligned. Indeed, the truthlikeness of a theory can be understood as a measure of its verisimilitude. Accordingly, and for the sake of simplicity, in what follows I will use truthlikeness as a general term encompassing both Popper's original usage and its later formal refinements.

when the new theory is more truthlike than the previous, meaning that it has more empirical content, a higher number of empirically corroborated, and a greater explanatory and predictive power than the theory it replaces.

Popper's view faces several problems. The truthlike content of scientific theories is assessed via empirical testing of some of their consequences. However, it is not possible to test *all* the consequences of a theory. This means that the following scenario is possible. At time t_1, some of the consequences of theories T_1 and T_2 are tested, and it is found that T_2 has more corroborated content than T_1. Therefore, T_2 is preferred over T_1 and this choice is regarded as progressive. However, other tests run at time t_2 could have shown that T_1 has more corroborate consequences than T_2. Furthermore, while empirical tests can adjudicate between the 'truth content' of two theories (that is, the set of empirically corroborated consequences of each theory), they say nothing about their 'falsity content' (that is, the set of consequences that would not be corroborated). Since theories have both truth and falsity content, one may end up accepting as progressive the acceptance of a theory that has a received a stronger empirical support so far, but that in reality possesses a greater falsity content than the previous one (see Tichý 1974; Miller 1974).

To solve these problems, Niiniluoto (1977, 1987) has proposed a new version of the truthlike, or 'verisimilatarian', approach that offers a quantitative account of how scientific theories can approximate the truth. Building on a semantic conception of truth, Niiniluoto defines truthlikeness as a function of how close the content of a theory is to the actual state of the world. In his framework, verisimilitude becomes a graded property that allows us to compare competing scientific theories not merely in terms of their truth or falsity contents, but in terms of their distance to the ideal of complete truth. It must be noticed that, in his conception, a theory may be more truthlike than others even if we do not realise it. Therefore, a change of theory may be regressive even though we believe it is progressive, and the other way round. Other difficulties revolve around the choice of the 'metric' to measure the distance between theories and the real world, as well as the complexity of the overall formal framework.

Kuhn's position escapes these issues. Generally, his conception of scientific progress is regarded as amounting to a *functionalist approach*, which regards progress as improvement of a function, rather than achievement of a goal. In *SSR*, Kuhn characterised science as a problem-solving activity. In the view proposed in that book, science does not make progress by getting closer to the truth but, rather, by improving its problem-solving power. Larry Laudan (1977, 1981) developed Kuhn's original view further. He specified that what he calls a 'research tradition' aims at solving two kinds of problems: empirical and theoretical. The 'problem-solving' effectiveness of a theory is then determined by how many problems it can solve, as well as by the importance or 'quality' of such problems. A transition from a theory to another is progressive if the new theory is more effective at solving problems than the old one. In this account, problem-solving effectiveness is assessed by the members of a scientific community (or by the followers of a research tradition). This account does not consider theories and their effectiveness 'in abstract', but as they are held and evaluated by groups of scientists. Therefore, it is consistent with Kuhn's collective approach (see Chap. 4).

The functionalist view, however, also faces some problems. Kuhn and Laudan can explain progress during periods of so-called normal science. As Kuhn maintained in *SSR*, however, after a scientific revolution, although the new paradigm often retains many of the successes of the old paradigm, some of them get lost, while some problems tackled by the old paradigm will be deemed as uninteresting or even unscientific. With the loss of some problem solutions and the change of problem-solving agenda, understanding how there can be progress through revolutions becomes difficult.

For a long time, philosophers regarded the truthlike and the functionalist views as the main (or the only) theories of scientific progress. In recent years, two novel views have emerged. For the *epistemic approach* (Bird 2007, 2022), scientific progress amounts to the accumulation of knowledge. While truth is necessary for the concept of knowledge, there is more to knowing (scientifically) than simply stumbling on some piece of truth. The epistemic view, therefore, poses a more stringent requirement to progress than the truthlikeness view. An interesting aspect of this view is that knowledge accumulates even when it is knowledge about previously held beliefs that turned out to be false. For example, we now know that the Sun does not go around the Earth. Knowledge has therefore been increased when we came to know that the proposition "the proposition 'the Sun goes around the Earth' is false" is true. Finally, Bird also maintains that some instances of increased problem-solving power are not considered to be examples of progress. Therefore, in his view, the epistemic conception is superior to both the truthlike and the functionalist conception.

For the *noetic approach*, scientific progress consists in increasing and refining understanding (Dellsén 2016). This view relies on the idea that 'knowledge' and 'understanding' are two distinct epistemological categories. If knowledge is defined as 'justified true belief', it is possible to argue there can be understanding that does not involve either true or justified beliefs (Dellsén 2017).[4]

Both the epistemic and the noetic view are widely discussed in contemporary philosophy. Analysing the arguments in favour or against them would go far beyond the present purposes. (For some discussions on the four conceptions of scientific progress, see Dellsén 2018; Shan 2023; Rowbottom 2023). For now, I consider whether the functionalist conception, attributed to Kuhn, is past redemption as many contemporary philosophers often assume. Recently, in fact, some philosophers have proposed to amend and rehabilitate it.

Douglas (2014) maintains that Kuhn did not manage to provide a satisfactory account of cross-paradigm progress because he defined problem-solving as being concerned with finding solutions to theoretical problems. For Douglas, Kuhn

[4]The philosophical literature in both epistemology and philosophy of science discusses several different definitions of 'understanding', some of which may not be necessarily distinct from knowledge. However, the noetic view on scientific progress developed in philosophy of science by Dellsén clearly relies on the distinction between understanding and knowledge.

presupposes the validity of the so-called *linear model*, which relies on the distinction between 'pure' and 'applied science'. In the linear model, progress in basic and theoretical research will (eventually) lead to successful applications. In this view, therefore, the progress of pure science is chronologically and logically prior to the progress of applied science. As Douglas explains, the pure/applied distinction is artificial. Furthermore, the engineering sciences or innovation research show how progress may work in the opposite direction then the one presupposed by the linear model, with advancements in applied science leading to new theoretical questions. Rejecting the linear model and the pure/applied distinction it presupposes permits to consider scientific activity as seeking more than just solutions to theoretical puzzle (solutions which, in turn, can be reinterpreted and even lost with theory change). Following Douglas, then, scientific progress can be defined as increased ability to predict, control, manipulate, and intervene in various (scientific as well as social) contexts. This ability, for Dougals, is not lost across scientific revolutions.[5]

Douglas appears not to hold a truthlike conception of scientific progress, which she implicitly rejects. Indeed, her attack on the pure/applied distinction can also be read as a criticism to the conception of progress as increased truthlikeness of scientific theories. Like Kuhn, then, she is interested in the problem-solving power of science. However, following on her rejection of Kuhn's 'isolationist' view of the scientific community (see Chap. 9), she does not regard scientific problems as being exclusively theoretical in nature, nor does she consider scientific progress as an issue that can be evaluated only by a scientific community that pursue its aims in isolation with the rest of society. In a sense, then, Douglas re-evaluates the functionalist account from the perspective of American Pragmatism and, especially, of James' philosophy.

This 'pragmatist-functionalist' view is a respectable philosophical position that deserves to be taken into consideration. However, I am not sure about the argument Douglas develops to criticise Kuhn. In short, she argues that the view of scientific progress as 'increased ability to predict, control, manipulate, and intervene in various contexts' requires (a) a removal of the linear model and of the pure/applied distinction in science, and (b) a non-isolationist conception of the scientific community. About (a), I agree with Douglas that the pure/applied distinction may be vague or artificial in many cases, though not necessarily in all of them. For instance, is the study of the properties of enzymes 'pure' or 'applied' research? Enzymes can be used for cancer therapies, so there are clear applied aspects to enzymology. At the same time, for such applications to work, fundamental theoretical research is

[5] "This is the kind of success that translates well across paradigms, that is rarely lost with theoretical change, and which matters greatly to both scientists and the public. While paradigm change can create losses in understanding or losses in explanatory unification as clear conceptual structures are swept away, what is not lost is the ability to predict phenomena and/or the ability to control aspects of the world. Theories or paradigms may come and go, but the ability to intervene in the world, or at least predict it, has staying power. We can think of explanatory frameworks and understandings lost in paradigm change (e.g. an intuitive grasp of what light is, a sense of place in the universe, a clear grasp of what makes something a species), but we are hard pressed to think of a predictive or manipulative capacity that has been lost" (Douglas 2014: 62).

required. There are also fields that start as 'pure' and then become 'applied'. For example, quantum mechanics started off as a purely fundamental science but later became applied for the construction of super quantum computers (or super bombs). The point is that many fields have both a pure and an applied side, but it is often possible to see which parts belong to one side or the other *grosso modo* (a distinction does not need to be neat: a 'fuzzy' distinction does not cease to be a distinction).

Therefore, I do not think that the removal of the pure/applied distinction is necessary. First, because it is still possible to conceive pure sciences as making progress in Douglas' sense of increasing explanatory power and predicting phenomena in the world. Of course, this does not solve the problem of how to measure theoretical progress after a revolution that has changed the theory. But problem-solving is not immune by this issue either since, after a revolution, even the list of problems that are regarded as scientific changes. Contrary to what Kuhn maintained, Douglas says the relevance of problems is assessed not by an isolated scientific community, but by the community at large. However, one may easily imagine social and cultural changes that, similarly to paradigm changes, alter the list of relevant problems. For example, we can imagine a society in which global warming is a serious issue to be tackled, that later decides to defund research aimed at solving that problem in favour of projects about other problems. The progress that research was making to solve the problem of global warming, and the problem itself, will be lost with the social and cultural change. Second, because by removing the distinction we may end up saying that, for example, quantum mechanics made progress only when it became applied, that is only when we started to use it to build super quantum computers (or super bombs). By accepting Douglas' argument, we may gain something (a form of pragmatic and socially oriented scientific progress that resists revolutionary changes), but we also seem to lose too much.

When it comes to (b), I also do not understand how openness and democratisation (which I take to be the opposite of an isolationist view) are necessary for science to improve the way Douglas claims it should. Everything one needs for defining scientific progress as an improvement of prediction, intervention, manipulation, and so on, is just a redefinition of the 'functions' of the functionalist approach. Such a redefinition, per se, does not require an open or democratised science. And indeed, in cases of social and cultural changes discussed above, it may even be problematic.

Redefining the functions of science is precisely the strategy followed by Shan (2019), who proposes to rehabilitate and expand Kuhn's functionalist account. To begin with, Shan argues that the 'Kuhn-Laudan' problem-solving view is only one possible version of the functionalist account. In his view, the key function science aims at fulfilling cannot be reduced to solving problems and puzzles. For Shan, in fact, science consists of a series of problem related activities, such as problem generation, problem specification, making hypotheses, reasoning, experimenting, and so on. His 'new functionalist account' considers scientific progress as improvement on these activities, and it is assessed holistically by the scientific community based on what Shan (2020) calls 'exemplary practice' (see Sect. 5.6). The scientific community, in other words, assesses the 'usefulness' of research based on its ability to provide a reliable and repeatable framework for posing problems, specifying them, and so on. All and all, it appears like Shan's new functionalist approach can account for many of the aspects of Douglas' conception of progress (that is, progress as improvement in prediction, experimentation, and so on), but without challenging the pure/applied distinction and the isolationist conception of the scientific community.[6]

The debate about scientific progress is intensifying. Supporters of the epistemic and noetic views may claim that improving problem solving and other functions is only a means to the end of accumulating knowledge or increasing understanding. Conversely, supporters of the functionalist account may respond that knowledge and understanding are useful in as much as they help solving problems. Kuhn's view on scientific progress is traditionally associated with functionalism, therefore one is obviously tempted to think that, in the contemporary debate, he would side with the contemporary supporters of functionalism. It is legitimate to wonder, however, whether his last writings may bring a different light on this canonical interpretation.

10.4 Knowledges

If one reads only *SSR*, interpreting Kuhn as supporting the functionalist conception of scientific progress seems plausible. However, in many of his later writings, Kuhn speaks about systems of beliefs. If scientific change involves the transition from one system of beliefs to another, and not just a change of problem-solving or exemplary practices, then a theory of scientific progress through revolutions should explain how the transition across systems of beliefs is progressive. One may wonder, then, whether Kuhn's view could also share at least some elements with the epistemic or the noetic conception.

For Bird (2010, 2022), scientific knowledge is possessed by a scientific community and not by any of its individual members in particular. Therefore, Bird's epistemic view does not conflict with Kuhn's collective approach (see Chap. 4). By contrast, it is not entirely clear whether understanding is a cognitive state possessed not only individuals, but also by groups. What it is clear, however, is that Dellsén (2023) criticises Bird's view on knowledge as possessed by a collective agent on the ground that scientific communities do not satisfy the 'criterion of commitment' of Gilbert's theory of group agency.[7] It must be said that Gilbert's position is only one

[6]To be clear, I am not claiming that the pure/applied distinction and the isolationist conception must *never* be challenged (although I do think that these issues require a deeper reflection rather than the plain acceptance or refusal of some ideas, or ideals—this was, in a sense, what Chap. 9 attempted to accomplish.) What I am saying is that one does not *need* to challenge such views when it comes to defining scientific progress as improvement over some key functions of science.

[7]"In Gilbert's influential terminology, the relevant groups must be 'established'. At a minimum, this requires that the members of the group have expressed a commitment to let the relevant propo-

of many theories of collective agency. On the one hand, such a view has been criticised by many and, *pace* Gilbert (and Dellsén), it is possible to conceive of a collective agent that is not characterised by the explicit commitment of its individual members (see Sect. 4.4). On the other hand, Dellsén (2023) claims that scientific progress should make public the results of science to increase understanding of both scientists and non-scientists, but he does not specify whether understanding will be then possessed by groups (i.e., the 'public') or by individuals. Indeed, his position with respect to the possibility of collective understanding is not defined. Therefore, it is not clear whether Dellsén's noetic approach conflicts with Kuhn's collective approach.

So far, then, it seems that only the epistemic and the functionalist account are clearly compatible with Kuhn's collective approach. As we know, Kuhn's view of progress is characterised as a form of functionalism, but the point now is: could it *also* be characterised in terms of an epistemic account?

Mayoral (2024) has recently analysed 'Does knowledge grow?', the *Foerster Lecture* that Kuhn delivered at the University of Berkley in 1976. In the lecture, Kuhn begins by rejecting the traditional definition of knowledge as 'justified true belief', on the ground that the formal conditions under which an epistemic agent makes knowledge claims, as opposed to just belief claims, are not clear. Indeed, the project of finding such formal conditions, like many other projects of the received view on science and knowledge, runs into difficulties. What Kuhn proposes is to substitute the view of knowledge as 'justified true belief' with a Quinean view in which a system of beliefs consists of two parts: a 'core', which is made of belief claims that have been proven to be reliable, and that are therefore presupposed as valid and considered *as if* they were knowledge claims; and other, less stabilised belief claims, more open to revisions.[8] The familiar problem is that a core can be revised too, leading to a new belief system. If that is the case, then knowledge (in the sense of 'belief claims considered *as if* they were knowledge claims') does not grow. Kuhn, however, explains that there is also a kind of non-propositional knowledge, a 'know-how', that is resistant to changes affecting the system of 'know-that'. Therefore, it is possible to regard know-how as growing through the accumulation of abilities and skills. If that is the case, then, we have always been right in attributing

sition stand as the collective attitude of the group, e.g. through voting on the proposition. [...] It is doubtful that anything like this requirement is always or even typically satisfied in paradigmatic cases of scientific progress. Scientific communities rarely have mechanisms for their members to express this type of commitment with regard to the theories or results obtained by its other members" (Dellsén 2023: 23).

[8] As discussed in previous chapters, Kuhn will further explain this view in his mature writings in which, on the one hand, he will keep using a terminology reminiscent of Quine's 'web of beliefs' but, on the other, he will also attempt not to relapse on Quine's 'extreme' holism by proposing a 'local holism'. It will also become clear that the 'core' of Kuhn's system of beliefs is that part of the lexicon of a community containing 'synthetic a priori' propositions which, if changed, will lead to a change of lexicon (see Chap. 7). The Foerster Lecture is evidence that Kuhn began to think about these issues as early as 1976, thus showing once again how his thought evolved through time, rather than undergoing a sudden 'turn'.

the functionalist view of progress to Kuhn. But one should not rush into conclusions, because Kuhn's picture is a bit more complicated.

Even though the two kinds of knowledge can be conceptually distinguished, for Kuhn, they are often inextricable in practice. Kuukkanen (2024) reads Kuhn as being contradictory or confused on this point. In his view, it is not possible to claim that know-that and know-how are both distinct and inextricable. He proposes to regard know-how as the fundamental sense of knowledge, which 'contains' know-that.[9] Therefore, if know-how grows, then knowledge grows. For Kuukkanen, in other words, Kuhn holds (or should have hold) a synthesis between the functionalist and the epistemic account, or an 'epistemological-functional' account, in which knowledge grows because know-how grows.

I am not entirely persuaded by this argument. To begin with, I do not think that Kuhn was confused regard the relation between know-that and know-how. The Foerster Lecture is not even the first time he makes a conceptual distinction of what is inextricable in practice. As I have already mentioned at the beginning of Chap. 6, for example, Kuhn explains how exemplars and symbolic generalisations are often inextricable in practice, even though it is possible to conceptually distinguish them.

Furthermore, it is possible to show instances in which, though they were inextricable in practice, the know-how grew but the know-that underwent a radical change, even though what counts as progress is attributed to the know-that. For example, the Chemical Revolution was driven by the development of new techniques for decomposing water into its constitutive elements and weighing them (Wise 1993). To be able to do so, surely Lavoisier had to solve several technical problems and increase his experimental know-how. The results of his experiments lead the community of chemists to abandon the phlogiston-centred system of beliefs and to endorse the new one. The transition between the two belief systems was a discontinuous change of know-that. Conceptually, we can distinguish Lavoisier's experimental know-how from his theoretical know-that. In practice, however, Lavoisier's know-how and know-that interacted in complex ways, to the point that it becomes difficult, and perhaps even pointless, to try to tell which one came first, or which one was more fundamental in his reasoning.

A not too often mentioned scientific revolution discussed in *SSR* is William Herschel's discovery of the planet Uranus in 1781. Uranus was the first planet observed with a telescope. From prehistoric times until Herschel's discovery, all the known astronomical entities had been observed with the naked eye. After the

[9] "When one knows how to do something, one has acquired a certain skill or capacity to perform an action. Consider chess. When does one know how to play chess? Knowing and being able to cite the rules of the game is not enough (and may not be necessary either); all that is needed is that the player is able to move the pieces appropriately. [...] And it may even happen that [the player] forgets the rules, or at least the ability to state the rules to a beginner, but is still able to play chess perfectly well [...]. Thus: knowing is doing intelligently. In a word, practice takes precedence over theorising. Naturally, even if 'knowing-how', thus practice, is prioritised over 'knowing-that', theory, it is not the case that one could not build a theory of playing chess. It is just that a theory of playing chess is based on playing, and any theory is an attempt to abstract and systematise the rules and the accepted moves of playing chess' (Kuukkanen 2024: 89–90).

discovery, more and new entities were observed. On the New Year's Day of 1801, Giuseppe Pazzi observed a 'minor planet' from the observatory of Palermo. In the following years, numerous 'minor planets' were observed, until a new class of celestial bodies, the *asteroids*, was established. The hypothesis of a 'missing planet' between Mars and Jupiter was discarded when it was observed that what was there, instead, was a 'belt' of asteroids. *Contra* Kuhn's own re-definition of revolutions in terms of radical taxonomic changes (see Chap. 6), neither Uranus nor asteroids are new kinds of entities that challenged the astronomic taxonomy by not respecting the no-overlap principles: Uranus was simply added to the existing class of planets, while asteroids were added as a new branch. Even though Uranus did not violate any expectations about planets, it did, in a sense, violate astronomers' expectations about their own object of study. After Herschel, the 'world of the astronomers' became much bigger and much more populated than it had been imagined before. Again, it is possible to distinguish the know-how necessary for the construction of new and more powerful telescopes, from astronomers' know-that. The simple and cumulative improvement of the know-how, however, had far reaching implications for the know-that, leading to a change in some of the core assumptions about the size of the universe and the numbers of entities that populate it, so much so that Kuhn classifies the discovery of Uranus as a scientific revolution. Therefore, it is possible to distinguish know-how and know-that, even though in practice they are often intertwined, if not inextricable.

Now, if I read Kuukkanen correctly, his epistemological-functional account of progress would characterise the Chemical Revolution and the discovery of Uranus as displaying a growth in the know-how (specifically, in the experimental techniques of chemistry and in the construction of more powerful and precise telescopes). Since the know-how is indeed knowledge, then these episodes display a growth of knowledge. However, the reason why we value these episodes in the history of science is precisely the impact they had on the systems of beliefs about chemistry and astronomy, not just because of how much the technical and experimental know-how improved. What is then one supposed to do with the revolutionary, non-cumulative changes of beliefs and know-that Kuhn was so interested about?

Maybe Kuukkanen could reply that non-cumulative changes in the know-that should not be part of a theory of scientific progress, which is first and foremost progress of know-how. This reply would be problematic in several respects. First, it assumes that know-that and know-how can be easily distinguished, even though, as already discussed, they are often strongly intertwined or even inextricable in practice. Second, even in case they were not inextricable after all, there is a clear risk of cherry-picking. One could simply see what grows in science and say that a theory of scientific progress should be about that, so that one could save our intuitions about scientific progress as growth. If scientific progress is growth and if only know-how grows, then a theory of scientific progress should explain only the growth of know-how. This, however, may be viewed as a rather *ad hoc* explanation of the progress of science. Third, Kuhn's dynamic perspective aims at explaining the rationality of changes of *beliefs*. Such changes, therefore, must be addressed by a theory of scientific progress.

Based on these considerations, I agree with Mayoral's reading of Kuhn's *Foerster Lecture*. Kuhn was neither confused nor mistaken when he distinguished the know-that and the know-how while also saying that they are inextricable. Therefore, we cannot simply conclude that the know-how grows while the know-that may go through discontinuities. Rather, as Mayoral suggests, Kuhn was attempting to *recast* the problem of the progress of science. I also agree with Mayoral on the point that, in doing so, Kuhn makes clear that scientific knowledge does not have only the instrumental value of solving problems and accomplishing specific tasks. In fact, for Kuhn, "Knowledge is also our source of insight into the world we live in, and [...] the world we live in *changes*. However, that change not only widens our world—it *transforms* it. Oftentimes, it turns it upside down" (Mayoral 2024: 314).

Contrary to the popular reading, one can interpret Kuhn as holding an epistemic conception: he is interested in how scientific knowledge grows (is so far as knowledge is defined as 'beliefs whose validity is taken for granted'). Unlike Bird's view, however, for Kuhn scientific knowledge does not accumulate through time (even the most reliable core of our belief system may be revised). Therefore, even though one could redefine Kuhn's view on scientific progress as somehow close to the epistemic conception, rather than to the functionalist one, one has still not solved the problem of how to explain progress through revolutions.

10.5 Aims

The debate about scientific progress could also be reframed as a debate about the aim of science, in the sense that the previous can be defined as an advancement towards the latter. In other words, one may simply say that, for the semantic approach, the aim of science is getting closer to the truth; for the noetic approach, it is increasing understanding; for the epistemic approach, it is the accumulation of knowledge; and for the functionalist approach, it is solving problems.

Bird (2019, 2022) specifies that the aim of science ought to be distinguished from the aim of scientists. Individual scientists may have all sorts of aims (increasing their own career prospects, getting public recognition, intellectual satisfaction, and so on). However, science as a collective enterprise has its own aim (which, in Bird's view, is the accumulation of knowledge). Bird, furthermore, stresses that science has indeed *one* aim. His ground for claiming so is the distinction between the 'constitutive' aim of science, the pursuit of which characterises science as such, and other 'contingent' aims. Science is dependent on several contingent factors, such as external funding. Patenting the result of research, writing grant applications, engaging with the public and other stakeholders, and so on, are all aims that promote science. Contrary to the constitutive aim, however, the contingent aims of science, such as those that promote it, are not constitutive of the 'scientificity' of science.

The idea that science has one constitutive aim conflicts with the view that science has essentially multiple aims, which may be both epistemic (such as understanding, prediction, and so on) and non-epistemic (Potochnik 2015). A similar line of argument claims that science does not have a 'global aim', which every scientific research everywhere at any time ought to pursue, but rather multiple and context-specific 'local aims' (Elliott and McKaughan 2014; Lusk and Elliott 2024). Talking about the aim of science, therefore, may shift the focus of the discussion on scientific progress.

What is Kuhn's position with respect to this issue? On the one hand, in *SSR* he defined science as a problem-solving activity, with problems, the way they are solved, as well as their significance, being context-dependent and varying from one paradigm to the other. Indeed, after a revolution, some of the central problems of the old paradigm may be dismissed as unimportant, or even as unscientific. Under this respect, Kuhn could be read as holding the view for which science has a constitutive aim, which is problem-solving, although what such an aim amounts to and what are the required actions to pursue it vary from paradigm to paradigm. As already discussed, this view would not explain why a revolutionary change of paradigm is indeed progressive. If not through the concept of truth (see Sect. 10.2), perhaps Kuhn may still escape relativism by characterising the aim of science as something different than the simple resolution of problems internal to a paradigm.

In his late years, Kuhn entertained such an idea. For instance, he pondered the 'amendment' to his view on values and theory choice suggested by H.G. Hempel.

> Science is widely conceived as seeking to formulate an increasingly comprehensive, systematically organised, world view that is explanatory and predictive. It seems to me that [Kuhn's desiderata for the goodness of scientific theories] may best be viewed as attempts to articulate this conception somewhat more fully and explicitly. And if the foals of pure scientific research are indicated by the desiderata, then it is obviously rational, in choosing between two competing theories, to opt for the one which satisfies the desiderata better. [...] [These considerations] might be viewed as justifying in a near-trivial way the choosing of theories in accordance with whatever constraints are imposed by the desiderata (Hempel 1983: 91).

Kuhn (1983) clearly endorses Hempel's view.

> [Hempel] points out that some of the difficulties with my accounts of theory choice would be avoided if desiderata like accuracy and scope, invoked when evaluating theories, were viewed not as a means to an independently specified end, like puzzle solving, but as themselves goals at which scientific inquiry aims. [...] Because it loosens the commitment to any particular prespecified goal like puzzle solving, Hempel's formulation is an improvement on mine; our points are otherwise the same. (Kuhn 1983, in *RSS*: 210)

Overall, Kuhn agrees with Hempel on regarding values (or 'desiderata') not as means to a 'prespecified' end (such as solving context-dependent problems). Rather, taken together, the values Kuhn speaks about allow to formulate the "increasingly comprehensive, systematically organised, world view that is explanatory and predictive", which is the actual constitutive goal of science.

The two philosophers disagree on one point. Kuhn does not think that Hempel's amendment to his theory would make the justification for theory choice a "nearly trivial" affair. For Kuhn, understanding that the satisfaction of some desiderata justifies choices in science corresponds to understanding what science is. To explain this point, he compares the acquisition of the meaning of 'science' to the acquisition of 'force' and 'mass' in classical mechanics. For Kuhn, one learns the meaning of 'science' by learning a cluster of interrelated terms, which also includes 'fine art', 'philosophy', 'theology', and so on. Kuhn also stresses that we need to keep in mind that 'science' is a relatively recent term.[10] Before its appearance, there was a different cluster of interrelated terms. Ancient Greeks, for example, did not practise science, but 'natural philosophy', which meant something different.[11] In the same way in which Newton's Second Law is constitutive of classical mechanics, in the sense that it allows the derivation of other empirical laws, while its change would correspond to a change of lexicon (see Sect. 7.3), so one cannot take off Kuhn's desiderata for scientific theories without modifying the contemporary lexicon in which 'science' has a meaning different from 'fine arts' or 'philosophy'.

> While we speak our own language, any activity that we label 'science', or 'philosophy' or 'art', and so on, must necessarily display pretty much the same characteristics as the activities to which we customarily apply those terms. Just as access to Newton's second law is required in order to pick out Newtonian forces and masses, so picking out the referents of the modern vocabulary of disciplines requires access to a semantic field that clusters activities with respect to such dimensions as accuracy, beauty, predictive power, normativeness, generality, and so on. Though a given sample of activity can be referred to under many descriptions, only those cast in this vocabulary of disciplinary characteristics permit its identification as, say, science; for that vocabulary alone can locate the activity close to other scientific disciplines and at a distance from disciplines other than science. That position, in turn, is a necessary property of all referents of the modern term 'science'. (Kuhn 1983, in *RSS*: 213–214)

Of course, many changes are possible *within* science (i.e., scientific revolutions, specialisation, theoretical development, and so on). Perhaps, one day the cluster of interrelated terms that contains 'science' will be modified, or the term itself may disappear, like it happened with 'natural philosophy'. Kuhn, however, is interested in *scientific* change and progress, that is in the changes and progress of an activity described by a term that is approximately 500 or 600 hundred years old. And this activity is distinguished by other activities in virtue of Kuhn's desiderata or values. The satisfaction of such desiderata is not just one of the aims of science (*via* Hempel's amendment), but it is its constitutive aim: what makes science scientific.

This aspect of Kuhn's philosophy has been discussed by Hoyningen-Huene (1992).[12] Indeed, his book-length treatment of the nature of science in terms of

---

[10] William Whewell was probably the first one to use the English word 'science' in its contemporary sense. The word itself, however, already existed in the so-called Middle-English, spoken between 1150 and 1500. Such a word was a borrowing from the French *science* (pronounced 'szjans'). Guy de Chaliac's *Grand Chirurgie,* published in 1425, contains one of the first French occurrences of the word used in its contemporary sense.

[11] Ancient Greeks did not even have the word 'science'. The etymology of the word, in fact, is the Latin *scientia,* which means 'knowledge'. This term derives from the Latin *scire* ('to know'), which in turns comes from *scindere* ('to divide', 'to distinguish'). The Ancient Greek word *épistème*, by contrast, means both 'knowledge' and 'understanding', and it is not related to terms indicating an activity (such as dividing or distinguishing); it is contrasted with *techne.*

[12] "Obviously, the fact that the decision can be justified with recourse to cognitive values, is not by itself sufficient for its rationality. Rather, the cognitive values themselves must be good reasons,

systematised knowledge (Hoyningen-Huene 2013) could be interpreted as a sophisticated development of the Kuhn-Hempel argument. It may also be interpreted as a version of the epistemic conception. While Bird characterises progress as 'accumulation of knowledge', Hoyningen-Huene characterises it as 'increasingly systematised knowledge': clearly, they both put knowledge at the centre of their accounts of scientific progress.

Other scholars, by contrast, argue that Kuhn does *not* need to postulate an aim for science. For example, Mladenović (2017) claims that Hoyningen-Huene's interpretation derives from his "residual faithfulness to the received view and its conception of rationality" (Mladenović 2017: 93). Characterising the aim of science as the satisfaction of Kuhn's values, for Mladenović, does not help escaping the charge of relativism, since the values are always open to interpretation and subject to different rankings. Mladenović, therefore, maintains that all Kuhn needs is a specification of "how to complete *specific* scientific tasks" (Mladenović 2017: 95), or 'mini tasks', rather than the postulation of an overarching goal. Such a goal, she claims, is not even needed to distinguish science from other activities.

I believe that Mladenović misunderstands both Kuhn and Hoyningen-Huene. Of course, values are open to different interpretations and rankings. However, they cannot be taken out of science without changing the meaning of 'science'. The fact that they can be applied in different ways does not make them any less constitutive of science than a set of 'fixed' principles. Indeed, Kuhn's dynamic perspective does not deny that there are things specific to science (i.e., scientific rationality), but only that such things can be universally interpreted and applied. Besides, if the postulation of a constitutive aim of science does not help with the charge of relativism, as Mladenović claims, it is not clear how the specification of 'mini goals' would solve such a problem either. When it comes to her explanation for why Kuhn would not need to postulate an overall aim for science to distinguish it from non-science, Mladenović quotes the parts where Kuhn (1983) speaks about the cluster of co-defined terms indicating disciplines and human activities. However, she does skip all the parts where Kuhn establishes a parallel between scientific values and Newton's Second Law, explaining how values are necessary to understand why 'science' is indeed different from 'fine art' or 'philosophy' in the same way in which the Second Law is constitutive of classical mechanics. In this way, she provides a reading of Kuhn that is at least partial, if not misleading.

In my view, Hoyningen-Huene is right in recovering the Kuhn-Hempel argument. Following this argument, one may conclude that Kuhn does indeed postulate a constitutive aim of science, in the form of systematised knowledge, that is as systematised established beliefs. From this perspective, solving problems would be functional to the achievement of this aim; or, in other words, solving problems

'rational' reasons which make the decision a rational one. [...] The justification of cognitive values has to have recourse to an ultimate goal of science: the commitment to cognitive values is then the means which guarantees that theory choice is made in accordance with this ultimate goal" (Hoyiningen-Huene 1992: 496–497). Such a goal, for Hoyningen-Huene, is the one indicated by Hempel (1983).

would promote science, but it would not constitute it. This interpretation would read Kuhn (or at least the 'late' Kuhn, who assimilated Hempel's amendment) as supporting once again a version of the epistemic view of scientific progress (assuming that progress as 'increasing the comprehensiveness and systematisation of an explanatory and predictive world view' amounts to a version of the epistemic view), but not a functionalist one.[13] Such an interpretation is supported by Kuhn's open endorsement of Hempel's amendment. More importantly, it is also supported by how the dynamic perspective is concerned with the comparison of different systems of beliefs, not of just problem solutions. And yet, something does not quite add up.

One of the perks of the functionalist/problem-solving approach is its consistency with an anti-teleological conception of scientific progress. Science keeps solving problems and puzzles in a process that is driven 'from behind'. This view is developed in contrast with the traditional received view which regards science as directed 'towards' the discovery of the ultimate truths of Nature. One may wonder whether Bird's epistemic conception is simply substituting 'truth' with 'knowledge'. In other words, one may wonder whether, for his epistemic conception, the accumulation or refinement of knowledge should be regarded as getting closer to an ultimate ideal state of 'maximal' knowledge about the world.

To sum up, the situation appears as follows. The functionalist view does not postulate an ultimate goal for science, but then it may run into difficulties when it comes to define scientific progress, due to the charge of relativism (i.e., if what counts as a scientific problem changes across paradigms, how can we say that a new paradigm improves on the previous?). The epistemic view postulates a constitutive goal (and Kuhn even agreed with Hempel on the need for such a goal), but it then gives up the evolutionary and anti-teleological conception Kuhn was so keen on. In the next section, I will explore the possibility that is possible to regard science as possessing a constitutive aim, though not an ultimate one.

---

[13] It would remain to be seen whether Hoyningen-Huene's view could also be read as a version of coherentism. Though he notices some similarities between his position and the one developed by Rescher (1979), Hoyningen-Huene maintains that the two views are substantially different (Hoyningen-Huene 2013: 170.175). First, there is a difference in the kind of projects pursued by the two philosophers: Rescher's project is epistemological, Hoyningen-Huene's project pertains to the philosophy of science. Second, Rescher is interested in 'systematicity' in the sense of being 'systematical', that is pertaining to a definite system; whereas Hoyningen-Huene is interested in what is 'systematic', in the sense of 'being ordered' (but without constituting a definite system). I do not find these differences to be significant enough. About the first point, many philosophers would not be so quick in separating epistemology from philosophy of science; and many philosophers of science draw from current debates in epistemology (a case in point is Bird 2022, whose philosophical view on science is inspired by the work in epistemology by Williamson 2002). When it comes to the second point, the difference between 'systematical' and 'systematic' appears to be of grade, rather than kind. Ultimately, my claim is that Hoyiningen-Huene's view on systematicity is inspired by the Hanson-Kuhn argument; however, Kuhn could also be interpreted as a coherentist (see Sect. 10.2), and so can Hoyningen-Huene's view. Contrary to Rescher and Kuukkanen (2007), however, I do not see coherentism to require convergent realism.

10.6 Niches

I want to begin this section by distinguishing between *constitutive* and *ultimate aims*. A constitutive aim characterises an activity as 'that activity'. An ultimate aim is the final objective that an activity is intended to reach, its *telos*. To my knowledge, not many philosophers of science distinguish between 'constitutive' and 'ultimate aim' of science. However, I find it of central importance when talking about scientific progress. One could say that, for example, the functional account defines solving problems, or improving other key functions, as the constitutive but not ultimate aim of science. In such an account, science aims at improving some of its functions, and pursuing such an aim constitutes science as scientific. However, the pursuit of the constitutive aim does not require the postulation of a final state of affairs to be reached down the road.

Bird (2022) agrees that, for the concept of progress in general, one does not need to postulate an ultimate end. For instance, progress in the arts or in politics is accomplished through gradual improvements, which however do not move towards an ideal final state. In other words, arts and politics do not have an ultimate aim. But that does not mean that they do not have any aim at all. Based on the distinction I propose, the constitutive aim of art would be, for example, the pursuit of significant artistic representations, while the constitutive aim of politics would be the pursuit of fair governance. In any case, Bird characterises the difference between the various conceptions of scientific progress being about the "ultimate or constitutive goals" of science (Bird 2022: 40). It is not always clear whether, for Bird, the accumulation of knowledge is *both* the constitutive *and* the ultimate aims of science.[14] By contrast, Hoyningen-Huene's (2013) position could be redescribed as holding that systematic knowledge is the constitutive but not ultimate aim of science. All Hoyningen-Huene needs for his theory to work is that, during a scientific revolution, the new paradigm will be shown to be more systematic than the old one, without the need of postulating that science progresses towards some sort of 'ultimate systematisation' (see Hoyningen-Huene 2013: 183–187).

Distinguishing between constitutive and ultimate aims brings some conceptual clarity, but it still does not help much in adjudicating between rival conceptions of scientific progress. Supporters of the functionalist conception may claim that the constitutive aim of science is improving some of its functions (i.e., solving problems), and that knowledge may help pursue such a constitutive aim. Supporters of the epistemic conception may claim that the accumulation of knowledge is the constitutive aim of science, though they may concede that it is not its ultimate aim, and that improving some functions may help pursuing such an aim. None of them, in my view, can satisfactorily deal with the episodes of belief change so central in Kuhn's

[14] Bird (2022) clearly rejects 'global arguments' about science. For instance, he thinks that arguments about the success of science 'as a whole' add little or nothing to the arguments of the success of theories in specific scientific fields. Such global arguments, in short, are unnecessary second-order arguments. However, there is still the chance that one may regard the accumulation of knowledge as the constitutive and ultimate aim of special sciences, even without ascending to the second-order level of global science.

dynamic perspective. The functionalist conception simply brackets such an issue: changes of beliefs notwithstanding, there is incremental improvement of functions and that is all we need for a theory of scientific progress. Changes of belief, therefore, are not even a central part of such a theory. For Kuhn, however, such changes must enter a theory of scientific progress, since know-that and know-how are inextricable. The epistemic conception, which regards progress as accumulation of knowledge, in a sense, denies that such changes pose any real threat. People used to believe that the Sun goes around the Earth, but we now know that that is false. Therefore, changes of beliefs are simply changes from false beliefs to knowledge, granted that truth is one of its necessary conditions. But not every previously held belief can be deemed as false so easily. We know that it is not the case that the Sun goes around the Earth but, for instance, is $F = ma$ false? Strictly speaking, it is not true (Cartwright 1983), but what are we supposed to make of its falsity? Questions like these make little sense within the Kuhnian framework, for which it is important to evaluate not '$F = ma$' in isolation, but as embedded in a system of belief. At this point I wonder: would it be possible to find *another* aim of science, which is constitutive but not ultimate, and which is consistent with Kuhn's dynamic perspective?

As already explained, Kuhn illustrates his views by drawing analogies between the development of science and biological evolution. Towards the end of *SSR*, he used a Darwinian analogy to explain how science evolves *from* a state to a better one, not *towards* an ultimate state while, in later writings, he compared the process of scientific specialisation to the phenomenon of biological speciation (see Sect. 8.5). In one of last interviews, he claimed:

> I tried at the end of [*SSR*] to say in what sense I thought there is progress. I largely squeezed out the answer to that, talked about the accumulation of puzzles, and I think I would now argue very strongly that the Darwinian metaphor at the end of the book is right, and should have been taken more seriously than it was; and *nobody* took it seriously. People passed it right by. (Kuhn 1995, in *RSS*: 307)

Based on the centrality that Kuhn himself accorded to his evolutionary analogies, some philosophers have begun to regard them as more than 'pedagogical tools' or communicative strategies that he used to convey philosophical theses in more immediate and intuitive ways. For example, Haufe (2022) attempts to "organise and clarify Thomas Kuhn's ideas about scientific knowledge under the evolutionary framework that he ultimately came to see as responsible for the epistemic phenomena that interested him" (Haufe 2022: 17). Effectively, Haufe uses the contemporary evolutionary framework to characterise science as a collective activity carried out by groups that, like biological species, survive to the evolutionary pressure of their environment, which in the case of science can be described as an 'epistemic niche'. Scientific communities and their niches co-evolve. Sometimes, a scientific community may break into small sub-communities, following a process that resembles biological speciation. Other times, some disciplines may disappear, following a process akin to mass extinction.

I agree with many of the things Haufe says. For instance, he considers revolutions and specialisations as being different in grade, rather than in kind, and as belonging to the same spectrum (Haufe 2022: Chap. 5–6; compare with my Politi

2018 and with Sect. 7.5). Curiously, however, Haufe is not always clear about *what* grows in science following evolutionary mechanisms. The title of his book promises to tell us 'How *knowledge* grows'. The subtitle, however, reads as 'the evolutionary development of scientific *practice*'. Effectively, his book is an evolutionary account of how scientific practice, consisting in posing and solving problems, experimenting, calibrating, optimising, and so on, survives to the selective pressure of an epistemic niche and it is passed down from one group of practitioners to its offspring. It seems, therefore, that Haufe favours practical know-how to propositional know-that, which would make his account a version of the functionalist approach, reinterpreted from an evolutionary perspective.

Kuhn maybe would have argued that Haufe's account could work for propositional knowledge too. In his view, a lexicon is the common possession of a community. Within the group, there are variations in the expectations individuals have about some kind terms and generalisations. A lexicon is passed through to the next generation: senior scientists transmit their lexicon to their 'offspring' (that is, to junior scientists), who will in turn pass it to the following generation, and so on. With each generation, the individual variations in the expectations associated with the lexicon may grow. Such pre-existing variations may help formulating a new lexicon if the existing one needs revision and change. Finally, know-how and know-that are inextricable: a lexicon is learned through practice, and practice refines and extends our lexicon. Therefore, I see no obstacle in extending Haufe's evolutionary reading to include propositional knowledge too. But not only that.

A genuinely evolutionary account of science helps highlight that the two kinds of knowledge are inextricable not only with respect to one another, but also with respect to the epistemic niche they are adapted too. As already explained in Chap. 7, the system of beliefs includes theoretical beliefs and beliefs driving practical activities. These beliefs, held by a community, shape the 'world' of that community. In this evolutionary account, the constitutive aim of science is not simply 'adequacy', in the empiricist sense of adequacy of statements and representations to empirical data. Rather, the aim is 'adaptation'. All the other aims of science (accumulating knowledge, solving problems, and increasing understanding) serve the constitutive aim of helping a population (the scientific community) to get more adapted to its environment (an epistemic niche). The more a population gets adapted to its environment, however, the more the population modifies the niche, which in turn will lead to mofications in the population.

Questions about the objective growth of science make sense only if there is a neutral way to 'measure' it, and the fixity of the world is believed to provide such a measure. Kuhn's dynamic perspective rejects the presupposition that the neutral and objective world is accessible to us and that, consequently, scientific progress can be measured from such a stable, transhistorical and transcultural Archimedean platform. The world of scientists, their epistemic niche, is *not* fixed, but changes with the changes of scientific beliefs and new data. In the transition from one paradigm to the other there is no 'growth', but there is only the acceptance of a paradigm that is more adapted to what the niche has become. So far, we are in a familiar Kuhnian territory, which however poses the problem of understanding what progress is if it is not growth. In the same way in which scientists do not have an Archimedean platform for evaluating different systems of beliefs, so philosophers and historians of science do not have an Archimedean platform for assessing scientific progress *outside* of the current epistemic niche. From our current perspective, from our current niche, present science is more 'evolved' than past science, because past science would not have been as adapted to our niche as the present science.

Adaptation to a niche is a constitutive but not an ultimate aim, because there is no ideal 'ending state' in a population has reached the maximal level of adaptation to a niche that will never change anymore. Rather, both the population and the niche are in an interactive relation, which leads to changes in both. The degree of adaptation of the scientific community's beliefs to a niche, and improvements over past beliefs, can be assessed only from *within* the niche itself, and, perhaps, this is all that one can ask from an evolutionary theory of scientific progress.

10.7 Conclusions

While we will never know whether Kuhn would have been able to develop his own theory of truth, the examination of some of his mature writings about the growth of scientific knowledge may have interesting, and perhaps unexpected, implications. Specifically, the widespread idea that Kuhn endorsed a functional theory of scientific progress may be questioned. In fact, it appears that his view on progress is closer to an epistemic theory. Kuhn does believe that science is a problem-solving activity, but the recognition of exemplary problems and of their solutions require a system of beliefs, whose core remains relatively stable through time, and which is inherited by following generations of scientists. In his view, there is no difference between what a community holds as knowledge and what a community believes. Furthermore, there is not even a neat distinction between know-how and know-that, which means that even solving problems may be reinterpreted within an epistemic theory of scientific progress. Finally, that Kuhn's theory of scientific progress does not postulate the existence of an ultimate aim for science does not imply that, in his perspective, science does not possess a constitutive aim. Whether Kuhn would have agreed with this interpretation is not known. In any case, examining the subtleties of his views and their possible implications—including those that Kuhn himself did not see—may help recast his philosophy under new lights.

References

Bird, Alexander. 2000. *Thomas Kuhn*. Chesham: Acumen and Princeton: Princeton University Press.

Bird, Alexander. 2007. What is scientific progress? *Nous* 41:64–89.

Bird, Alexander. 2010. Social knowing: The social sense of scientific knowledge. *Philosophical Perspectives* 24:23–56.

Bird, Alexander. 2019. The aim of belief and the aim of science. *Theoria* 34:171–193.

Bird, Alexander. 2022. *Knowing Science*. Oxford: Oxford University Press.

Cartwright, Nancy. 1983. *How the Laws of Physics Lie.* Oxford: Oxford University Press.
Dellsén, Finnur. 2016. Scientific progress: Knowledge versus understanding. *Studies in History and Philosophy of Science* 56:72–83.
Dellsén, Finnur. 2017. Understanding without justification or belief. *Ratio* 30:239–254.
Dellsén, Finnur. 2018. Scientific progress: Four accounts. *Philosophy Compass* 13:e12525. https://doi.org/10.1111/phc3.12525.
Dellsén, Finnur. 2023. Scientific progress: by-whom or for-whom? *Studies in History and Philosophy of Science* 97:20–28.
Dimitrakos, Theodore. 2023. Do Kuhnians have to be anti-realists? Towards a reconception of Kuhn's historiography. *Synthese* 202:21. https://doi.org/10.1007/s11229-023-04225-z.
Douglas, Heather. 2014. Pure science and the problem of progress. *Studies in History and Philosophy of Science* 46:55–63.
Elliott, Kevin, and Daniel J. McKaughan. 2014. Non-epistemic values and the multiple goals of science. *Philosophy of Science* 81:1–21.
Haufe, Chris. 2022. *How knowledge grows: the evolutionary development of scientific practice.* Cambridge: MIT Press.
Hoyningen-Huene, Paul. 1992. The interrelations between the philosophy, history and sociology of science in Thomas Kuhn's theory of scientific development. *British Journal for the Philosophy of Science* 43:487–501.
Hoyningen-Huene, Paul. 2013. *Systematicity: The Nature of Science.* Oxford: Oxford University press.
Hempel, Carl G. 1983. Valuation and objectivity in science. In *Physics, Philosophy and Psychoanalysis: Essays in Honor of Adolf Grünbaum*, ed. Robert Cohen and Larry Laudan, 73–100. Boston: Reidel.
Kindi, Vasso. 2011. The challenge of scientific revolutions: VanFrassen's and Friedman's responses. *International Studies in the Philosophy of Science* 24:327–349.
Kuhn, Thomas S. 1979. Metaphor in science. In *Metaphor and Thought*, ed. Andrew Ortony, 409–419. Cambridge: Cambridge University Press. Reprinted in *RSS*:197–215.
Kuhn, Thomas S. 1983. Rationality and theory choice. *The Journal of Philosophy* 80:563–570. Reprinted in *RSS*:208–215.
Kuhn, Thomas S. 1991. The Road since Structure. *PSA: Proceedings of the Biennial Meeting of the Philosophy of Science Association 1990*: 2–13. Reprinted in *RSS*:90–104.
Kuhn, Thomas S. 1995. A discussion with Thomas S. Kuhn. Interview with Aristides Baltas, Kostas Gavroglu, and Vasso Kindi. Published in *RRS*:255–323.
Kuhn, Thomas S. 2022. The Last Writings of Thomas S. *Kuhn*, ed. Bojana Mladenović. Chicago: Chicago University Press.
Kuukkanen, Jouni-Matti. 2007. Kuhn, the correspondence theory of truth and coherentist epistemology. *Studies in History and Philosophy of Science* 38:555–566.
Kuukkanen, Jouni-Matti. 2009. Closing the door to cloud-cuckoo land: A reply to Šešelja and Straßer. *Studies in History and Philosophy of Science* 40:328–331.
Kuukkanen, Jouni-Matti. 2024. Kuhn, progress and knowing-how: an epistemological-functional account of scientific progress. In *Rethinking Thomas Kuhn's Legacy*, ed. Yafeng Shan, 77–101. Cham: Springer.
Laudan, Larry. 1977. *Progress and Its Problems*. Berkeley: University of California Press.
Laudan, Larry. 1981. A problem-solving approach to scientific progress. In *Scientific Revolutions*, ed. Ian Hacking, 144–155. Oxford: Oxford University Press.
Lusk, Greg, and Kevin Elliott. 2024. Against global aims for science: Values, epistemic priority, and a local aims approach. *Synthese* 204:47. https://doi.org/10.1007/s11229-024-04633-9.
Mayoral, Juan V. 2024. Kuhn's 'Does knowledge "grow"?': Second thoughts on the nature of epistemic progress. In *Rethinking Thomas Kuhn's Legacy*, ed. Yafeng Shan, 291–319. Cham: Springer.
Miller, David. 1974. Popper's qualitative theory of verisimilitude. *British Journal for the Philosophy of Science* 25:166–177.

Mladenović, Bojana. 2017. *Kuhn's Legacy. Epistemology, metaphilosophy, and pragmatism.* New York: Columbia University Press.
Nickles, Thomas. 1989. Heuristic appraisal: A proposal. *Social Epistemology* 3:175–188.
Nickels, Thomas. 2006. Heuristic appraisal: Context of discovery or of justification? In *Revisiting Discovery and Justification: Historical and Philosophical Perspectives on the Context Distinction*, ed. Jutta Schickore and Friedrich Steinle, 159–182, London: Springer.
Niiniluoto, Ikka. 1977. On the truthlikeness of generalisations. In *Basic Problems in Methodology and Linguistics*, ed. Robert Butts and Jaakko Hintikka, 121–147. Dordrecht: Reidel.
Niiniluoto, Ikka. 1987. *Truthlikeness*. Dordrecht: Reidel.
Popper, Karl. 1972. *Objective Knowledge*. Oxford: Clarendon Press.
Potochnik, Angela. 2015. The diverse aims of science. *Studies in History and Philosophy of Science* 53:71–80.
Rescher, Nicholas. 1979. *Cognitive Systematisation: A Systems-Theoretic Approach to a Coherentist Theory of Knowledge*. London: Rowman and Littlefield.
Rowbottom, Darrell. 2023. *Scientific Progress*. Cambridge: Cambridge University Press.
Shan, Yafeng. 2019. A new functionalist approach to scientific progress. *Philosophy of Science* 86:739–758.
Shan, Yafeng. 2020. Kuhn's "wrong turning" and legacy today. *Synthese* 197:381–406.
Shan, Yafeng, ed. 2023. *New Philosophical Perspectives on Scientific Progress*. London: Routledge.
Šešelja, Dunja, and Christian Straßer. 2009. Kuhn and coherentist epistemology. *Studies in History and Philosophy of Science* 48:19–49.
Tichý, Pavel. 1974. On Popper's definition of verisimilitude. *British Journal for the Philosophy of Science* 25:155–150.
Williamson, Timothy. 2002. *Knowledge and its Limits*. Oxford: Oxford University Press.
Wise, Norton. 1993. Mediations: Enlightenment balancing acts, or the technologies of rationalism. In *World Changes: Thomas Kuhn and the Nature of Science*, ed. Paul Horwich, 207–256. Cambridge: MIT Press.

Chapter 11
Outro: Is a Distinctive Kuhnian School Even Possible?

Rather than with an analysis of some of the most talked about concepts of *SSR*, such as incommensurability or scientific revolutions, in this book I began with an analysis of Kuhn's general philosophical perspective, which he developed as a remedy to some of the difficulties of the static perspective. From the rejection of the unwarranted presupposition concerning the accessibility of an absolute and neutral basis for scientific evaluation, the dynamic perspective unfolds through 'Socratic juxtaposition' to the static perspective. Kuhn did not draw his dynamic perspective from the study of the history of science. Rather, he studied the history of science because of his philosophical perspective on science. Based on the history of science, he then developed a descriptive-normative account of the progress of science through periods of normal science and revolutions. Such an account can be adjusted, challenged, or even rejected by new or better historical evidence. The dynamic perspective per se, however, cannot can be challenged only by philosophical arguments. It may however bring more clarity on the often-discussed history-based model of science, that Kuhn sketched in *QPT*, developed in *SSR* and then adjusted in many post-*SSR* writings.

One of the main implications of the dynamic perspective is the abandonment of an algorithmic theory of scientific rationality, which is in turn grounded on a rather individualistic conception. In fact, if there are universal rules of rationality, every rational individual will apply them in the same way. By dropping the algorithmic theory of rationality, Kuhn develops a collective approach to the study of science, which in turn requires the development of a theory of collective rationality. This shift poses the methodological problem of how to study a community of practising scientists. A collective approach requires a study of the intersubjective dynamics of the group itself, as well as of the 'common shared culture and practice' that constitutes a collection of individuals as a collective agent. I contend that Kuhn wrestled with the concepts of paradigm, DM, exemplar, taxonomy, and lexicon precisely in an effort to do so. The choice of a 'linguistic concept' such as lexicon is driven by methodological reasons. Even though, in practice, the collective beliefs, practices, and artefacts of a community are all be, the ethnographers of science (as well as the 'quasi-ethnographers', in the case of historians of science) may conveniently begin from the study of the language shared by the members of a scientific community, which is therefore akin to a linguistic community.

V. Politi, *Thomas Kuhn's Philosophical Perspective*, Boston Studies in the Philosophy and History of Science 348,
https://doi.org/10.1007/978-3-032-13540-7_11

When properly understood, Kuhn's dynamic perspective and collective approach allow not only to clarify some aspects of his history-based model of scientific progress. In fact, it can also be extended to explore issues that Kuhn himself did not consider with enough depth. In my view, for example, he was not entirely correct in his characterisation of specialisation. The dynamic perspective and the theory of collective rationality may also bring some new light on issues that Kuhn did not address directly. For example, it may help to see the value-freedom of science and the autonomy of the scientific community as not necessarily related, since an autonomous scientific community could still be value-laden after all. Even though the concepts of truth and progress remain open problems, my interpretation, without developing a brand-new theory on scientific progress, shows why Kuhn attempted to reframe these concepts.

This is, in a nutshell, what I have argued in this book. In concluding it, I want to return to its beginning, with Alexander Bird's claim that, since he underwent a 'linguistic turn', Kuhn did not manage to establish a distinctive philosophical legacy. I have already argued that Kuhn did not undergo a 'linguistic turn'. However, the problem of the lack of a distinctive Kuhnian legacy remains. I therefore propose three ways to interpret it.

First, Kuhn was always a bit 'off' with respect to his contemporaries. When he was talking about paradigms, DMs, and exemplars, philosophers were talking about theories and theory change. When he chose scientific lexicons to study the 'common culture' and the practices of a scientific community, philosophers decided to undergo a 'practical turn'—often without examining the methodological challenges of studying a collective practice—and then accused him of being too old-fashioned. When he was developing his theory of collective rationality, he was blamed for giving an image of science as driven by irrational factors. However, now that philosophers have caught up and have become interested in the social aspects of science, instead of rescuing the concept of a DM, they simply dismiss it or ignore it. In a sense, Kuhn was 'the right philosopher at the wrong time'. The very things for which he was criticised in a decade became reasons for praise and inspiration in another, and vice versa. Maybe the right moment for establishing a proper 'Kuhnian school of thought' is yet to come.

Second, it may be the case that such a Kuhnian school has always been there. Philosophers who disagree with some of the historical or conceptual details of Kuhn's history-based model of science regard themselves as 'post-Kuhnians' (or, at least, as non-Kuhnians). However, how many of them would disagree with his dynamic perspective? Do they challenge the philosophical arguments that support such a perspective, and do they develop a different general philosophical perspective of their own? I suspect that at least some of the philosophers who urge us to overcome or overthrow Kuhn's philosophy may actually agree with his anti-foundationalism, his blurring of the discovery/justification distinction, his theory of collective rationality, and so on.

Finally, could it be that a Kuhnian school of thought is just not possible? Kuhn developed a new philosophical perspective on science as well as a very general history-based model of the scientific progress. To do so, he relied on philosophical arguments, the history of science, metaphilosophical considerations about the relations between history and philosophy, experimental studies on the nature of perception, developmental psychology, linguistics, philosophy of language and cognition, and so on. He contributed to a range of topics far too wide for a single philosopher to handle. Indeed, it was far too wide for him as well, so much so that he did not always manage the numerous threads he was trying to keep together in the best possible way. Therefore, *one* Kuhnian school is simply impossible. Philosophers, however, can choose and pick some of those threads and develop them further. The use of scientometrics or ethnographic methods in philosophy of science, the establishment of HPS, the study of scientific practice, the development of model-based epistemologies, and so on, all come from a common Kuhnian ancestry. Going back to Kuhn's philosophy may even lead to the discovery of new threads, with the consequent development of new methods and approaches to be incorporated into philosophy of science.

That the work of Thomas Kuhn keeps being fertile and inspiring is, at least in my view, far more valuable than the existence of *one* distinctive, crystalised and perhaps dogmatic Kuhnian school of thought.

Index

V. Politi, *Thomas Kuhn's Philosophical Perspective*, Boston Studies in the Philosophy and History of Science 348,
https://doi.org/10.1007/978-3-032-13540-7

MIX
Papier aus verantwortungsvollen Quellen
Paper from responsible sources
FSC® C105338

If you have any concerns about our products,
you can contact us on
ProductSafety@springernature.com

In case Publisher is established outside the EU,
the EU authorized representative is:
Springer Nature Customer Service Center GmbH
Europaplatz 3, 69115 Heidelberg, Germany

Printed by Libri Plureos GmbH
in Hamburg, Germany